FUNDAMENTALS
OF
AGRONOMY

Fundamentals of
Agronomy

Dr. Gopal Chandra De

Palli Siksha Bhavana
(Institute of Agriculture)
Visva-Bharati, Sriniketan
West Bengal, India

Oxford & IBH Publishing Co. Pvt. Ltd.

New Delhi

(A Unit of CBS Publishers & Distributors Pvt Ltd)

CBS Publishers & Distributors Pvt Ltd

New Delhi • Bengaluru • Chennai • Kochi • Kolkata • Lucknow • Mumbai
Hyderabad • Jharkhand • Nagpur • Patna • Pune • Uttarakhand

Fundamentals of

Agronomy

ISBN-13: 978-81-204-0416-8
ISBN-10: 81-204-0416-5

© 2013, 1989 Gopal Chandra De

CBS Reprint: 2017, 2019, 2023

OXFORD & IBH

New Delhi
(A Unit of CBS Publishers & Distributors Pvt Ltd)

Published by Satish Kumar Jain and produced by Varun Jain for
CBS Publishers & Distributors Pvt Ltd
4819/XI Prahlad Street, 24 Ansari Road, Daryaganj, New Delhi 110 002, India
Ph: 011-23289259, 23266861, 23266867   Website: www.cbspd.com
Fax: 011-23243014   e-mail: delhi@cbspd.com

Corporate Office: 204 FIE, Industrial Area, Patparganj, Delhi 110 092, India
Ph: 011-4934 4934   Fax: 011-4934 4935   e-mail: publishing@cbspd.com;
                                          publicity@cbspd.com

Branches

- **Bengaluru:** Seema House 2975, 17th Cross, KR Road, Banasankari 2nd Stage, Bengaluru 560 070, Karnataka, India
  Ph: +91-80-26771678/79   Fax: +91-80-26771680   e-mail: bangalore@cbspd.com
- **Chennai:** 7, Subbaraya Street, Shenoy Nagar, Chennai 600 030, Tamil Nadu, India
  Ph: +91-44-26680620, 26681266   Fax: +91-44-42032115   e-mail: chennai@cbspd.com
- **Kochi:** 42/1325, 1326, Power House Road, Opp KSEB, Power House, Ernakulam Kochi 682 018, Kerala, India
  Ph: +91-484-4059061-65,67   Fax: +91-484-4059065   e-mail: kochi@cbspd.com
- **Kolkata:** 147, Hind Ceramics Compound, 1st Floor, Nilgunj Road, Belghoria, Kolkata-700056, West Bengal, India
  Ph: +033-25633055, 033-25633056   e-mail: kolkata@cbspd.com
- **Lucknow:** Basement, Khushnuma Complex, 7 Meerabai Marg (Behind Jawahar Bhawan),Lucknow-226001, UP, India
  Ph: +0522-4000032   e-mail: tiwari.lucknow@cbspd.com
- **Mumbai:** PWD Shed, Gala no 25/26, Ramchandra Bhatt Marg, Next to JJ Hospital Gate no. 2, Opp. Union Bank of India,
  Noorbaug, Mumbai-400009, Maharashtra, India
  Ph: 022-66661880/89   e-mail: mumbai@cbspd.com

Representatives

• Hyderabad	0-9885175004	• Jharkhand	0-9811541605	• Nagpur	0-9421945513
• Patna	0-9334159340	• Pune	0-9623451994	• Uttarakhand	0-9716462459

Printed at Chaman Enterprises, Daryaganj, New Delhi, India

Preface

"If you give a fish to a person he will be contented for the day and if you teach him how to fish he can maintain himself throughout his life and his generations."

Agriculture has been a mainstay of human being since time immemorial. Agriculture in any form has been practised in those areas where men have been living permanently. They have been restrained from plundering and/or indiscriminate hunting rather domesticated a number of facultative organisms as pursuivants. They have been utilizing their diligence and intelligence to integrate natural germplasms of plants and animals with natural resources like land, light, water, air and other biotic and abiotic factors of production to produce usable products directly or after processing by their own muscle power or any aid found in nature or developed by them for ease, cheap and swiftness.

Agricultural development is multidirectional having galloping speed and rapid spread with respect to time and space.

After the introduction of modern varieties of various durations with improved plant types along with improvement of input resources, farmers started using improved cultural practices in intensive cropping systems with labour-intensive programmes to improve production potential per unit land, time and input. A large number of alternate genotypes are available to the farmers. Agronomists provided suitable environment to these genotypes to foster them and to manifest their yield potential in newer areas and seasons.

Dealing with various types of soils and climatic conditions is a prerequisite to grow crops. Understanding of relationships between crop plants and natural and manmade factors of production is a guideline to provide better environment to crop plants and to increase use-efficiency of these factors. A wide spectrum knowledge of soil, water, nutrients, weeds etc. helps to manage them efficiently in favour of crop production for years. A comprehensive idea on seeds and their handling, sowing and planting of different crop varieties under different agroclimatic conditions, cropping pattern and cropping systems helps to raise crops together or one after another considering prevailing conditions and available facilities.

Modern concept on improved cropping system is found to be successful in shifting the direction of agricultural practices from crop oriented one to system oriented approach. Modern agrotechniques on crop production, protection and preservation have opened the opportunities to absorb, adapt and apply new knowledge and technology to the farmers. Recently introduced methods of cropping also approach diversified farming systems with components complementary to each other. Non-traditional crops with respect to zone and season showed their promise. All these have dwindled the earlier traditional agriculture and established the scientific agribusiness.

Modern mechanization, chemicalization etc. are widening the scope of utilizing the genetical, hydrological, meteorological as well as technological advances even in the farmers' level for commercialization. Scientific cropping system therefore, has the highest potential even to the marginal and small farmers. It also helps to improve use-efficiency of local resources like land, light, livestock, labour force etc. and the costly and scarce inputs like fertiliser, water etc., improve the physical, chemical and biological properties of soil as well as provide opportunity of on-farm seed production, storage, processing etc.

Very often we ignore or miss the basic knowledge or information on which the multi-storeyed, multidisciplinary approach of crop production stands. It is a strenuous strive to keep pace with the progress of such a vast subject like Agronomy which is in practice throughout the world. An effort has been made to collect and collate new concepts of Agronomy into a somewhat single unit understandable to the undergraduate students and useful as remedial course to the workers and scientists of Agricultural and allied disciplines.

All my strains will be worthy if this treatise containing the past and present forms of Agriculture, basic concepts of

Agronomy, Agronomic principles, seeds, sowing and planting, soil and soil management, tillage, types of tillage operations, nutrient management, dry farming, weed management, methods of cropping and cropping pattern concisely dealt in separate chapters becomes helpful to them for whom this is made.

GOPAL CHANDRA DE

Contents

Introduction

"A man without food for three days will quarrel, for a week will fight and for a month or so will die."

Agriculture is a branch of applied science. The term agriculture has been derived from the Latin word 'ager' meaning land or field and 'cultura' meaning cultivation i.e., the science and art of farming including the work of cultivating the soil, producing crops and raising livestock. The Greek 'geoponic' (cultivation in earth), 'hydroponic' (cultivation in water) and 'aeroponic' (cultivation in air) refer to the three main spheres of agriculture.

Agriculture is the most important enterprise in the world. In a true sense it is a productive unit where the free gifts of nature namely, land, light, air, temperature, rain water, humidity etc., are integrated into a single primary unit (crop plants or their usable parts) indispensable for human beings. Secondary productive units namely, animals including livestock, birds, and insects, feed on these primary units and provide concentrated products such as meat, milk, wool, hide, eggs, honey, silk and lac.

The important cultural energies utilised for the production and protection of agricultural commodities are irrigation and drainage, organic, biological and mineral fertilisers, chemicals, farm equipment and draft power. These are used to maximise productivity per unit time, water, land, labour and rupee investment.

Agriculture provides us food, feed, fibre, fuel, furniture, raw materials and feed-back materials for and from factories, funds, flood control, a free, fair and fresh environment, abundant food driving out famine and friendship eliminating fights. It also considers employment generation, enonomics, education, ecology, energy consumption, use of equipment and earning for production, protection, processing, consumption, preservation and war against waste, transport and trade. Even though agricultural commodities are mostly seasonal, bulky and perishable in nature, they help the nation to earn and conserve a greater amount of foreign exchange and to build up the national economy. Satisfactory agricultural production brings peace, prosperity, harmony, health and wealth to individuals of a nation by driving away distrust, discord and anarchy. It helps to elevate the community consisting of different castes and communities to a better social, cultural, political and economic life. Agriculture consists of growing plants and rearing animals in order to yield, produce and, thus it helps to maintain a biological equilibrium in nature. It congregates the interaction of all environmental factors namely, water, heat, light, air and soil distributed in the different spheres such as the lithosphere, pedosphere, atmosphere, hydrosphere and photosphere. Agriculture helps to meet the basic needs of humans and their civilisation by providing food, clothing, shelter, medicine and recreation. Their eyes and minds are soothed by dynamic changes from grey (bare soil) to green (growing crop) to golden (mature crop) and gay harvests.

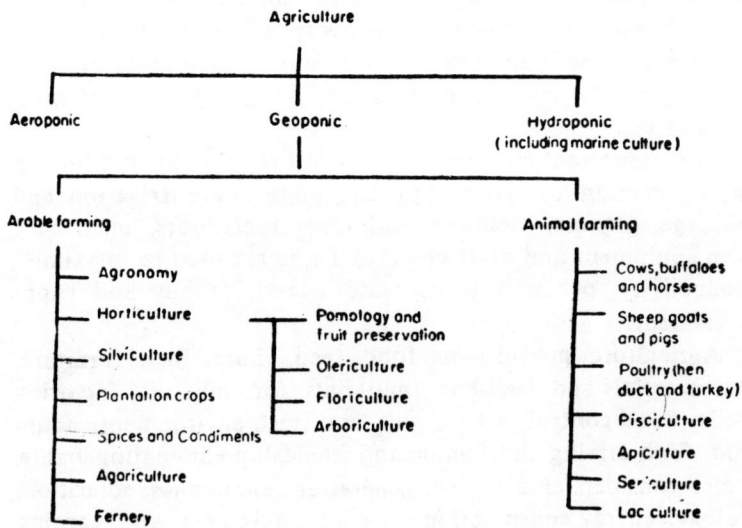

In a wide sense agriculture implies the effective use of land, water, light and other resources of environment through the production of field crops, forage crops, fruit crops, farm animals, fisheries and forestry.

The word *agriculture*, thus may be expanded as Activities on the Ground for Raising Intended Crops fo · Uplifting Livelihood Through the Use of Rechargeable Energies.

Agriculture deals with the enterprises which have been shown graphically in p. 2.

Agriculture encompasses the applied aspects of the following basic sciences:

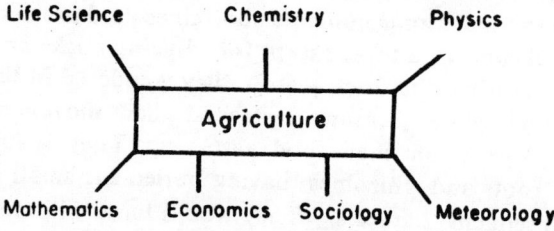

India as a subcontinent has the following land utilisation statistics:

1.	Total geographical area	328.848	million hectare
2.	Total reporting area	304.3	,, ,,
3.	Area under cultivation	143.0	,, ,,
4.	Total cropped area	179.75	,, ,,
5.	Area sown more than once	36.75	,, ,,
6.	Area not available for cultivation	161.3	,, ,,
7.	Forest	66.4	,, ,,

CHAPTER 1

The Past and Present Forms of Agriculture

It appears from excavation, legends and remote sensing tests that about ten thousand years ago man invented agriculture. It is believed that women were the pioneers for cultivating useful plants from the wild flora, while men went out for hunting and gathering food. Some nomadic aborigines used fire and weapons. They did not have tools except for digging sticks and mollusc shells. During the drier season they wandered in bands. The wetter and colder seasons restricted their movements. They would depend on their food gathering. They would dig ov' edible roots and rhizomes having buried the small ones there for subsequent harvests. They would hunt wild animals and use their flesh as food and skin as clothing. Thereafter they entered upon a more secure stage of living. Probably woman by her intrinsic insight, first observed that plants come up from seeds. Man might have consciously planted seeds and nurtured the sprouts to harvest. They used fire and digging sticks to prepare garden plots in which they would grow small grains. Once the seeds were planted the band would move on and return later for the harvest.

A primitive form of agriculture was shifting cultivation, in which people working with the crudest of tools, cut down a part of the forest, burnt the undergrowth and started new garden sites. After a few years when these plots lost fertility, choked with grass or brush or became heavily infested with soil-borne pests, cultivators would shift to a new site. This system would support a sparse population where abundant land was available.

Man invented the use of live animals as beasts of burden. They started domesticating wild animals facultative in nature.

People with more members in a group selected somewhat better areas (near a stream or river) and settled down in permanent village sites and started cultivating the same land more continuously. However, the tools, crops and cropping methods might have been primitive. This system was probably the rudimentary system of settled farming. Since then agricultural activity became subsidiary to gathering and hunting.

Thereafter an advanced form of primitive agriculture was the subsistence form of agriculture. This was based on the principle "grow it and and eat it" instead of on a commercial basis. In some cases it was based on crop production and in other cases on a more intimate blend of crops and livestock. Hunting and gathering were subsidiary occupations. Men in meaningful numbers began to settle down to sedentary life.

A sparse population of nomadic tribesmen followed their flocks and herds in the seasonal pursuit of water, forage and shelter. Commercial ranching gradually blended with subsistence farming.

Under shifting cultivation land abandoned became degraded. Subsequent subsistence farming followed naked fallow rotation. Thereafter legume rotation was adopted. Under mixed farming comprising field crops and animals, field grass husbandry was adopted and the same fields were used for cropping and grazing.

The concept of property in land evolved and the individuals held the privilege of autonomous action. The area over which a band wandered and carried on their food-gathering was considered to be for the joint and mutual exploitation of the group. They held a territory assuming they could defend it from other bands by the right of prior occupance. Any individual property in land was right to a certain garden plot and domestic site, earned by virtue of having cleaned, improved and used it.

During their movement from one location to another they would come in contact with a large number of plant and animal types. They selected the most useful ones for domestication and improved them.

Battles, fights on rivalries among different groups and individuals for property or passion helped to blend newer ideas for the components, commodities and comforts of life. Mutual exchanges, gifts, offerings etc., also blended the style of living, food habits and nature of earning.

This right to use the land with the progress of time was recognised on a continuing basis year after year. With still

further evolution land was held to be inheritable and the ownership title was granted and defended by group authority.   These
progressive stages of security in mobile and immobile properties,
contributed proportionately to the greater stability and well-
being of the society.   They concentrated on the welfare of their
individual families, and the produce of their land and animals.
They devoted themselves more intensively to utilise their land to
the best of their ability.   However, they held together to defend
any unwanted enemy of their society.   This was still the transi-
tional stage from food gathering to food growing.

The self-sufficient agricultural village was the least common
denominator of a permanent human society.   It developed and
perpetuated itself without many tools or other capital and without dependence on the outside world.   Several crafts also developed for and around agricultural production.   Agriculture
then experienced extreme adversity as long as the minimum of
food, feed and fibre were raised. The standard of living was low.
With the span of time there was the periodic and geographic
rise of different civilisations.   Agricultural production expanded
and towns and cities developed, trade and arts flourished.

Downtrodden people, slaves and bonded labour were the
principal sources of human labour.   Draft animals were the
plodding oxen.   Farming equipment was animal drawn and
consisted largely of hand tools.   It was organised on a commercial basis and management was developed to a fine art.   The
skills developed were inherited through generations of cultivators who became a major section of the then civilised society.
Advanced farming practices included seed selection, green
manuring with legumes, rotation of crops, use of animal and
crop refuse as manure, irrigation, pasture management, rearing
of bullocks and milch animals, meat and wool production and
rearing of birds.

In subsequent stages the feudal manor emerged, and was
essentially an agricultural village.   It consisted of a nucleated
village built around the manor-house and surrounded by the
village land.   The best lands were laid out in large arable fields,
the medium lands were used as grasslands for mowing and grazing and the least productive areas were used for grazing, woodlands, funeral grounds and other common activities.   Being a

self-contained the village included a worship altar and a meeting place.

Arable lands were divided into different plots according to topography and use. During the favourable season crops were grown in the fields and livestock was kept away from the crops but after the harvest cattle would graze in these areas under the supervision of herdsmen. During the cropping season livestock were allowed to graze in common grazing lands and in adjoining woodlands and forests. By-products of crops were used to feed the animals during stall feeding. Quack remedies based on locally available plants with therapeutic values and minerals were the major method of treatment for common ailments of animals. Infestation of insect-pests was also combated by indigenous methods.

There were ponds and tanks in the villages. Some of these water reservoirs were reserved for drinking water, washing purposes, fisheries, or small-scale irrigation. The banks of water systems were used for arboriculture and fruit growing.

Emperors, monarchs, kings, zamindars, governors, heads of states etc., would claim authority over their respective geographical land areas and rule over them. Cultivable areas would be distributed to lease holders, tenants or share croppers directly or through intermediaries. Such cultivators or tillers would enjoy little benefit except for mere subsistence.

The enclosure movement began to be consolidated in some countries during the 14th century. Open and scattered fields of a tenant were compacted into a farm unit under an individual owner. These consolidated, fenced and individually controlled farm units were managed by the farmers with commercial intent. Land as an inheritable and transferable property was subdivided and fragmented to an indefinite extent without the consideration of its economic viability to the tenants who depended on subsidiary occupations to supplement their earnings subsequently. In long and medium term cycles, natural calamities affected agricultural production and the farming community and the kingdom or nation as a whole would fall in distress. Facing such famine was a futile effort except by large-scale emigration. The toll of human and animal lives at such times, was considerable. No other occupation could compensate for these disasters.

Land, though an immobile property was capable of producing an enormous quantity of mobile short to long-term crops meant for the livelihood of a large number of people. The helping hands of the peasantry and ploughmen were usually family members. There were also different types of human labourers and work animals. The then tools and implements were a slightly improved form of primitive ones. Bullock or horse drawn tools were limited to the iron banded wooden plough, bullock cart etc. Thereafter wooden and iron implements and equipment were developed. All of them were manually operated or drawn by animal power.

There were many ups and downs, shortages and disasters in this expanded (both spatial and temporal) development of agriculture in spite of the diligent endeavours of humans for a greater dependence on agricultural production.

Modern agriculture started during the early eighteenth century. Crops were grown in sequence without allowing the field to remain fallow. Crop rotation was introduced. Organic recycling was intensified. A number of exotic crop plants and animals were introduced. The crops selected were clean, stable in yield potential and well adapted to the local agro-ecological systems. Land improvement was introduced. Farm implements and equipment were improved. The care and management of crops were advanced and specialised, crop protection practices were introduced and preservation and processing improved. Gradually livestock and flock breeding and rearing and crop farming became diversified. Sheep meant for wool were bred for the production of better and more efficient meat animals. Dairying became a profession. Fish rearing and fishing became an occupation. Greater emphasis was laid on feeding, breeding, caring for and weeding out of different livestock and birds meant to meet various requirements more efficiently and economically. Crop plants also drew due attention and they were grown in different categories as field crops, plantation crops and garden crops.

In the nineteenth century the real science of agriculture started. Nutrition, physiology, ecology etc., developed. Research and development in fundamental and basic sciences were brought under applied aspects in agriculture. It took the shape of a teaching science. The most neglected occupation of a downtrodden and dull community drew the attention of aristrocrats, scientists, researchers and others in almost all walks of life. Research on specialised branches of several disciplines were developed and scientific findings were translated into production practices. Laboratories, farms, stations, centres and institutes for research, teaching, training and demonstration were developed. Books, journals and scientific and popular literature were introduced. News media and audio-visual aids were developed to disseminate new findings and information to the rural masses engaged in agriculture.

At present a lot of developments on hydrological, mechanical chemical, genetical, and technological aspects of agriculture are in progress. Governments are apportioning a greater share of development for agricultural improvement. Central and State Departments of Agriculture, Agricultural Universities and Institutions are flourishing. International Research Institutes are being set up. Farmers specially small and marginal ones of the underdeveloped and developing countries are supplied with improved seeds, fertiliser, irrigation, equipment, electrical energy, pesticides, improved genetic stocks of cows, chicks, kids, pigs, fingerlings, silk worms and their improved feed. Even funds for both recurring and non-recurring expenditure with reasonable cost or rate of interest, or free of cost, as well as technology services are being provided freely. Preserving, processing, pricing, marketing, insuring, distributing, consuming and exporting-importing policies are strengthening. Agro-based industries and crafts are developing. Need-based planning, programming and executing are in progress. Improved use-efficiency of non-monetary, less-monetary and monetary inputs is in the process of adoption. From specialised farming, diversified and integrated farming is emerging. Land reforms, development and diversion from degradation are going on. As the size of operational holdings are rapidly decreasing to marginal and sub-marginal levels, farming by individuals is going to be recast to co-operative, community or contractual farming.

Agriculture now is not merely production oriented but is becoming an agribusiness consisting of enterprises such as livestock, poultry, fishery, silviculture, sericulture, apiculture and lac culture along with field and forage crops and plantation crops. These are the stepping stones of the green, white and yellow revolutions of today and simultaneously of the multi-colour revolution with manifold quantitative, qualitative and generative yields.

It is surprising that even in the era of nuclear energy in industrial development and expeditions to space, man retains his animality and primitiveness in food habits and is a competitor with his animals for foodgrains.

Even though agriculture was, is and will have tremendous potential for life and longevity in this luminous world, yet it is far behind the technologies developed in other sectors such as industry, space expeditions and defence.

CHAPTER 2

Basic Concepts of Agronomy

The word agronomy has been derived from two Greek words—'agros' meaning field and 'nomos' meaning to manage. Literally it means the art of managing fields and technically it means the science and economics of crop production by management of farm land. In other words it is the art and underlying science in production and improvement of field crops with the efficient use of soil fertility, water, labour and other factors related to crop production. Agronomy is the field of study and practice of ways and means of production of food, feed and fibre crops. Thus agronomy as a branch of agricultural science deals with principles and practices of field management for the production of field crops. Among all the branches of agriculture, agronomy occupies a pivotal position and is regarded as the mother or primary branch. Like agriculture, it is nothing but an integrated and applied aspect of different disciplines of pure sciences. It has three clear branches: (1) crop science (mainly field crops); (2) soil science; and (3) environmental science, that deals only with applied aspects.

The central theme of agronomy is of soil-crop-environment relationships. Field crops without soil cannot be considered and soil without crops is barren. The core of agronomy is the field of crop plants with the theme of controlling the environment (microclimate) and the nature of agronomy is based on soil-plant-environment relationships. The study of the magnitude of variation in yield, cause and effect relationships, internal and external factors and their interrelationships, techniques of increasing use-efficiency of inputs, evolving technologies for better management practices of soil, water, plant nutrients, weeds and crop plants are the major aspects of agronomy to boost production and its usable products per unit land, time and input. Scientific crop production includes crop improvement, improved agro-techniques, the ameliorating agro-climate of the locality and other aspects of the surrounding area for the entire field duration of the crop concerned.

AGRONOMY thus denotes Activities on the Ground to Raise Outspread and Noble crops to Obtain Massive Yields.

2.1 BASIC PRINCIPLES OF AGRONOMY

Science deals with five w's—they are what, when, where, why and how ? As an applied science agronomy deals a little more in detail in most cases, i.e., which, whether, who, whom and how many, how much and how long.

A principle means a scientific law that explains natural action and agronomic principles are the ways and means for the better management of soil, plants and environment for economically maximum returns per unit area for years.

Some of the factors of crop production are controlled by the environment and the rest are man-made factors. The environmentally controlling factors are modifiable to a limited extent, for instance drought is mitigable by irrigation or gales are retrievable with shelterbelts.

Principles of crop management depend largely on the type of farming namely, specialised, diversified, mixed and integrated and also on the physical and technological facilities available— irrigated farming, dry farming and rain fed (*barani*) farming.

The fundamental principles of agronomy may be listed as:

a) Planning, programming and executing measures for maximum utilisation of land, labour, capital, sunshine, rain-water, temperature, humidity, transport and marketing facilities;

b) choice of crop varieties adaptable to the particular agro-climate, land situation, soil fertility, season and method of cultivation and befitting to the cropping system;

c) proper field management by tillage, preparing field channels and bunds for irrigation and drainage, checking soil erosion, levelling and adopting other suitable land improvement practices;

d) adoption of multiple cropping and also mixed or inter-cropping to ensure harvest even under adverse environmental conditions;

e) timely application of proper and balanced nutrients to the crop or crops in sequence and improvement of soil fertility and productivity. Correction of ill-effects of soil reactions and conditions and increasing soil organic matter through the application of green manure, farm yard manure, organic wastes, biofertilisers and profitable recycling of organic wastes;

f) choice of quality seed or seed material and maintenance of requisite plant density per unit area with healthy and uniform seedlings;

g) proper water management with respect to crop, soil and environment through conservation and utilisation of soil moisture as well as by water that is available in excess. Efficient use of this water for life-saving or protective irrigation, scheduling irrigation at critical stages of crop growth and also growing crops in non-traditional seasons or belts with irrigation or rain or residual moisture;

h) adoption of adequate, need-based, timely and exacting plant protection measures against weeds, insect-pests, pathogens, as well as climatic hazards and correction of deficiencies and disorders;

i) adoption of suitable and exacting management practices including intercultural operations to get maximum benefit from dearer and difficult to get, low-monetary and non-monetary inputs. Adequate input supply should not be a substitute for deficiencies in management practices;

j) adoption of suitable method and time of harvesting of crop to reduce field-damage and to release land for succeeding crop(s) and efficient utilisation of residual moisture, plant nutrients and other management practices. Adoption of suitable post-harvest technologies.

CHAPTER 3

Agronomic Principles

Before going to any agronomic field operations some basic knowledge about the crop plants must be acquired. Some basic information about the crop plants and how wisely they can be included as components in a particular farming system are discussed below.

3.1 CLASSIFICATION OF CROPS

Classification is an acceptable product of scientific studies and is obviously based on well demarcated characteristics of the things or concepts so classified.

A crop is an organism grown or harvested for obtaining yield. Crop plants do not belong to any particular portion of the plant kingdom. Several crop plants are alike with respect to ontology (development of an organism), morphology, anatomy, physiology and requirement of particular type of ecology at different stages of their growth and development. All these help to generalise similar crop plants as a class for attaining a better understanding of them. Of course, while considering some crops as a group, some exceptions must always be expected.

Crops are grouped in several ways namely, according to the range of cultivation, place of origin and distribution, different characters, uses, cultivation requirements and other common behaviour.

Crop plants are grouped into three main classes according to the range of cultivation. They are :

a) *Garden crop*: Crop plants that are grown on a small-scale in gardens such as in kitchen gardens, flower gardens and backyard gardens for instance onion, brinjal.

b) *Plantation crop*: Crop plants that are grown on a larger scale on estates. They are permanent (need not

be replanted after each harvest) in nature i.e., harvesting continues for a prolonged period from a single planting, for instance tea, coffee, cocoa.
c) *Field crops*: Crop plants that are grown on a vast scale. They are mostly seasonal—such as rice, wheat, cotton.

Among these three classes agronomy deals with field crops only. Field crops are classified in the following ways:
1. According to the place of origin;
2. Botanical classification;
3. Commercial classification;
4. Economic classification;
5. Seasonal classification;
6. Classification according to ontogeny;
7. Agronomic classification;
8. Classification based on leaf morphology;
9. Classification based on serving special purposes.

3.1.1 Classification According to the Place of Origin

A number of crops may be grown elsewhere but each crop has its native place where there are a large number of wild relatives even today. The agro-climatic situation of the place of origin may provide relevant information about the requisite or ideal environment for its adaptability. The wild relatives may be used as a gene pool for any further improvement of the crop concerned.

According to the place of origin crops are grouped into:
a) Native: which are grown within the geographical limits of their origin. Crops of Indian origin are rice, barley, black gram, green gram, sarson, castor, sugar-cane and cotton.
b) Exotic or alien or introduced: which are grown even beyond their site of origin. Some of the crops which are now grown in India but introduced from other countries, are tobacco, potato, jute and maize.

3.1.2 Botanical or Taxonomic Classification

Crop plants are dealt with under the natural orders or the

families under which they come, in a systematic arrangement. This classification has certain advantages in the understanding of the morphological character of any particular family of crop plants with different agro-botanical or morpho-agronomical characteristics or peculiarities. Crop plants are grouped into families:

a) *Poaceae (Graminae)* or grass family: rice, wheat, oats, barley, maize, sugar cane, napier, sorghum, para grass.

b) *Papilionaceae* (legumes) or pea family: peas, cow-pea, pigeon-pea, Bengal gram, black gram, green gram, soybean, velvet bean, *guar*, groundnut, berseem, sunn-hemp, lentil, *dhaincha*.

c) *Cruciferae* or mustard family: mustard, *toria*, radish, cabbage, cauliflower, knolkhol.

d) *Cucurbitaceae* or gourd family: sweet gourd, ash gourd, bitter gourd, ridged gourd, cucumber, pumpkin.

e) *Malvaceae* or cotton family: cotton, lady's finger, roselle.

f) *Solanaceae* or brinjal family: potato, tomato, tobacco, brinjal, chillies.

g) *Tiliaceae* or jute family: jute.

h) *Asteraceae (Compositae)* or sunflower family: sunflower, safflower, niger.

i) *Chenopodiaceae* or spinach family: spinach, beet, sugar beet.

j) *Pedaliaceae* or sesame family: sesame.

k) *Euphorbiaceae* or castor family: castor, tapioca.

l) *Convolvulaceae* or sweet potato family: sweet potato.

m) *Umbelliferae* or coriander family: coriander, anise, cumin, carrot.

n) *Liliaceae* or onion family: onion, garlic.

o) *Zingiberaceae* or ginger family: ginger, turmeric.

3.1.3 Commercial Classification

The plant products (usable products) which come into the commercial field go under popular groupings such as

a) Food crops: rice, wheat, green gram, soybean, groundnut.

b) Feed crops: oats, napier, sorghum, maize, stylo, rice bean, berseem, lucerne.
c) Industrial or commercial crops: cotton, sugar cane, sugar beet, tobacco, jute.
d) Food adjuncts: turmeric, cumin, garlic.

3.1.4 Economic or Agrarian or Agricnltnral Classification or Classification According to Use of Crop Plants and Their Products

a) Cereals: *Ceres* (a) Roman word) was the name of a Roman goddess who was the 'giver of grain'. Cereals are the cultivated grasses grown for their edible starchy grains (one seeded fruit known as caryopsis). In gener-al, the larger grains used as staple food are cereals such as rice, wheat, maize, barley, sorghum.

b) Millets: Millets are the small grained cereals which are of minor importance as food and they have a single cover (due to fusion of testa to the pericarp) such as *bajra, ragi, kauni, shyama.* Millets are also used as staple food in drier regions of the developing countries.

c) Oil seeds: Crop seeds that are rich in fatty acids, are used to extract vegetable oil to meet various require-ments, for instance mustard, rape, *rat*, sesame, sun-flower, safflower, castor, linseed, groundnut, soybean.

d) Pulses: Seeds of leguminous crop plants used as food. On splitting they produce *dal* which are rich in protein such as green gram, black gram, Bengal gram, peas, cow-pea, pigeon-pea, soybean, lentil, lathyrus

e) Feed or forage: It refers to vegetative matter, fresh or preserved, utilised as feed for animals. It includes hay, silage, soilage, pasturage and fodder. Forage crops are anjan grass, berseem, centro, cow-pea, Dinanath grass, field bean, Guinea grass, *jowar*, lucerne, maizes marvel grass, napier, oats, para grass, rice bean, Rhodes, grass, sarson, Setaria, stylo, Sudan grass, teosinte, turnips and velvet bean.

f) Fibre crops: Crop plants grown for fibre yield. Fibre may be obtained from seed, such as cotton; stem or bark such as jute, mesta, rosellc, sunnhemp and flax;

leaf such as agave, pineapple.

g) Sugar and starch crops: Crops grown for the production of sugar and starch are sugar cane, sugar beet, potato, sweet potato, tapioca and asparagus.

h) Spices and condiments: Crop plants or their products are used to season, flavour, taste and add zest and sometimes colour the fresh or preserved food, for instance ginger, garlic, fenugreek, cumin, turmeric, chillies, onion, coriander, anise and asafoetida.

i) Drug crops: Crops that are used for the preparation of medicines, for instance tobacco, mint.

j) Narcotics, fumitories and masticatories: Crop plants or their products that are used for stimulating, numbing, drowsing or relishing effects such as tobacco, *ganja*, opium poppy, anise.

k) Beverages: Products of crops used for mild, agreeable and stimulating liquors meant for drinking such as tea, coffee, cocoa.

3.1.5 Seasonal Classification

Different annual crops are grown in different seasons of the year. Such seasons are most prominent in the plains in contrast to coastal and mountainous ranges and in areas receiving monsoonal climate. Crops are grouped under the seasons in which are their major field duration:

a) *Kharif* or South-West monsoon crops: Crop plants grown during June-July to September-October such as rice, maize, castor, groundnut. They require warm-wet weather during their major period of growth and shorter day-length for flowering.

b) *Rabi* crops: Crop plants grown during October-November to January-February such as wheat, mustard, barley, oats, potato, Bengal gram, peas, linseed, lentil, berseem, cabbage and cauliflower. They require cold-dry weather for their major growth period and longer day-length for flowering.

c) *Zaid* or summer crops: Crop plants grown during February-March to May-June such as black gram, green gram, sesame and cow-pea. They require warm-dry weather for their major growth period and longer day-length for flowering.

With respect to rice crops grown during different seasons *aus, aman* and *boro* seasons are prevalent. The word *aus* has been derived from the, Sanskrit *ashu* meaning quick or early. It is designated to the rice varieties grown during the pre-monsoon period i.e., harvested in August-September. It is also known as autumn rice with respect to the time of harvest. The word *aman* has been derived from Arabic meaning safety which indicates stability of the crop. It is synonymous to *kharif* or winter crop of rice. *Boro* refers to the rice crop grown in submerged land lower in elevation. It is grown during January-February to April-May. It is also known as summer rice.

This seasonal classification is not a universal one. It only indicates the period of the year when a particular crop has been grown, for instance *kharif* rice, *kharif* maize, *rabi* maize, *boro* rice, *rabi* pulse, summer pulse.

Some crops overlap the seasons such as jute which may be sown in February-March and harvested in July-August. Sesame is an adaptable crop grown during summer in the southern plains of West Bengal whereas it is a suitable crop during the monsoon period in the north and both in the *kharif* and *rabi* seasons in south India. Some photoneutral or weakly photo- and thermo-sensitive crop varieties may be grown throughout the year provided irrigation is available, for instance rice, maize, groundnut. Again some long duration temporary crops such as sugar-cane, cotton, pigeon-pea etc., do not follow the regular pattern except that low temperatures in winter provide a triggering factor contributing towards the maturity of the crops.

3.1.6 Classification According to Ontogeny

Each crop plant completes its life cycle after passing through progressive stages of development. Crop plants are grouped according to their life cycles: a) Annual crops: Crop plants that complete their life cycles within a season or year such as rice, wheat, maize, mustard and tobacco. They produce a crop of seed and die. Some of these crop plants may produce tillers. If such rooted tillers are separated from the main shoot and planted, each tiller will survive that season as a new plant, but will not live until another season.

b) Biennial crops: Crop plants having a life span of two consecutive years. The first year's growth is purely vegetative with top growth usually confined to rosette of leaves. The tap root is often fleshy and serves as a food storage organ. During the second year it produces a flower stalk from the crown. After producing seed the plant dies, for instance sugar beet, beet, cabbage, radish, carrot. Due to unfavourable weather conditions, a floral stalk may come out (bolting) even in the first year.

c) Perennial crops: Crop plants that live for three or more years. They may be seed-bearing (spermophytes) or non-seed-bearing. They may propagate by seeds and/or by vegetative propagules that may be developed in aerial or subaerial or in both parts, such as sugar cane, napier, ginger, sweet potato and garlic. In general, perennial crops occupy land for more than 30 months.

3.1.7 Agronomic Classification or Classification According to Cultural Requirements

A group of crop plants are alike in cultural requirements due to their similar agro-botanical or morpho-agronomical character:

A. According to suitability of toposequence:

1) Crops grown on high land such as *arhar*, groundnut, sunnhemp, maize, *jowar*, *bajra*, cotton, sesame, napier and Dinanath grass. They cannot tolerate water stagnation.

2) Crops grown on medium land such as jute, potato, sugarcane, upland rice, *ragi*, wheat, Bengal gram, black gram, barley, oats and mustard. They require sufficient soil moisture but cannot tolerate water stagnation,

3) Crops grown on low land such as rice, *dhaincha*, para grass and jute (*capsularis*). They require an abundant supply of water and can stand prolonged waterlogged conditions.

B. According to the suitability of the textural groups of soils:

1) Crops of sandy to sandy loam (light) soils: potato, sweet potato, sugar-beet, onion, carrot, turnip, green gram, black gram, sunflower, *jowar*, *bajra*.

2) Crops of silty to silty loam (medium) soils: jute, sugar cane, maize, cotton, mustard, tobacco, Bengal gram, peas, pigeon-pea, berseem, lucerne, cow-pea.

3) Crops of clay to clay loam (heavy) soils: rice, wheat, barley, linseed, lentil, chickling-pea, para grass, Guinea grass, marvel grass.

C. According to the tolerance to the problem soils:

1) Crops tolerant to acidic soils: wet-rice, potato, mustard.

2) Crops tolerant to saline soils: chillies, cucurbits, wheat, *raya*, *bajra*, *jowar*, sesamum, *guar*, barley.

3) Crops tolerant to alkaline soils: barley, cotton, chick-pea, peas, beans, berseem, groundnut, maize.

4) Crops tolerant to waterlogged soils: wet-rice, *dhaincha*, para grass, napier, Guinea grass.

5) Crops tolerant to soil erosions: marvel grass, *dhaulu*, groundnut, black gram, *kudzu*, rice bean, moth bean.

D. According to tillage requirements:

1) Arable crops: Crops requiring preparatory tillage: potato, radish, berseem, sugar cane, rice, tobacco, maize.

2) Non-arable crops: Crops that may not be requiring preparatory tillage: black gram, para grass, cow-pea.

E. According to the depth of root system:

1) Shallow-rooted crops: rice, potato, *toria*, onion.

2) Moderately deep-rooted crops: wheat, groundnut, castor, tobacco, barley, chick-pea, cow-pea, berseem.

3) Deep-rooted crops: maize, cotton, *jowar*, sugar beet, sweet potato.

4) Very deep-rooted crops: sugar cane, safflower, lucerne, pigeon-pea.

F. According to the tolerance to hazardous weather conditions:

1) Frost tolerant crops: sugar beet, beet.

2) Cold tolerent crops: potato, cabbage, mustard.

3) Drought tolerant crops: *bajra*, *jowar*, barley, safflower, castor, cotton, chillies, tetrakalai, sesame.

G. According to water supply:

1) Irrigated crops: potato, *boro* rice, cabbage, berseem, wheat.

2) Rain fed upland crops: jute, maize, *ragi*, upland rice,

pigeon-pea, cotton, castor, sesame, cow-pea, Guinea grass, Dinanath grass.

3) Rain fed but partial irrigated crops: Bengal gram, wheat, *jowar, bajra*, mustard.

4) Residual or conserved soil moisture crops: *toria, bajra*, barley, safflower, lentil, linseed.

5) Rain fed crops with supplemental irrigation: *kharif* rice, sugar cane, black gram, green gram.

6) Rain fed plus flooded crops: deep water rice, tidal wet land rice, jute (*capsularis*), sugar cane, *dhaincha*.

H. According to method of sowing or planting:

1) Direct seeded crops: upland rice, wheat, mustard, maize, *jowar, bajra*, groundnut, peas, beans, grams, berseem, Dinanath grass, rice bean, stylo.

2) Planted crops: sugar cane, potato, sweet potato, napier, Guinea grass, para grass.

3) Transplanted crops (after raising seedlings in the nursery): transplanted rice, tobacco, onion, *ragi*, brinjal, cabbage, cauliflower, *bajra*.

I. According to requirements of intertillage specially earthing up:

1) Intertilled crops: potato, sweet potato, groundnut, maize, napier, sugar cane, turmeric, arum, ginger.

2) Non-intertilled crops: *jowar* (fodder), Dinanath grass, para grass, oats, lentil, linseed, chickling-pea.

J. According to length of field duration:

1) Very short (up to 75 days) duration crops: green gram, black gram, *toria*, maize, radish, spinach, beet.

2) Short (75 to 100 days) duration crops: upland rice, potato (early), lady's finger, cauliflower, sunnhemp, sunflower.

3) Medium (100 to 125 days) duration crops: rape, wheat, *jowar, bajra*, barley, potato (main), groundnut, Bengal gram, soybean, sesame, cabbage, tomato, *guar*, tobacco (*rustica*), onion, *boro* rice, jute.

4) Long (125 to 150 days) duration crops: mustard, coriander, tobacco (*tabacum*), anise, cumin, *kharif* rice, *mesta*, roselle, cotton, rice bean, berseem.

5) Very long (150 days and above) duration crops: sugarcane, sugar beet, castor, pigeon-pea, napier, para grass, Guinea grass.

K. According to the method of harvesting:

Different crops are harvested by different methods: by

1) Reaping: rice, wheat, mustard, sugar cane.

2) Uprooting by pulling: *toria*, lentil, linseed, Bengal gram, *khesari*, black gram, radish, cabbage.

3) Uprooting by digging: potato, onion, garlic, ginger, sugar beet, sweet potato, colocasia, turmeric, groundnut, carrot.

4) Picking: cotton, brinjal, lady's finger, gourds, maize, cow-pea, green gram, chillies.

5) Priming: tobacco.

6) Cutting: oats, napier, berseem, spinach, amaranth.

7) Grazing: para grass, pasture grasses.

L. According to post-harvest but pre-processing requirements:

Some crops require special care before they are used or sold, for instance:

1) Curing: tobacco, mustard, sesame.

2) Stripping: jute, *mesta*, roselle, sunnhemp, flax.

3) Shelling: groundnut, maize, sunflower.

4) Ginning: cotton.

5) Seasoning: turmeric, chillies.

6) Grading and sorting: potato, rice, wheat, fibre crops, tobacco.

3.1.8 Classification Based on Leaf Morphology

Crop plants have two types of foliage leaves:
a)   Broad leaves: mustard, peas, potato, tobacco, cotton, jute, berseem, cow-pea.
b)   Narrow leaves: rice, wheat, sugar cane, napier, para grass, onion, garlic, ginger, turmeric, asparagus.

3.1.9 Classification Based on Serving Special Purposes or Special Purpose Crops

A. Crop plants which are grown with the intention of serving purposes in addition to obtaining yield or to meet a particular purpose are grouped into special purpose crops:

1) Catch crops or emergency crops or contingent crops: These are crops cultivated to 'catch' the forthcoming season.

They replace a main crop that has failed due to biotic or climatic or management hazards and utilise the remaining period of the season. They provide livestock feed at a difficult time of year, check weed growth, conserve soil, utilise added fertiliser and moisture. They are generally of very short duration, quick growing, fast bulking, harvestable or usable at any time of their field duration and adaptable to the season, soil and cultural practices, for instance green gram, black gram, cow-pea, *bajra*, spinach, radish, coriander, onion.

2) Restorative crops: These are crops which provide a good harvest along with enrichment or restoration or amelioration of the soil, such as legumes. They fix atmospheric nitrogen in root nodules, shed their leaves during ripening and thus restore soil conditions.

3) Exhaustive crops: These are crop plants which on growing leave the field exhausted because of a more aggressive nature, for instance sesame, brinjal and linseed.

4) *Paira* crops: These are crop plants which are sown a few days or weeks before the harvesting of the standing mature crop. These crops are grown on residual moisture without preparatory tillage. The standing crop and the later sown (*paira*) crop become simultaneous (forming a pair) for a brief period and for the rest of the duration they remain as single crops: for instance black gram in *aus* paddy, *khesari* in *aman* paddy, sweet guard in potato. *Paira* crops in succession may constitute relay cropping.

5) Smother crops: These crop plants are able to smother or suppress the population and growth of weeds by providing suffocation (curtailing movement of air) and obscuration (of the incidental radiation) by their dense foliage developed due to quick growing ability with heavy tillering or branching, planophyllic or procumbent or trailing habits, for instance barley, mustard, cow-pea, sweet gourd.

6) Cover crops: These crop plants are able to protect the soil surface from erosion (wind, water or both) through their ground covering foliage and/or root mats: for instance groundnut, marvel grass, black gram, rice bean, tetrakalai, *khesari*, *kudzu*, sweet potato, para grass.

7) Nurse crops: These crop plants help in the nourishment of other crops by providing shade and acting as climbing sticks

such as *rai* in peas, *jowar* in cow-pea, *Tephrosia*, *Glyricidia*, *Crotalaria* in tea. Leguminous or deciduous plants shed their leaves and thus enrich the surface soil for better growth of the crops grown underneath. Shade loving plants (sciophytes) such as turmeric, ginger and corm are nourished by tall crops such as pigeon-pea, castor, and climbing vegetables such as cucumber, little gourd, ash gourd and bean.

8) Guard or barrier crops: These crop plants help to protect another crop from trespassing or restrict the speed of wind and thus crop damage such as safflower in gram, *Saccharum munjo* (*Sar*) around the crop fields situated on the banks of torrential rivers.

9) Brake crops: These crops are grown to break the continuity of the agro-ecological situation of the field under multiple cropping systems. The inclusion of such crops in the cropping system helps to reduce the inoculum of soil-borne harmful biotic agents such as weeds, pests, pathogens and parasites and improves soil conditions for crop growth. Growing potato, pulses or oil seeds in continuous cereal cropping systems, for instance rice-rice, rice-wheat, rice-maize breaks the continuity of a large number of pests due to variations in host ranges and changing of agro-ecological situations.

Brake crops are also used to designate guard crops particularly those which help to brake the wind speed and protect crop plants from wind hazards.

10) Trap crops: These crop plants are grown to trap soil-borne harmful biotic agents such as parasitic weeds, *Orobanche* and *Striga* that are trapped by Solanaceous and sorghum crops respectively. These weed seeds germinate when they come in contact with roots of these crop plants. Thereafter the destruction of these crops reduces the inoculum of such parasitic weeds. Similarly, nematodes are trapped by some Solanaceous crops. On uprooting crop plants, nematodes are pulled out from the field soil.

11) Mulch crops: These crop plants are grown to conserve soil moisture from bare ground by their thick and multilayered foliage, trailing habit and sometimes, self-seeding nature, for instance cow-pea, *Atylosia*.

12) Sod or turf crops: These crop plants belong to the grass family and have sod type tillers with matted foliage and

roots close to the soil surface. They are grown to conserve soil from erosion particularly in non-arable areas, for instance marvel grass, *Digitaria sanguinalis, Cynodon dactylon.*

13) Cleaning crops: These are crop plants whose agronomic practices make the field clean from weeds and stubble, for instance potato, groundnut, ginger, turmeric, and colocasia which reqire considerable earthwork such as earthing up, preparation of irrigation and drainage channels, placement of topdressed fertilisers and harvesting by digging which helps to disturb the soil surface, the site of weed growth.

14) Fouling crops: These are crop plants whose cultural practices allow the infestation of weeds intensively, for instance maize, cotton and direct seeded upland rice as they have low coverage over the ground at their earlier stages of growth and wider spacing and slow growth at the beginning.

15 Cash crops: These crop plants are grown for sale to earn hard cash. The processing of such crops after harvest is beyond the means of individual farmers, for instance jute, tobacco, cotton and sugar cane.

16) Ware crops: These crop plants are grown for temporary storing as intact in ware-houses for future use or sale, for instance potato, beet, carrot. They are in general, fleshy fruits or roots of higher values.

17) Truck crops: These crop plants are grown to market fresh, for instance lady's finger, brinjal, berseem, spinach, basella, radish.

18) Cole crops: These crop plants are essentially cold weather crops belonging to the Cruciferae family capable of withstanding considerable frost such as cabbage, cauliflower and brussels sprouts. Colewarts is the ancestor of wild cabbage (cliff cabbage) from which the name cole has been derived

19) Silage crops: These crop plants are grown to preserve in pits (silo pits) in a succulent condition by a process of natural fermentation or acidification for feeding livestock during lean months or off seasons, for instance maize, cow-pea, *jowar, bajra,* berseem.

20) Soiling crops: These crop plants are grown to harvest while they are still green and fed fresh to livestock in stalls, for instance berseem, oats, cow-pea, napier, rice bean, teosinte.

21) Ley crops: These consist of any crop or combination of

crop plants that is grown for grazing or harvesting for immediate or future feeding to livestock, for instance marvel grass+
berseem, Dinanath grass+cow-pea, berseem+mustard.

22) Green manuring crops: These crop plants are grown
to be incorporated into the soil fresh to increase the fertility of
the soil. Green manuring may be (a) green leaf manuring or (b)
green manuring *in situ*, such as *dhaincha*, sunnhemp, *Glyricidia*,
subabool.

23) Mixed crops: These consist of two or more crops that
are grown simultaneously in the same field without preserving
their identity with respect to field area. Seeds of these crops
may be mixed together before sowing and broadcasted irregularly or drilled in rows or may be sown at the same time and
grown with the same management practices. They may be harvested together or separately, for instance peas+*rai*, wheat+
mustard, gram+safflower, maize+cow-pea, *jowar*+cow-pea,
linseed+*toria*, berseem+mustard.

24) Intercrops: These consist of two or more crops that
are grown simultaneously in alternate rows in the same field.
The crops are not necessarily sown at exactly the same time and
their harvest times may be quite different but they are usually
simultaneous for a significant part of their growing periods.
These crops are grown with the same agronomic practices without separate identities except for the respective rows. Here there
is one main crop while the others are subsidiary crops, for
instance sesame+black gram, pigeon-pea+groundnut, sorghum
+pigeon-pea, wheat+mustard, rice+pigeon-pea, berseem+
mustard.

25) Companion crops: In intercrop situations when subsidiary crops are usually of a shorter duration in a long duration
main crop i.e., the main crop gets the company of a short duration crop for a certain period at the early stages of its growth,
for instance potato, onion and spinach in autumn planted sugar
cane, basella, lady's finger and amaranths in spring planted
sugar-cane; radish in potato; cow-pea, rice bean, green gram in
napier.

26) Parallel crops: These consist of two or more crops
which are grown simultaneously but do not have inter-competi-
tive effects i.e., they are parallel with respect to competition, for
instance maize+cow-pea, sorghum+pigeon-pea.

27) Mono or whole or sole or pure crops: These crop plants are grown as pure or solid stands, for instance transplanted rice, jute, tobacco, oats and rice bean.

28) Plant crop or first harvest crop or stub crop or first cycle crop: In perennial or multicut crop plants this refers to the first harvest after sowing or planting the crop, for instance in sugar cane, napier, cotton, pigeon-pea, berseem, oats, para grass, Guinea grass and marvel grass.

29) Ratoon crop or stubble crop or coppice or providence or regenerated or second or third-cycle crop: In perennial or multicut crop plants this refers to the subsequent harvests taken from the regrowth of the root stocks, stubble and stumps, after the first harvest such as for sugar cane, napier, berseem, rice, *jowar*, *bajra*, oats, Guinea grass and rice bean. Such crops reduce the cost of cultivation, do not need seed and sowing, reduce the crop cycle period, produce a higher yield per unit time and often require less input.

30) Seed crops: These crop plants are grown for seed production (generative yield), for instance the seed crop of jute, cabbage, cauliflower, potato, tobacco, oats, berseem, cow-pea and maize. Such crops require special care and management to maintain the satisfactory genetic identity and purity of the crop variety (agro-ecotype) and a healthy crop.

31) Leaf crops: These crop plants are grown to harvest leaves for economic yield, for instance tobacco, sisal, spinach, basella and greens.

32) Fuel crops: These crop plants are grown to obtain fire wood or solid fuel as a by-product along with their economic yield, for instance jute, sugar cane, pigeon-pea, cotton, mustard and sesame.

33) Energy crops: These crop plants are grown to obtain liquid energy such as ethanol and alcohol, for instance sugar cane, potato, maize and tapioca which are used for fermentation and distillation.

34) Outer or border crops: Crop plants that are grown on the border areas of the plot or field of another crop, for instance safflower on the border of potato; radish or barley or onion on the border of cabbage or cauliflower plots; amaranths, lady's finger and basella in the outer ridges of cucumber, sweet

gourd, bitter gourd, ridged gourd plots; *mesta* in the outer ridges of sugar-cane.

35) Riparian crops: These crop plants are grown along irrigation and drainage channels or water bodies such as water-bind weed (*kalmi sak*), pepperwort (*sushni sak*), para grass, *dhaincha*, *Bhabhar* or *Babui* (*Elaliopsis binate*). They help to protect the soil from erosion.

36) Avenue crops: These crop plants are grown along the farm roads and fences such as pigeon-pea, sisal, *Glyricidia* and *Tephrosia*.

37) Contour crops: These crop plants are grown on or along the contour lines to protect the land from erosion, for instance marvel grass, Dinanath grass, *Setaria* and anjan grass.

38) Strip crops: They are of six types:

a) Erosion permitting crops: These crops are grown to permit soil erosion in large sloping (where vertical distance is more as compared to horizontal distance, say, more than ten per cent) lands which require land shaping for successful arable farming. Such crops have lower foliage cover and holding capacity of roots to surface soil, for instance *jowar*, *bajra* and maize. They are sown along the slope in strips.

b) Erosion restricting or resisting crops: These crop plants are grown to restrict soil erosion. They are grown across the slope in strips alternately with erosion permitting strip crops to introduce land shaping by cultural means, for instance marvel grass, groundnut, *Atylosia*, *kudzu*, *shyama*, horse gram and kidney bean.

c) Contour strip crops: These crop plants are grown in strips of suitable width across the slopes (bi- or multi-directional) on the level or contour lines alternating with erosion restricting crops particularly where land shaping is not needed i.e., in low gradient grounds, for instance *jowar*, pigeon-pea, groundnut and marvel grass.

d) Field strip crops: The crop plants that are grown on farm lands in more or less parallel strips across fairly uniform slopes but not on exact contours, for instance maize, pigeon-pea, *guar*, *jowar*, *bajra*, cow-pea, sunn-hemp, *mesta*, green gram, sesame.

e) Wind strip crops: The crop plants (mainly tail crops such as *jowar*, *bajra*, maize, pigeon-pea, safflower and ustard that are grown with low crops such as potato, peas, groundnut, black gram and green gram in alternately arranged straight and long but relatively narrow, parallel strips laid right across the direction of the prevailing wind regardless of the land contour. Sometimes bi-directional rows of wind strip crops are also sown to combat wind from any direction.

f) Permanent buffer strip crops: These are crop plants such as permanent legume, grass or brush or shrubs that are grown on a permanent or temporary basis on the strips that are established to take care of critically steep or highly eroded slopes in fields under contour strip cropping, for instance subabool, babul, *Cassia, Prosopis, Glyricidia, Tephrosia, sar (Saccharum munjo)*, Guinea grass, *Panicum repens*, lemon grass, *Cymbopogon naraus, C. martini, Andropogon squarrosus, Imperata cylindrica* (thatch grass).

39) Augmenting crops: When subcrops are sown to supplement the yield of the main crops, the subcrops are known as augmenting crops, for instance Japanese mustard with berseem, Chinese cabbage with mustard. Here the mustard or cabbage helps in getting a higher yield of fodder in spite of the fact that berseem gives a poor yield in the first cutting.

40) Alley crops: When arable crops are grown in alleys formed by trees or shrubs, established mainly to hasten soil fertility restoration, enhance soil productivity and reduce soil erosion they are known as alley crops. Such crops should have slight shade tolerance and should be non-trailing, for instance sweet potato, black gram, turmeric and ginger in between the rows of Eucalyptus, Subabool, and *Cassia*.

CHAPTER 4

Seeds

Introduction

Seeds may be defined in many ways: (1) a seed is a fertilised ripened ovule; (2) the part of a flowering plant that contains- the embryo and develops into a new plant if sown; (3) a seed is a young plant packed and ready for transport to wherever it may be wanted to start growing; (4) anatomically a seed is an embryo plant consisting of a rudimentary stem and root together with a supply of food sufficient to establish a plant in a new location, all encased in a protective coat; (5) physiologically, a seed is essentially a meristematic axis with storage organs covered with membranes and the outer one is the seed-coat; (6) a seed is a mature ovule in a dormant state, of a plant where the metabolic, synthetic and morphogenetic activities are suspended; (7) a seed is a unit of reproduction of flowering plants in general; (8) agronomically, a seed or seed material or propagule is the living organ(s) of the crop in rudimentary from that is used for propagation, or in other words, any part of the crop from which a new crop will grow.

A seed is the initial capital of a plant and has the functions of perpetuation, multiplication and dissemination, which are vital to the plant species lacking vegetative propagation. All flowering plants may not bear seeds, all seed-bearing plants may not produce new plants from their seeds and all seeds which bear on mother plant may not produce new plants. Some plants may bear seeds in a specific season and at a specific site. Seeds, as dormant and minute structures may survive where and when mother plants cannot. A plant may bear seeds once or many times in its life. Seeds in some plants may ripen synchronously while in others they may not. Some crop plants may bear seeds or fruits in their apex as in rice or wheat, below the apex as in maize, or even below the ground as in groundnut. A crop plant can produce seed varying from a few dozen as in peas and lentil to thousands or even lakhs as in tobacco. The weight of each seed varies from less than a milligram as in tobacco to a few grams as in beans or castor. Seeds may be formed in single seeded (rice) or multiseeded (tomato) fruits which may be dry or fleshy. The size, shape and weight of individual seeds is extremely variable. Sizes depend mostly on the form of the ovary, the location of the seed and fruit on the mother plant, the conditions under which the mother plant is growing, specially during the developmental stage of the seed and the genetic character of the plant.

Seeds arise as a result of self or cross fertilisation, apomixis or non-sexual processes (as in *Dandelion*). Soil, as a source of infinite life, is a natural site for seed population reserves which are most important both for the plant not to become extinct, particularly when the mother plant is not capable of surviving and for the animals who live on these plants. Seeds may remain viable both in soil and in the storehouse for years. This is most important for man and his agriculture. It has been found that the population of germinable weed seeds in the plough layer of cropped land is three and a quarter thousand per square metre (De et al. 1986).

A seed consists of three main parts: (1) embryo which in due course gives rise to the new plant; (2) endosperm or the storage tissues which contain the substances that nourish the embryo during its development prior to and for sometime after germination; and (3) seed-coat or a protective covering which shields the embryo and endosperm.

Seeds may be monoembryonic (maize) or polyembryonic (sometimes in citrus, mango, etc.). Seeds may be dormant or non-dormant and may germinate by the following process hypogeal, epigeal and vivipary.

In some seeds the endosperm is greatly reduced and food materials are reserved in the cotyledons and the endosperm of different seeds are rich in a variety of storage materials e.g. starch (as in cereals), hemicellulose (as in palm-seeds, coffee etc.), protein (as in pulses) and oil (as in oil seeds).

Seed-coats consist of two layers, united or free (as in grams and castor) or with membranous layers made of the seed-coat and wall of the grain that are fused together (as in rice and maize) and is provided with micropyle or caruncle (the out-growth formed at the micropyle e.g., in castor). Seed-coats of cotton and silk cotton have seed fibres as appendages for seed dispersal. These appendages have a fibre valuable for humans.

Hard seed-coats prevent the attack of insect-pests and pathogens during storage, compared to soft and thin seed-coats. This structure or combination of structures functions not only as a mechanical protection for the seed, but is also important in many cases as a germinating regulator by controlling moisture and gas exchange between the seed and its environment. In some cases it also carries chemical substances which are active germination inhibitors.

Germination is the process by which the embryo wakes from its dormant state, emerges and develops from the seed-embryo of those essential structures which indicate its ability to produce a normal plant under favourable conditions. Germination is also a phenomenon of expression of life, a symbol of awaken-ing from sleep and from the lifelessness of seeds. In germi-nation seeds begin their heterotrophic life from the autotrophic stage and resume growth. For crop seeds there should be absolute germination and the production of synchro-nous and healthy seedlings. Agronomically, germination means the capacity of seeds to give rise to normal sprouts within a definite period fixed for each crop under optimum field conditions.

Primary environmental factors which influence germination in field conditions are water, oxygen, temperature, light, soil structure and micro-organisms. The soil structure and micro-flora are constantly in a state of flux as a result of the changes in temperature and in the oxygen and water supply. Tempera-ture, water and oxygen supply change with time; these environ-mental conditions and changes can vary greatly from one locality in the soil to another. These changing and variable environmental conditions act on seeds.

Although the soil microflora must be considered as a dyna-mic, ever-changing part of the seed's environment, the seed cannot be considered as a static object waiting to be attacked. Seeds lose large amounts of such organic nutrients such as carbohydrates, aminoacids and coenzymes to the medium in which they germinate. These nutrients alter the activity of the soil microflora and may determine the population density of selected organisms in the vicinity of the seed. This may influence the pathogenicity of such organisms as *Pythium* which causes pre-emergence damping off (Pollock, 1972).

When a dry seed is placed in a moist environment, it absorbs water in three stages: (1) an initial period of rapid uptake; (2) a lag period in which little water is absorbed; and (3) a second uptake stage which is associated with embryo growth. This uptake is considered to be the result of physical absorption of water by colloidal materials in seeds and thus it occurs in both living and dead seeds. In the hydration process the seed absorbs water and swells up. The seed-coat becomes more permeable to oxygen and carbon dioxide. The seed-coat is often ruptured. With increased hydration of the cells during germination the enzymes already present in the seeds become activated and the zymogens are changed into enzymes. This hydration may result in specific enzymic activities that hydrolyse storage materials and synthesise new substances for growth and development. These reactions are programmed by RNA and are controlled by various enzymes which may respond to environmental conditions. Water sensitivity, oxygen availability and light quality influence the rate and direction of the reactions. Energy for these reactions is supplied through respiration. The normal progress of seed metabolism during germination results in the growth of the embryo into a vigorous seedling, capable of establishing itself in the given environment and of developing further into a normal plant. Natural deterioration of the seeds or injury through external factors may cause physiological disorders that disrupt germination and seedling growth. These biochemical changes associated with weakened germinability and loss of vigour are characterised by a decline in metabolic activities (lower respiration, reduced germination and slower growth rate). There is greater total activity of hydrolytic enzymes—phytase, protease and phosphatases, and reduced activity of respiratory enzymes—catalase, peroxidase, dehydrogenase, cytochrome oxidase and glutamic acid decarboxylase. Cell membrane permeability is increased and leakage of sugars, amino acids, inorganic solutes and fatty acids results (Anderson, 1970; Maguire, 1972).

It is an established fact that field germination is always lower than the laboratory germination tests of seeds. Though the mortality of seedlings after germination in the field frequently depends upon entomological, phytopathological, edaphological and meteorological factors as well as toxic effects of organic secretions and applied chemicals including herbicides, pesticides, fertilisers, etc., yet there is the influence of the storage period (ageing) of the seed. Seed quality is determined in the period of its formation (setting) and subsequently modified by the environment including the man-made treatment it experiences, until it produces a new plant.

Emergence is the process of the coming out of the tender seedling from the seed through a column of soil or similar substratum in the field. A seed may germinate but may not be able to emerge because of various reasons. The important factors affecting the emergence of seedlings are (1) deep sowing or depth of soil cover over the seed; (2) inadequate or excess soil moisture; (3) poor aeration; (4) higher soil compaction or impermeable layer of soil or other materials; (5) low temperature; (6) rapid desiccation of the soil (7) longer time period; (8) injurious level of salt content; (9) poor seed capacity; and (10) detrimental physical, chemical and biological soil conditions.

Under favourable conditions most of the crop seedlings emerge within three to five days while Guinea grass, tobacco, onion, coriander, spinach, carrot, chillies and brinjal emerge within six to ten days after sowing under normal conditions.

4.1 CHARACTERISTICS OF A GOOD OR QUALITY SEED

Quality seed ensures a uniform crop establishment with uniform vigour and population of seedlings per unit area of the field. The weaker and late-emerging seedlings are poor in vigour, subsequent growth and productivity both in quality and quantity as compared to vigorous and early emerging ones. Therefore, the selection of good seeds is of prime importance for raising crops and reaping a rich harvest. A quality seed should possess the following characteristics.

a) The seeds should be of adaptable crop variety or

hybrid and their duration should be according to the agro-climate and cropping systems of the locality;

b) The seeds should be pure (true to type), with high sowing quality (viable seeds that germinate rapidly to give rise to strong and vigorous seedlings under normal conditions), good yield potential, evenness in growth pattern and maturity and should meet the purpose of cultivation;

c) The seeds should be free from seed-borne diseases and physiological disorders due to deficiency of plant nutrients or blonding or bleaching due to adverse weather conditions during seed development or in the post-maturity period respectively;

d) The seed should be large (according to the variety), plump, bold, uniform in size, shape, colour, texture, development and of proper test weight (weight of 1000, bold seeds);

e) seed should be clean and free from inert matter such as crushed rock, dirt, grit, chaff and trash, as well as from adhered soil, sticky substances such as pulp or the juice of the fruit;

f) The seed should be free from noxious or objectionable or satellite weed seeds;

g) The seed should be free from insects, insect eggs, disease spores etc., in or on the seeds;

h) The seed should be whole, not broken, crushed, peeled off, half filled, half rotten or affected with damp;

i) The seed should be as fresh as possible or of the proper age;

j) The seed should contain a required amount of moisture.

Seed quality index is the vigour of seed germination i.e., the percentage of germinated seeds with normal seedlings to their total number on that day for which germination has been tentatively fixed.

Sowing of poor seeds results in irregular bare patches in the field and it becomes beyond the compensation capacity of the plants to meet the adequate plant density uniformly distributed per unit area. Thin crop density is associated with poor quality of produce, more weed growth and low produce from the crop. Irregular and thin crop density is usually beyond the scope of correction because of late detection and may require resowing which in turn causes delay, wastage of seed, soil moisture, nutrients, cultivation practices, expense, energy, effort and enthusiasm.

There is a close relationship beetween seed quality and its germination. A significant role is played not only by seed size and weight, but also by the place of seed setting on the plant. The location of the seeds in the cobs, earheads, panicles and capsules as well as the location- of fruiting bodies such as capsules and pods on the mother plant are the most important factors in the formation of different quality seeds. This is related to the different effects of external conditions such as the intensity and quality of light, duration of illumination and effect of temperature and moisture on seed formation, as well as to the differences in supply of nutritional matter and other substances. A wheat seed weighing 35 mg produces seeds with weights ranging from 24 to 35 mg. Seedling germination vigour of both root and shoot also varies among seeds produced in different locations of the plant, though it is partially controlled by the management practices followed and other prevailing conditions.

Seedling emergence is either delayed or preponed in seeds collected from different locations of the plant. Wheat seedlings appear three to four days later in seeds taken from the upper part of the spike, compared with seeds obtained from the middle part. The seed of the fodder bean collected from the lower tier which has formed earlier, germinates more rapidly than that from the upper tier. In rice, the mother tiller produces the highest number with heavier seeds compared with primary and secondary tillers. Seeds formed at the tip of the panicle are heavier than at the middle and base of the panicle and the heavier seeds germinate early with vigorous seedlings.

In general, larger seeds produce seedlings one to two days earlier. Larger and heavier seeds produce stronger and healthier seedlings capable of maintaining subsequent growth vigorously and also produce a higher yield than the smaller and lighter seeds. Fully developed fruits or fruiting bodies such as cob, spike and bolls, produce larger and heavier seeds which have good potential as quality seeds.

The thicker the seeds the higher their sowing quality. The larger seeds absorb water at a lower rate and germinate with lower soil moisture. Sometimes the germination of smaller seeds of wheat, oat, and barley is higher than that of larger seeds. This is probably due to the more rapid swelling of smaller seeds in which the ratio of the surface area to volume is greater than that in larger seeds. Those seeds which swell more intensly do not always germinate more rapidly.

Larger seeds have a higher germination per cent at all depths than smaller seeds. Plants from larger seeds produce more tillers or branches and accumulate a larger quantity of dry matter. Larger seeds on sowing can withstand unfavour-able soil moisture conditions over a long period of time, while smaller seeds under the same conditions deplete their reserves in the process of respiration and physiological rearrangement. When conditions improve, the seedlings emerging from these are either unable to change over to autotrophic feeding and die or their growth and development are strongly retarded.

By grading seeds by the flotation method in water with brine solution, it is possible to eliminate immature, injured and infect-ed seeds and, as a result, not only is germination improved but the underground yield significantly increases. Better-filled, larger seeds transfer more nitrogen from the endosperm to the embryo after sowing, than do small seeds. The sowing quality of the seed is reduced not only through its morphological peculiarities, but also by the chemical composition and physio-logical stages of the endosperm and embryo. Biochemical vari-ability does not depend upon the period of harvesting alone, nor upon the seeds in different stages of physiological maturity. This is frequently observed in fully mature seeds forming in different parts of the generative organs. In the panicle of oats there are three categories of seeds, large, stub and slim. The protein content is at the minimum in wheat grains developed from the third flower of the glumes. Seeds formed in the ray florets of the capitulum of the sunflower are larger and heavier than that of seeds formed in the disc florets. In safflower the seeds of primary anthodia have a greater seed weight as com-pared with seeds formed at the tertiary.

Grains of typical colour of the variety of wheat have an earlier and higher per cent of germination than red grained seeds. The germination of red seeds of white colour is lower than that of yellow. In linseed, the sowing quality of brown seed is better than that of yellow. The plants developing from brown seeds are slightly taller and produce larger seeds with a higher degree of vitality.

The germination of seeds depends upon their natural peculiarities and biological conditions. Germination is fairly high in some seeds even in the period of their formation, while in others it reaches a reasonable level only at the time of ripeness. For the purpose of seed production, sunflower, mustard and flax can be harvested 15 to 20 days earlier and coriander ten days earlier than for market purposes.

Some crop seeds collected at the wax stage have a better sowing quality compared to mature ones and the dead-ripe stage. This is associated with the completion of structural formation, accumulation of vitally important compounds and presence of inhibitors in them.

External environmental conditions exert a direct as well as an indirect influence on seeds. A significant role in the supply of vitally important materials to developing seeds is played by the redistribution of metabolites between the vegetative and reproductive organs and between the seeds forming in the spikes and capsules.

During grain formation and harvest, cold, cloudy and rainy weather prevents the maturation of seeds. Seed formation under conditions of stress results in reduced weight as well as in a reduction in the number of seeds.

Plant nutrition has a significant effect on seed quality. Foliar feeding of nitrogen in time (at the grain filling stage) to grain crops, improves germination, initial growth and growth vigour of seedlings and weakens dormancy.

A normal moisture supply to formed seeds is an essential condition for their high sowing quality. Wrinkling in seeds, which results from rapid moisture loss, for instance in the case of drought during seed maturation or with early harvesting of the crop, is one of the indicators of low seed quality. Due to rainy weather the moisture content of grain is quite high at the time of harvesting. Any delay in drying the seeds leads to a significant loss in their sowing quality. On the contrary, under the conditions of high air temperature and lower rainfall, the duration of post-harvest maturation of seeds is reduced.

With the application of chemical ripener e.g., magnesium. chlorate at the rate of 20 kg/ha 40 days after the flowering of the sunflower, seed drying while in the plant was satisfactory. With this chemical, the moisture content of the seeds is twice reduced without affecting the sowing quality. The application of defoliants such as calcium cynamide and sodium pentachlormate, lowers the seed moisture content at harvest time, of lupine and other legume crops. In this case seed maturation significantly improved, its growth vigour increased and its germination per cent and 1000 seed weight were higher. Spraying legumes with reglon (at 1.5-3.0 lit/ha) three to five days before harvesting led to an increase in seed harvest, the germination of which was normal. This was also observed when treating legumes with magnesium chlorate, butafen and other desiccants. The defoliation of cotton with magnesium chlorate and butyphos at the optimum time resulted in an increased accumulation of nitrogenous substances in the seeds and increased germination.

Seeds (grains) of many of the modern rice varieties sprout while they are on the mother plant when the ripe crop is exposed to water in cyclonic weather conditions. To prevent the sprouting of seeds in such conditions, submerging rice sheaves and immersing or dipping the panicles in five per cent salt solution and heaping the sheaves as per convenience are recommended. The seed germination is prevented temporarily for ten to twelve days. There would not be any discolouration of grains or mould formation for ten to twelve days (AICRIP, 1984).

For many types of seeds additional time is required to complete the post-harvest maturation process. Seed quality is increased with storage for one to two years. With a more prolonged dry storage, seed germination falls sharply and in some cases no germination occurs. Many crops pass through a period of post-harvest maturation when they are left in bundles for a week or so in the field or in a closed space for curing. Many seeds need no additional maturation and with storage their chemical values and sowing qualities deteriorate. Storage conditions namely, dry storage can give a shorter period of longevity but moist and cold storage particularly cryogenic storage produce longevity for centuries.

Seed viability in the course of storage is greatly influenced by its hereditary character. In general, hybrid seeds of tomato have a higher percentage of germination after a prolonged period of storage compared with normal seeds. There is also varietal difference with respect to viability, germination rate and vigour after prolonged storage. Seed fractions of large and small seeds of a single variety retain their germination capacity after storage to a different extent.

The location and time of seed setting on the plant are significant with respect to germination capacity. Seeds collected from earlier formed capsules bear a higher germination capacity than in later formed capsules even after a considerable period of storage.

Storage temperature, moisture and duration have a considerable effect on seed germination. When seeds are piled in the store, variation in seed quality occurs according to location of the seed in the pile. Seeds located at the bottom lose their viability more rapidly than seeds located at a shallower depth or at the surface. This is particularly due to the difference in ambient temperature experienced by the seeds.

Under conditions of low temperature not only is the consumption of reserve material reduced through a reduction in enzymatic activity t also there is an accumulation of inhibitory substances for germination. In such conditions the structural formation of the cells, which ensures normal seed germination, is better preserved.

The germination capacity of seeds is lost very rapidly with storage under conditions of a high moisture content and a high temperature. Such conditions influence the intensity of seed respiration and reduce dry matter.

During storage the intensity of respiration drops and the activity of dehydrogenase is reduced. The composition of many dehydrogenases includes vitamins such as PP, B, B_2 and others. The vitamin content of a seed with a low germination capacity is significantly less than that of normally germinating seeds. The loss of seed germination is related to the toxic effect of the high concentration of lactic acid on embryonic tissues and a drop of other acids, for instance citric, malic, tartaric and succinic acids. The quantity of free fatty acids increases in oil seeds after prolonged storage. This is the result of hydrolysis in the lipid part of the seed by lipase. During storage the iodine quantity and tocopherol content decrease, while the non-saponifiable substances increase. In the process of storage, glyceride of the oil changes and the quantities of other substances in the oil also undergo a change. All the factors which facilitate oxidation and hydrolysis in the lipid fraction also facilitate a reduction in seed germination. A large quantity of glyceride in the highly unsaturated fatty acids, linoleic and linolenic causes a more intensive oxidation of the lipid fraction and a more rapid reduction in seed viability. The quantity of tocopherol is also reduced in these seeds. Reduction of tocopherol, which is a natural antioxidant, expedites the oxidation of carotinoides and fatty acids in the embryo, which may be one of the causes for a reduction in seed viability.

During storage the carbohydrate content of seeds also changes significantly. Sucrose, raffinose and stachyose are present in different quantities in different seeds. The consumption of sugar for respiration as well as for the formation of organic acids and other vitally important compounds, leads to a reduction in its quantity. After a prolonged period of storage starch totally disappears in the seeds, and the quantity of hemicellulose and other matters that could have been converted into sucrose are sharply reduced.

The quantity of nitrogenous substances in seeds also changes significantly. The protein content in many cereal seeds is reduced during storage. A slow disintegration of protein and the accumulation of asparagine, aspartic acid, serine, glutamic acid, cystine and other amino acids in the embryos of cotton seeds are observed in the process of their storage. The asparatic and glutamic acid content increase twice compared to freshly harvested seeds.

The quantitative composition of amino-acids also changes. Leucine and phenylalanine are absent in fresh seeds, while in seeds which have passed through a dormant period these amino-acids accumulate significantly. Prolonged storage results in the degradation of protein or its excessive amination. The reduction in the viability of seeds is associated with the coagulation of protein in the embryo.

A significant rearrangement of chromosomes is also observed with the prolonged storage of seeds. During storage inhibitors accumulate in some seeds in their different parts.

Germination is also reduced during storage as a result of the action of microflora, micro and macrofauna including insects and rodents which develop vigorously under conditions of high humidity and high temperatures.

Fungi develop intensively on the embryo rather than on the endosperm. A slight injury to the embryo leads to a sharp reduction in seed germination. During seed storage the most significant disruption in metabolism is observed in the embryo, while in the endosperm disruption occurs to a lesser extent. The sequence of death in the individual parts of seeds occurs in the course of prolonged storage. In maize seeds, the scutellum dies first, followed by the radicle and plumule.

Abnormal seed germination (those which do not show the capacity for continued development into normal plants when grown in good quality soil under favourable conditions of water supply, temperature and light) such as damaged, deformed, decayed or tree seedlings depends greatly on the mechanical damage that occurs at the time of harvesting, threshing, cleaning, grading or even transporting and sowing the seeds. Seeds are liable to severe damage when their moisture level is high. The physiological processes change markedly in injured seeds; the intensity and nature of these changes depends upon the type of injury. A micro-injury to the embryo of the dry or soaked seeds causes a sharp increase in the intensity of respiration. With damage to the seed-coat, the injured seed absorbs water and oxygen more intensively, which sometimes enhances the germination of such seeds but the life processes are subsequently retarded resulting in the reduction of growth vigour and germination.

Which part of the endosperm is injured is an important consideration. The removal of equal quantities of endosperm from the sides of the pappus causes an insignificant drop in laboratory germination and less suppression in the growth process compared to the removal of endosperm from the dorsal side near the embryo. The seedlings from such grains often lose their geotrophic and phototrophic orientation; the leaves are etiolated and become smaller in size. Injured endosperms also invite microbial infection resulting in decayed seedlings.

Damage to or death of the scutellum leads to a loss in seed germination. The scutellum is not only that part of the seed which connects the endosperm with the embryo, it is also the active biochemical system in which important metabolic processes take place, which play a significant role during seed germination.

4.2 TYPES OF PURE SEEDS

The seeds are evolved or identified and tested. If they are found to be good they are multiplied and released for commercial production. Four classes of pure seeds are recognised by the International Crop Improvement Association. They are:

a) *Breeder or Nucleus seed*: Breeder seed or vegetative propagative material is directly controlled by the originating or in certain cases the sponsoring plant breeder or institution, and provides for the initial or recurring increase of the foundation seed. This is the seed that is produced directly under the supervision of the plant breeder.

b) *Foundation seed*: This includes elite seed, would be seed stocks (seeds, tubers, corms, bulbs, plants or other propagating materials) that are so handled as to maintain *specific genetic identity and purity*, and that may be designated or distributed by representatives of an experimental station. Foundation seed is the source of all other certified seed classes, either directly or through registered seed. It is also known as the mother seed.

c) *Registered seed*: This is the progeny of the foundation or registered seed that is so handled as to maintain *satisfactory genetic identity and purity*, and that ·has been approved and certified by the certifying agency. This class of seed is of a quality suitable for the production of certified seed. It can be produced by the farmers and other growers under special contract with the certifying agency.

d) *Certified seed*: This is the progeny of the foundation, registered or certified seed that is so handled as to maintain *satisfactory genetic identity and purity*, and that is approved and certified by the certifying agency. This is seed designed for use by farmers in sowing/planting their commercial crops. Two classes of certified seeds are produced: they are F_1 and F_2; recertification is not permitted from F_2 generation of seeds.

The multiplication of seeds to obtain certified seed is done under ideal situations.

Separate tags are used to identify the various seed classes. The foundation tag is white, the registered tag is purple and the certified tag is blue.

4.3 OTHER TYPES OF SEEDS IN AGRONOMIC USE

a) *Improved seed*: The better seed substituted for one which is not so good with respect to genetic and physical factors, is an improved seed. Such seeds have at least ten to fifteen per cent more genetic potential and are resistant to pests and diseases, well adapted to agro-climatic conditions of the locality, high in response to better conditions of growth. They have a wide range of adaptability, tolerance to adverse conditions of environment such as drought, flood and frost. Their quality is acceptable to the local market and consumers. Such seeds are in general superior to land races (an ancient, local, usually heterogenous population. Traditional land races and cultivars have advantages over the subsistence farmers whose strategy is not so much for better performance in the best years as for adequate performance in the worst. Old cultivars and land races lag behind the yield but may excel in adaptability to such extreme conditions as drought, flood, heat and cold and in resistance to pests and diseases. Such land races are in general superior in quality).

b) *Hybrid seed*: Hybrid seed is the seed produced by hybridisation i.e., by crossing between two or more homozygous inbred lines to obtain a desirable type having high yield potential. Only the F_1 generation of hybrids is recommended for use as seed for commercial production. To obtain such F_1 hybrid seeds, parents are to be maintained and freshly bred each time particularly if the same vigour and known desired qualities are to be maintained.

Hybrid seeds may be the product of single or double cross or multiple cross.

c) *Composite seed*: Composite seeds are produced by inter-crossing a number of selected varieties by making germplasm complexes. Such composites possess the genetic potential for high levels of production and are comparatively more stable than hybrids. Thus they need not be replaced after F_1 generation for commercial cultivation.

d) *Mutant seed*: Mutant seeds are the seeds produced by mutation breeding with the help of mutagenic agents. Mutant seeds have greater potential in yield with superior quality and are stable compared to hybrid and composite seeds.

4.4 AGRONOMIC SIGNIFICANCES IF THE SEED IS NOT PURE

Modern agriculture demands uniformity of type, ripening, time, responses to agricultural practices and of the consumers' quality within a single field for efficient treatment, easy harvesting and good marketing. The processing industry demands large, uniform batches for ease and cheapness of processing and constant consumer qualit.

If the seed is not pure (a) it may be unsuitable for the tract or season or land situation; (b) it may cause difficulty in scheduling field operations and adopting suitable cultural practices as there will be differences in growth and development patterns, harvesting time and method, of different plant types in the same crop field; (c) it may be difficult to protect crop plants from insects, pests, diseases, weeds, climatic hazards etc.; (d) it may differ in quality and quantity of economic and biological yields; (e) it may not be suitable for multiplication or preservation as seed for future cropping; and (f) it may cause difficulty in grading, storage, consumption and marketing.

Vegetative propagules such as tubers, setts, stem cuttings and rhizomes do not deteriorate genetically as easily as seeds, particularly when the seeds are developed by cross pollination.

4.5 PROCUREMENT OF QUALITY SEED

Seed may be purchased or procured or home preserved after multiplication in the owner's field.

Certified or pure seeds of assured cultigen and cultivar should be procured from (a) selected growers, (b) a locality where the only variety is grown, (c) a reliable farmer, (d) Government seed multiplication farm, (e) Government sponsored organisations, (f) a registered, reliable, well reputed dealer.

4.6 REAL VALUE OF SEED

It is the per cent of germinability of per cent of purity of the seed lot of a crop plant:

$$RV = \frac{\text{Purity } \% \times \text{Germination } \%}{100}$$

and expressed in per cent.

This is essential to determine the actual quantity of seed that will be required to obtain a desirable plant density per unit area of land and to fix prices or to judge whether the seed can be used for sowing or not. Conditions affecting the real value of seeds are the method of production, the method of handling and the method of storage.

A crop can be grown for its quantitative and qualitative yield where it is adapted, but for seed production (generative yield) it may require a special site, environment, season, protection and care. For example, cabbage and cauliflower can be grown as vegetables in plains even with short bracing winters whereas seed production is easy in areas having a prolonged and severe winter as in Kalimpong. Seed production of jute favours rainless, low humid and high temperatures particularly during the phase of the seeds. Maharashtra state is the supplier of jute seeds for fibre production in eastern India.

From the beginning of cultivation until today the best plants, the best fruits or fruiting body of the plants, have been collected and preserved for seed with the greatest care. In modern days seed selection, extraction, processing, preservation and distribution have been established on a scientific basis instead of by mere intuition and tradition. In International Research

Institutions germ plasm banks have been establi.shed **as** genetic resource laboratories.

4.7 SEED DISTRIBUTION

Besides the selling of seeds of modern cultivars through the National Seed Corporation, the Tarai Seeds Development Corporation, the State Farm Corporation etc., small packages (minikits) of certified seeds of different cultivars are distributed free of cost and with some incentives through Panchayats. The selection by the farmer is mainly based on his interest in growing the crop in his field.

4.8 SEED DORMANCY

Dormancy is the arrested development and reversible rest period of plant organs, either of a seed or of any vegetative part. Dormancy is not an absolute block to germination but an adaptation to evade unfavourable environmental conditions for a per:od of time. This is mostly a plastic character of the seed. Failure or delaying of germination of a fully mature and viable seed of a species under conditions of moisture, temperature, light and oxygen supply which are normally favourable for the later stages of germination and growth of that species, is seed dormancy.

Practically all existing plant species, at least once during their life cycle produce specialised cells or multicellular bodies that exhibit dormancy. This may be to initiate an entirely new life cycle or new phase of the existing cycle. The formation of dormant structures is commonly associated with the suspension of metabolic, synthetic and morphogenetic activities that are associated with the minimum physiological activities and a minimum moisture content. During this period there is very poor or total suspension of respiration or rather anaerobic respiration with higher respiration quotient.

4.8.1 Importance of Seed Dormancy

In seed bearing plants the dormant seeds are much more resistant to unfavourable conditions than the mother plant and act as a solution to the periodic as well as non-periodic changes in the environment. Seeds under dormant conditions may easily be carried over a distance and passage of time from its site and time of production. Dormancy of weed seeds is a characteristic that enables weeds to persist for decades, even centuries, in the soil as serious infestations despite the soil disturbances that attend crop production. Many annual weeds with dormancy in their seeds germinate under a narrow range of environmental conditions. They are specialists in utilising opportunities under such conditions. Conversely, the generally rapid, predictable germination of non-dormant crop seeds over a wide range of temperatures and soil moistures, is an adaptation man utilises in ensuring the successful establishment and subsequent growth of crop plants.

In temperate zones, dormancy is a survival mechanism that prevents the fall germination of species that are not winter hardy. In tropical zones, many weed seeds have no dormancy and germinate soon after falling to the ground. Some weed species have periodicity of germination—they show a definite pattern of peaks of germination at regular intervals, other species have a less regular pattern and some species germinate freely throughout the year. These patterns are synonymous with the quasi-simultaneous, continuous and intermittent germination as suggested by Salisbury, 1961 under constant conditions, and could be the result of seed polymorphism. Dormancy is perhaps the single most important characteristic of weeds that enables them to persist and flourish (NAS, 1971).

Dormant seeds possess the highest potential for resuming growth and forming fresh individuals and are nature's general insurance for survival (perpetuation) of the species. This is particularly important for weed seeds in crop fields as some, but not all germinate at one time. Therefore dormancy is (a) a form of plant dispersal and multiplication; (b) a limit to the time at which seeds germinate for completion of their life cycle under favourable seasonal conditions; (c) the capability of extending the longevity of the seeds; (d) the means of survival with a high level of tolerance to adverse conditions by over summering and/or over wintering; to escape unseasonable growth, drought, frost, flood, fire, soil erosion, digestion in the alimentary duct of animals and birds as well as man-made modifica-

tions of the habitat or disruptive forces such as repeated cultivation, irrigation or drainage; (e) determining the location of
germination of seeds. This is particularly important for parasitic weeds which germinate with the stimuli from their hosts.
Obligate weeds which occur only in cultivated or disturbed land,
determine their site of germination. In some field crops where
the economic yield is that of the seed itself, dormancy helps
man to harvest, process, store and use as seed for future crop
and/or as food, feed, oil seed, etc. A limited dormancy in crop
seeds or kernels or seed like structures is a useful attribute that
prevents sprouting, when the mature crop or mature part of the
crop gets wet due to rain or flood or when harvesting operations
are delayed by wet weather.

4.8.2 Types of Dormancy

Based on the behaviour of seeds three categories of seed
dormancy have been recognised (Harper, 1957). They are:

a) *Innate* (*or primary or natural or inherent or physiological
or endogenous*) *dormancy:* Some seeds that are *born* dormant
show this type of dormancy when they are attached to the
parent plant. This may last for a few weeks to one or more
years. This is considered an inherent property to the mature
seed as it leaves the parent plant. The dormancy levels of seeds
depends on the weather conditions during their maturation on
the plant. Such dormancy prevents the seed from germinating
viviparously and also usually for some time after the ripe seed
is shed or harvested (Roberts, 1972 a). In some plants achloro-
phylous seeds bear innate dormancy.

Such dormancy may be due to the presence of some germination inhibitors, such as a physiologically immature embryo,
rudimentary embryo and mechanically resistant seed-coats preventing entry of water and or gases.

To break such dormancy seeds may require an exposure to
an after-ripening treatment, alternate freezing and thawing in
temperate regions, leaching of natural inhibitors or other
similar methods.

b) *Induced* (*or secondary*) *dormancy:* Some seeds may
achieve dormancy due to the interaction of the seed with
the environment for example, with warm, dry conditions.

Dormancy develops within the seed when a non-dormant seed is exposed to certain external conditions and continues even after the conditions are changed or the seeds are removed from them. When non-dormant imbibed seeds are buried deep in soil or any such medium and lack the required amount of light or oxygen or temperature or under conditions of excess carbon dioxide or a combination of all these, dormancy is induced. Such dormancy lasts until the seed experiences the condition necessary to break it (Fryer and Makepeace, 1977).

c) *Enforced (or environmental) dormancy:* Some seeds have dormancy *thrust* upon them due to conditions of deficient oxygen, excess carbon dioxide, cold temperature, ethylene etc., induced by man or in nature. In such dormancy, limitations of the environment or habitat prevent non-dormant viable seeds from germinating and germination occurs as soon as seeds are freed from the limiting conditions. When non-dormant seeds are stored under dry conditions, the seeds are forced to remain in a dormant state. Weed seeds buried deep in the soil are put under enforced dormancy and bringing seeds to the surface with a favourable environment may terminate enforced dormancy.

Such dormancy may last for a century or more.

Seeds from a single plant species may exhibit one or more types of dormancy or all three in succession over a period of time. Beside the above three types of dormancy some seeds show special types of dormancy. They are:

a) *Relative dormancy:* This is found in certain varieties of lettuce which germinate well at 20°C or below but which are dormant at high temperatures (thermal dormancy) and require light for germination (Toole et al. 1956).

b) *Epicotyl dormancy:* Some species of *Paenoy* showed epicotyl dormancy and a combination of root and epicotyl dormancy or double dormancy in *Sanguinaria* sp.

This can be overcome by moist storage at 1 to 10°C for two or three months after the seed has germinated to form a root.

c) *Multiple dormancy:* Seeds of *Trillium* sp. require the passage of one cold period for the after-ripening of the radicle and of the embryo; during the next summer, the radicle emerges and becomes established, but the epicotyl remains dormant until the passage of another cold period.

4.8.3 Mechanisms of Seed Dormancy

Seed dormancy is controlled by four relatively distinct developmental phases e.g., (a) inductive, (b) maintenance, (c) trigger, and (d) germination (Amen, 1968).

a) *Inductive phase:* Seed dormancy or the delayed germination character of the seed begins with an inductive phenomenon. Certain events during the maturation of seeds may lead to the onset of dormancy. These events may be triggered off environmentally by light and temperature, or chemically and they are termed photoinduction, thermoinduction and chromoinduction respectively.

Dormancy and the germination of seeds are regulated by a critical balance of inhibitor-promoter complexes. During seed maturation, this balance is shifted in favour of the inhibitor, imposing dormancy. This may happen either through curtailment of synthesis of promoter substances (for example gibberellic acid, cytokinin, kinetin, auxin, amino acid, vitamin, nucleic acid, enzyme, fusicoccin) or through a build up of inhibitory components (abscisic acid, coumarin and its derivatives, parasorbic acid, ammonia, phthalids, ferulic acid, caffin, caffeinic acid, protocatecholic acid, salicylic acid, 4-oxibenzoic, vanillic, 4-oxicorric acids, citral, benzaldehyde, cinnamic acid, aldehyde, scopoletin, protoanemonin dehydroacetic acid, cyanides, vitamin P, 2-thiouracyl, phosphone, phosphone-S) which do not reduce the viability of the seed nor produce any growth abnormalities in the seedling after germination.

Growth inhibitors may appear as antagonists of vitamins, gibberellins, auxins, aminoacids and other widely essential compounds. By activating these metabolites, inhibitors disrupt other reactions of metabolism as well. Growth inhibitors suppress the activity of enzymes which synthesize IAA and of enzymes containing SH groups, which leads to a change in the nature of the transformation of essential compounds. The suppression of nucleic acid synthesis is particularly significant (Ovcharov, 1977).

Such natural inhibitors are not confined to any particular

part of the seed and may even be found in structures covering the seed. Inhibitors of germination have been found in the pulp or juice of the fruit containing the seed, seed-coat, endosperm or embryo (Devlin, 1975), or by a direct antagonism among them. Abseisic acid (ABA) is antagonistic to gibberellin (GA). Endogenous growth promoters are in high concentration in the early maturation of seeds but they decrease with the cessation of embryo growth and ripening of fruit. Dormancy may also be induced by the formation of impermeable seed-coats which impose anaerobic conditions on the seed resulting in the build up of growth retardants (coat-imposed dormancy).

b) *Maintenance phase:* During the maintenance phase of dormancy the metabolic activity of the seed is very much reduced by blocks at the specific metabolic sites. These metabolic blocks are due to the presence of endogenous inhibitors (like ABA) which are either antagonistic with endogenous promoters (like GA) or interfere with their synthesis. The hormone promoters help in the synthesis and activation of hydrolytic enzymes (a-amylase, protease, lipase etc.). Thus, dormancy is due to lack of or inactivity of hydrolytic enzymes.

c) *Trigger phase:* In a dormant seed, germination is triggered off by a factor or an agent that elicits germination. This triggering agent should continue to be present for the completion of germination. This agent may be a photochemical one involving phytochrome-red light mechanism, thermochemical one involving stratification (i.e., exposure to moisture near freezing temperature), the inhibitor-removal mechanism involving scarification (i.e. scratching or rupturing or partial decaying of the seed-coats), and leaching of the inhibitor-promoter complex favouring the promoter. During this phase, the inhibitor is either removed or its effect is counteracted by activation or synthesis of the promoter.

d) *Germination phase:* This process involves the breaking of the rest period or the suspension of growth period or the dormancy of the seed and subsequent germination with a resumption of growth. It is under hormonal control and the naturally occurring hormones e.g., auxins, gibberellins and cytokinins function as germination agents via the inhibitor-promoter complexes. Of these gibberellins are the predominant germination agents early in the germination phase during the food reserve degradation stage. Cytokinins exert their influence later on the initiation of cell proliferation and expansion (Rao, 1983).

4.8.4 Forms of Dormancy

The germination of seeds may be blocked by the absence of some external factors considered necessary for this process to occur. However, some seeds do not germinate even though they are exposed to a favourable environment considered adequate for germination. This may be because of some internal factors. Such externally and internally regulated blocks are associated with peculiarities of structure or behaviour or formation of special situations in the seeds.

A: Properties of the seed-coats:

1) Physical impermeability of seed-coat to water:

Many seeds of legumes, for instance peas, gram and beans have seed-coats that are impermeable to water. The coat may be removed by the long leaching action of soil-water, the activities of soil animals, biochemical changes of seed-coats during ageing, activities of soil microbes at the microsites or inhibition through the hygroscopically activated valve in the hilum.

Some of the agro-techniques to break dormancy of such seeds are filing, rolling over sand or any abrasive substance and treating with commercial sulphuric acid.

2) Low permeability of seed-coats to gases:

Heteromorphic seeds produced in Asteraceae, Chenopodiaceae etc., do not permit gaseous exchange freely through their seed-coats.

The removal of seed-coats, exposure to 40°C or chloroform, acetone or alcohol and the drying and washing of seeds improve permeability to gases.

3) Mechanical resistance by the seed-coats:

Chromatin containing seed-coats of lettuce seeds show such resistance in absence of light. Some seeds of *Amaranthus* sp. also show such resistance.

Light treatment induces the formation of enzymes such as pectinase, cellulase etc., which degrade the seed-coats. Chemical or mechanical scarification may break dormancy in the seeds of *Amaranthus* sp.

4) Inhibitory action of substances within the seed-coats:

Most of the seeds of succulent fruits do not normally germinate in the maternal tissue. Testa of peach seeds contain ABA, an inhibitor. Seeds of *Acer* sp. contain an inhibitor other than ABA in their seed-coats. Some desert 'rain gauge' seeds contain large amounts of osmotic materials—the exposure of such seeds to ample amounts of rainwater (25 mm) elutes the osmotic potential.

A seed as a whole or different parts of it may contain or produce larger amounts of ABA, phenolics—specially flavonoides and may restrict germination.

Soaking and discarding water at regular intervals promotes leaching of inhibitors.

B: Under-development of the embryo:

1) Immature embryo:

After-ripening in dry storage or stratification may break such dormancy.

2) Physiological condition of the embryo itself:

a) Low-temperature requirement: Temperate seeds such as apple require chilling treatment for a considerable period to break their dormancy. Some seeds require treatment with $-2°C$.

b) Alternate temperature requirement: Varying temperatures between $5°C$ to $22°C$ favoured the germination of most species as compared to constant temperature (20 to $22°C$). In nature, the diurnal flux in temperature, potassium nitrate or red light (phytochrome system) treatment may nullify the varying temperature requirements.

c) High-temperature requirement: Seeds of oil-palm require as high as 50 to $60°C$ to break their dormancy.

d) Specific light requirement: Some seeds have an absolute requirement for light whereas in other seeds, exposure to light is inhibitory to germination and in still others, germination is associated with a photoperiodic response i.e., an alternation of light and dark periods.

Many seeds have been found to be light sensitive and they are only responsive to the stimulus when imbibed with water. The germination responses of the seeds *Lepidium virginicum*, *Nicotiana tabacum* and some varieties of *Lactuca sativa* are promoted by light (positively photoblastic) while others such as *Phacelia tenacetifolia* and *Nemophila insignis* are inhibited by light (negatively photoblastic). Some seeds have been shown to exhibit a true photoperiodic effect (Villiers, 1972).

Red light (660 m μ wl) promotes and blue, specially far-red light (735 m μ wl) inhibits germination. The effects of red and far-red light on dormancy have been found to be reversible. The last treatment determines the response of the seed.

Removal of seed-coats, exposure to specific photoperiod, imbibition period and higher temperatures may reduce the light requirement whereas lactone, coumarin, rhamulosin and patulin induce light requirement.

e) Specific water requirement: It acts to induce elution of growth inhibitors of the embryo itself, osmotic material, drying seeds of fleshy fruits to specific water content (e.g., tomato seeds to 60 per cent moisture) induces germination, permits gaseous exchange and the dispersal of ribosomes away from the endoplasmic reticulum.

f) Specific temperature requirement: Pre-chilling requirement may be nullified by ageing, light treatment and drying.

C: Presence or formation of germination inhibitors within the seed:

Seed inhibitors need to be washed out or detoxified by washing seeds, soaking and replacing water at regular intervals.

D: Seeds requiring stimulus from host plants:

Parasitic (root or shoot) weeds need some stimuli from their host plants for their germination. In the absence of the host species they remain dormant. Growing trap crop induces their germination and thereafter destruction before the formation of generative organs is the way to escape from them.

E: Seeds also remain dormant under the conditions of allelopathy:

Some plants or their parts either living or dead release some chemicals or substances which seriously affect the vital activities of plants or their organs of the same type or others. Leachate of some seeds even prevents the germination of other seeds of the same seed-lot. Washings, exudates or even products of decay prevent the germination of seeds.

The harmful effects of seed germination in soil may also be caused by substances located in the post-harvest residue of field crops. Extracts from the stubble and roots of wheat, rye, barley and maize with a dilution of 1 : 20, delay seed germination and growth in wheat and vegetable plants. Extracts from the underground and above ground parts of lucerne, clover, and vetch show a strong effect (Mitrieva and Pavlov, 1963). They noted that extracts prepared from young plants have a stronger inhibitory effect compared to extracts from immature plants. Compounds such as agropyren, tryptophan, arginine, alanine, leucine, glutamic acid, transcinnamic acid, phloridzin B-caryophyllene, bisabolin, prussic acid, angustion, phenolic compounds, saponin, glucobrassicin, ferulic acid, vanillic acid, quinones and hydrogen sulphide which are of biotic origin are found in soil or soil water and inhibit germination as well as root or stem growth.

4.9 VIABILITY OF SEEDS

Viability is the capacity of the seed to germinate or to maintain its normal life process or capability of growing or maintaining its living state. Sometimes when a seed is given suitable conditions for germination it fails to germinate. This inability may be due to dormancy which is a temporary condition in living seeds which can often be removed artificially, or to loss of viability—a degenerative change which is irreversible and generally considered to represent the death of the seed. A nonviable seed will be considered to be one which could not germinate when given near optimal conditions, even when it is nondormant. Thus a viable seed is one which can germinate under favourable conditions, provided any dormancy that may be present is removed (Roberts, 1972 b). A viable seed may be dormant or non-dormant. However, to a farmer, viability of a seed refers to the capacity of the seed to develop into an acceptable seedling even in the field while non-viability refers not only to dead or diseased seeds but also to those that produce or are expected to produce abnormal or diseased seedlings (Moore, 1972). Under optimum conditions a viable weed seed germinates within ten days (Timson, 1965) while a crop seed germinates within seven days (Takahashi, 1984).

The problems of seed viability are important in a number of applied fields. The factors affecting viability before harvest are a special concern of seed producers and the problems encountered after sowing are important to farmers, agronomists and horticulturists. The problems of maintaining viability in storage have always been an important concern of seedsmen including plant breeders (Roberts, 1972 a).

In addition to maintaining modern genetic resources, preventing genetic erosion by collecting and conserving the diverse genetic materials distributed in different natural centres, and upgrading the existing ones or evolving new ones complemented by reconstitution with the existing ones, are gaining overriding importance in the increase of the longevity of seeds or seed materials. The viability of the seed is the best estimate of planting value and determines the profit or loss of the farmer, trader and seed producer.

4.9.1 Onset of Viability

During the formation of the seed it acquires the capacity to germinate after it reaches the phase of ripening and prior to attaining the phase of maturation. Seeds of many species of Asteraceae become viable before they are fully developed. If plants of *Phalaris minor* are pulled out while in flower, they can set about 50 per cent viable seeds and if panicles of *Avena* sp. are cut or pulled at the milk stage of seed development, 30 per cent of the seeds are already viable but not yet dormant (Crafts and Robbins, 1973).

4.9.2 Longevity or Duration of Viability of Seeds

Longevity depends on storage conditions and the genetic character. A given storage condition may lengthen the life span of one species and shorten that of another. Even while in the parent plant, seeds deteriorate by weathering or by biotic factors. Hot, dry or sunny weather at the time of seed ripening is found to be favourable for good seeds. High soil moisture, humidity, rainfall, temperature, wind speed, desiccating wind, scorching sunlight, infestation of insects and diseases can adversely affect the seed before harvest. The state of maturity of seed crops when harvested is a major factor responsible for part of the variation in the viability and size of the seed. The method of harvesting, field drying, curing or threshing and weather conditions during the period have a great influence on seed viability. Cleaning and drying to storable conditions, if improperly done, further injure the seeds. If the seeds are held at a high moisture content awaiting cleaning at ambient high temperature conditions, a decline in viability will occur. Even the period between storage and sowing can be deleterious to the seeds. The time during transit in the retail store and in the farmers' shed is also critical. Proper drying and packaging in moisture proof containers and subsequent seasoning prior to sowing can minimise seed deterioration during this time. Aerial, edaphic and biotic environmental conditions after sowing also determine the viability of seeds for producing desirable seedlings.

Seeds of some weeds can remain viable in soil burial deep underneath for years without deterioration. Exposure to light, alternating temperatures, mechanical disturbances, microbial activity on the seed-coats and some other unknown factors either alone or in combination may serve as a stimulus to bring about the germination of these viable burial seeds.

On the basis of the life span of seeds under optimum conditions Quick (1961) defines three biological classes: (a) Microbiotic—less than three years; (b) Mesobiotic—three to fifteen years; (c) Macrobiotic—more than 15 years. Most of the cultivated species are either microbiotic or mesobiotic and weed and wild plant species are either mesobiotic or macrobiotic. This situation causes 'one year's seeding: seven years' weeding'.

A seed may be non-viable for various reasons. The most important ones are:

A. *Mechanical injury:* Hard coated, large and flat seeds with brittle embryos are susceptible to mechanical damage. Destructive forces encountered in harvesting, threshing, cleaning, grading and imperfect handling result in mechanical damage. Injury may be external or internal. Injury invites infection resulting in further deterioration of viability.

Mechanical damage may include detached seed structures,

breaks within structures, abnormally shaped structures, scar tissues, infections, restricted growth, uneven placement of cotyledons, unnatural shrinkage of cotyledons and splits or otherwise abnormally developed hypocotyls and primary roots.

The most intensive injuries reduce viability immediately. The small injuries beyond the level of healing often do not cause an immediate loss in viability, but become increasingly critical as ageing occurs. In seeds that are extremely dry and brittle, fracturing is the predominant type of injury. Bruising is usually prevalent in seeds with sufficient moisture for toughness.

If an initial injury is non-critical in that it has no immediate effect on viability, but is located on or near an essential part of an embryo structure, a seed can easily become non-viable with only a minor amount of additional deterioration. Injuries near the point of attachment of cotyledons to the embryonic axis, or on most other vital parts of the embryonic axis (radicle, epicotyl and plumules), usually bring about a more rapid loss of viability during storage than with injuries of similar size located in less important areas of the seed.

B. *Water damage*: Such damages are associated with rapid and uneven uptake or loss of water of a seed or a seed-lot during the process of handling or storage. Seeds processed at a higher moisture content usually tend to show a greater degree of damage than that of seeds with a lower moisture content. The high level of moisture associated with seeds that bruise, greatly hastens the rate of deterioration.

Water damage may cause 'natural crushing' (involvement of plasmolysis-deplasmolysis injuries). This may occur both internally and externally. There may be 'drought damage' of seeds which gives poor emergence or produces weakly-developed seedlings in field plantings (Moore, 1972).

C. *Biotic damage*: A variety of organisms cause damage to seeds in various forms and intensity. Almost every crop has one or more seed-borne disease which can be serious in their effect on seed viability.

D. *Ageing*: In general, low moisture content, cool temperatures and low oxygen tension increase the longevity of seeds in storage. Many seeds retain their viability longer when buried in moist soil than when kept in dry storage.

Results of ageing may be due to (1) the accumulation of deleterious products of metabolism; (2) wear and tear of organelles to become inefficient; (3) somatic mutation (Curtis, 1963).

4.9.3 Loss of Viability and Crop Yields

Loss of viability in seeds can affect the yield of a crop in two ways: (a) decreased germination can lead to a sub-optimal plant population per unit area; (b) poorer performance by the surviving plants. The first problem can be overcome by increasing seed rates but the necessary adjustment may be greater than that indicated by the germination test, because of the decreased vigour of the surviving seedlings, particularly in those species which are unable to compensate for a reduced density (by tillering or branching). In many seeds if there is a loss of viability, decreased yields are obtained from the plants produced from the surviving seeds.

The actual age (chronological age) of the seed is much less important than the environment in which it is held. Between slow ageing (long storage at a lower temperature) and rapid ageing (short period of storage at a higher temperature) of seeds, slow ageing results in a severe loss of yield even with a small loss of viability (Roberts, 1972 a).

With decreased growth rates at an early stage the crop becomes more susceptible to adverse conditions during emergence and early establishment and is expected to be more susceptible to soil capping, pests, diseases and weed competition.

Seed stock having a higher loss of viability should not be used for seed production as it would contain a large amount of genetic mutation.

4.10 SEED TREATMENT

Considering the value of seeds in crop production they are cared for even more intensively than foodgrains as seeds are to be kept viable and healthy under intact conditions from the time when they are on the mother plant until they produce new plants. For protecting the seeds both in the store and the field as well as to have a higher rate of germination, healthy seedling vigour and higher production they are processed through or dressed with mechanical, physical, chemical and biological procedures or substances. Seeds are treated to meet several objectives; they are:

a) to prevent the infestation of insect-pests and diseases as well as the deleterious effects of moisture and temperature both in the store and field by sun-drying, sieving, winnowing, flotation technique and dressing with inert dust such as sand, clay, ash, calcium phosphate, magnesium phosphate, camphor, napthalene, para dichloro benzene, sodium chloride, sodium carbonate, sodium bicarbonate, sodium barium flurosilicate, sodium sulphate, lime, barium oxide, aluminum oxide, mercuric compounds, dried neem leaves, tobacco, pyrethrum dust and derris root dust;

b) to make the seeds free from insect-pests or disease germs that have already infested or are yet to infest the seeds, by dipping, dressing or coating seeds with disinfecting chemicals in liquid, dust, slurry or paste form or by fumigation (application of toxic fumes/vapours/gases of a substance to the infested seeds for a certain period in a reasonably air tight container/chamber/fumatorium) with ethylene dibromide (EDB), ethylene dibromide and carbon tetrachloride, ethylene dichloride and carbon tetrachloride (EDCT), aluminum phosphide and by sun-drying, hot water and hot air treatment;

c) to enforce dormancy with ethylene, deficiency of oxygen, excess of carbon dioxide or a warm temperature;

d) to inoculate microbial culture (e.g., *Rhizobium*) around the seed;

e) to break dormancy whenever it is needed. This is done by scarification, stratification, treatment with heat, light and chemicals such as thiourea, potassium nitrate, gibberellin and cytokinin;

f) to induce higher germination per cent by pre-soaking seeds in warm water or sulphuric acid or placing them in a warm temperature; to promote early seedling vigour by dipping in fertiliser solution or coating or pelleting with nutrition slurry and early rooting and fruiting by IBA, GA etc.;

g) to harden the seeds by inducing tolerance to adverse weather and soil conditions such as drought, frost, cold, heat and salt by dipping in a solution of potassium nitrate, calcium chloride, boron, agrosan, cycocel, sodium chloride, sodium sulphate, magnesium sulphate, citric, fumaric, succinic, malic acid, purines, pyrimidines, caffein, uracil, xanthine, uridine diphosphate and by imposed drought i.e., partial imbibition followed by drying back;

h) to obtain polyploids, disease and pest resistance, improved plant type and plant behaviour including acclimitisation through genetic amelioration by treating seeds with mutagenic agents such as colchicine, X-ray or Y-ray;

i) to protect the seeds from herbicidal injury with herbicide antidotes and crop protectants, for instance with activated charcoal, NA (1,8-Napththalic anhydride), R 25788 (N, N' diallyl-2, 2 dichloroacetamide), CGA 43089, CGA 92194, Mon 4606, AD 67, CDAA, R-28725 (AD-2), R 29148, S 499, AC 222293, benzil hydrazone, carboxin (5, 6-dihydro-2-methyl-1, 4-oxathion-3-carboxanilide) and ferrous sulphate;

j) to prevent the viviparous germination of rice when still on the mother plant standing in the field or on sheaves by dipping panicles in five per cent salt solution; to prevent picking of seeds after sowing by frugivorous birds and rodents. This can be achieved by mixing seeds with malathion, BHC, dithane M-45 and other insecticides.

For treating seeds prior to storage under dry conditions organomercurials are preferred in view of their greater effectiveness over others. For treating seeds before sowing, seeds are dressed with disinfectants in dry, wet or slurry conditions. For treating leguminous seeds it is better to use non-mercurials, for instance captan and thiram.

Any seed treatment should not affect the viability and germinability of the seed. Therefore, treatment of seeds must be done with accuracy so that the objective is met with properly. Treated seeds should never be used for consumption or for feeding livestock or birds. It is also to be noted that no one recipe is applicable to all types of seeds.

CHAPTER 5

Sowing and Planting

Sowing is the placing of a specified quantity of seeds in the soil in the optimum position for germination and growth while planting is the putting of plant propagules (may be seeds, seedlings, cuttings, tubers, rhizomes, clones) into the ground to grow as crop plants.

In modern farming practice the establishment of plant populations at predetermined densities, without the intervention of gap filling or thinning, is sought in order to obtain at the lowest cost, maximum yields of produce of the precise quality demanded by the industries using such produce as their raw material.

Seeds are sown either directly on the field (seedbed) or in the nursery (nursery bed) where seedlings are raised and transplanted later.

5.1 DIRECT SEEDING

Direct seeding may be done by:
a) broadcasting,
b) dibbling, and
c) drilling.

a) *Broadcasting*: Broadcasting is the scattering or spreading of seeds on the soil which may or may not be incorporated into the soil or covered with soil or similar other materials. Broadcasting of seeds may be done by hand, mechanical spreader or aeroplane.

This method is suitable for close planted crops which do not require a specific geographic area for the optimum expression of their morphogenesis and development and are more plastic to compensate their morphological structures according to prevailing conditions. Under situations where the number of plants per unit area (per m^2) is more important than that of definite spacing from plant to plant, broadcasting is the usual method of sowing. Crop plants not requiring special types of cultural practices e.g. earthing up or picking may be sown by broadcasting. Crops such as upland and flooded rice, wheat, millets, mustard, jute, black gram, greens, fodder crops such as Dinanath grass, berseem, lucerne and *jowar*, spices such as coriander and cumin are generally sown by this method. For mixed cropping, broadcasting is the usual practice of sowing seeds.

Though it is an easy, quick and cheap method of seeding there are difficulties in uniform distribution, placing in optimum and uniform depth of soil and in providing soil cover and compaction. As all seeds are not placed in uniform depth there is no evenness of germination and uniformity in seedling vigour.

Broadcasting of seeds is done in dry, semi-dry and wet fields. For sowing in wet fields, seeds must be soaked with water for eight to twelve hours and incubated for a few hours so that the radicle just begins to emerge from the seed.

b) *Dibbling*: This is a method of putting a seed or a few seeds or seed materials in a hole or pit or pocket, made at predetermined spacing and depth with a dibble or planter or very often by hand.

This method is suitable for wider spaced planted crops requiring a specific geometric area for their canopy development or cultural practices such as weeding, earthing up and irrigation in furrows. Seeds may be dibbled in level fields or on ridges or the sides of the ridges or in localised pits or pockets, that form hills, clumps, rings, or stations distinctly separated from each other. For such methods of seeding the entire field need not be prepared for the seedbed, only the seeding zones.

This method is suitable for planting maize, cotton, castor, potato, groundnut, pigeon-pea, cow-pea, soybean, sunflower, sugar cane, sweet potato, onion, garlic, turmeric, ginger, gourds. napier and Guinea grass.

Dibbling is more laborious, time consuming and expensive compared with broadcasting but it requires less seeds and it gives rapid and uniform germination with good seedling vigour.

Seeds are sown in dry or semi-dry soil conditions and

manures and fertilisers including pesticides and soil conditioners may be applied simultaneously.

c) *Drilling*: Drilling is a practice of dropping seeds in furrows. Furrows of predetermind dimensions are made, seeds are dropped at a definite depth and distance, covered with soil and compacted. Sowing implements such as the seed drill or seed-cum-ferti drill are used. The use of the seeding funnel (*pora* and *kera*) and sowing behind the plough (plough planting, wheel track planting, strip planting) are also practised. After seeding the field may be levelled or ridged.

During seeding other operations such as the drilling of manures and fertilisers, pesticides and soil amendments are also done simultaneously.

To reduce the risk of the irregular dropping of seeds from the seed box due to the irregular shape and stickiness of the seeds, they are pelleted. Pelleting with synthetic covers which holds the chemicals such as mineral nutrients, growth regulators, rhizobium culture, pesticides, soil conditioners and other inert substances on seeds, improve seed germination and subsequent growth. Pelleting or coating is the process of enclosing a seed with a small quantity of exogenous material just large enough to produce a globular unit of standard size to facilitate precision drilling. It is essential that pellets should be stable until drilled but should then easily disintegrate to release the seed or do not affect the imbibition and subsequent establishment of seedlings.

During drilling seeds may be sown continuously or at regular intervals in the rows. These rows may be straight and parallel or staggered and irregular. Rows may be arranged as pairs (paired row planting), uni- or bi-directional (cross row planting), or with a skip row after a few rows (border method). Drilling may be adopted for both pure cropping and intercropping situations.

Drilling requires more time, energy and cost but it maintains a uniform plant population per unit area or per running metre in the rows. In addition the rows are set according to the requirements. Seeds are placed at uniform depth and covered and compacted uniformly in this method.

Crops such as wheat, mustard, upland rice, barley, *jowar*, black gram, green gram, Bengal gram, berseem, lucerne, jute, safflower, sesame, niger, taramira, cumin, anise and coriander are sown by drilling.

5.1.1 Depth of Sowing

Placing the seed lower than the true soil surface and the amount of soil above the seed (cover) determines the favourable conditions around the seed. The depth of sowing varies with seed capacity (size and weight), soil texture, structure, degree of capping, soil moisture content, depth of water table, season of cultivation, chemical composition of soil and agro-techniques followed.

In general the range of sowing depth varies from soil surface to 10 cm below the surface. In certain conditions it may be deeper. Deeper sowing delays field emergence and thus delays crop duration. Deeper sowing sometimes ensures crop survival under adverse weather and soil conditions.

Seeds are covered with loose and friable soil. In soil prone to crust formation, seeds are covered with particle organic mulch or anticrusting material such as vermiculite.

5.1.2 Plant Density

The number of harvestable plants per unit area depends on various factors. The important factors are plant character and the duration, time and method of sowing, fertility status of soil, purpose of cultivation, management practices and method of harvesting.

5.1.3 Field Conditions for Sowing

Seeds may be sown when the soil is dry, when the soil moisture around the seeding zone is tco low to cause hydration of seeds. Seeds such as sesame, jute, *jowar* and *bajra* may be sown under such soil moisture conditions awaiting rainfall which induce germination and the establishment of crop plants.

Seeds may be sown when the soil has the optimum moisture content for crop germination and establishment. Such soil moisture conditions can be achieved after rainfall, irrigation and drainage or from conserved water. Crop seeds such as wheat, mustard, castor, barley, rice, potato and jute may be sown under such soil moisture conditions.

Seeds may be sown when the seeding zone is in wet (saturated) conditions. Pre-germinated or pre-sprouted seeds or seed materials are sown under such soil moisture conditions. Crops such as rice, sugar cane, berseem, lathyrus and black gram are sown in such conditions.

Seeds may be sown in dry or semi-dry soil conditions and immediately after sowing the seeding zone is irrigated. Crops such as onion, garlic, sugar cane, napier, para grass, sweet potato and Guinea grass are sown or planted following this method.

During seeding the field should be free from established weeds, readily decomposable organic matter and toxic substances.

a) *Soil structure*: The ploughed layer of an arable soil has an artificial structure caused by preparatory tillage and other work involved in growing a crop. The natural peds (the coherent microporous units of soil) are distributed to become the clods of the newly ploughed field and these in turn, during subsequent cultivations, are broken down to become the crumbs of the seedbed (tilth). The larger inter-crumb pores permit surplus water to drain quickly from the soil and admit air. The water required by the plant is retained within the finer pores of the crumb. Plant roots can move freely through the inter-crumb pores to all parts of the soil, in search of nutrients and water. Underground modified stems and roots which have storage organs make room for themselves by pushing crumbs outward. Such a soil is also easily cultivated. However, the strength of the forces holding the crumb together determines its stability and hence the durability of the crumb structure. Organic matter enhances structure and increases stability either because coarser fractions hold the crumbs apart or because decomposed material cements or coheres the smaller particles together.

Soil structure is important for placing seeds in the desirable soil depth, covering and compacting and thus providing intimate soil-seed contact. When sown, many seeds fall initially into the smaller inter-crumb pores where they germinate. Those which fall into macropores or void spaces do not germinate until they come in close contact with a microporous matrix which may be formed subsequently due to the impact of rain or irrigation or traffic and effectively incarcerated by the collapse of the unstable crumbs or clods.

b) *Soil water*: During the process of imbibition the seeds get wetter and the soil gets drier if additional water is not available. Thus water content in the soil, across the soil-seed interface and in the seed are important. In saturated soil, the water movement is entirely within the liquid phase which is continuous. As the soil dries, the liquid phase decreases in volume and the liquid-filled paths become more tortuous with a corresponding decrease in the cross-section. In clay soils, fissures form first and clefts, crevices and deep cracks form on further drying. In dry soils, water films are discontinuous and water moves by molecular diffusion in the vapour phase.

A newly sown seed rests within an air-filled pore-space surrounded by soil. At least a few points of the seed come in contact with the soil at which water moving in the liquid phase, passes into the seed. Rough seed-coats of many seeds decrease the efficiency of physical contact between seed and soil.

Crusting (capping), which often occurs when flooded soil surfaces dry rapidly, results in more of a problem for the reliable production of stands than an excess or deficiency of water in many soils. Due to surface crusting the gas regime in the soil changes, oxygen content decreases and carbon dioxide concentration increases. Crusting changes the nature of the conversion of nutritional matter. Crusting compacts the soil and restricts the infiltration of water thus inducing a run off. Firming induces greater water stability. Unfirmed soil has large enough pores to permit unstable water-borne particles to re-sort themselves to form a crust when the water evaporates after irrigation or rain.

Keeping the soil surface continuously moist even by very fine spray irrigation until the seedlings emerge or by the application of anticrusting materials including mulches, reduces the deleterious effect of capping.

The longer a seed spends in the ground, the longer it is at risk. Staggered germination has low uniformity of seedlings. Earlier emerging seedlings will progressively restrict the supply of light, water and nutrients to the later emerging ones.

A high degree of uniformity of a population is an advantage

which minimises the tendency for the late emergers to be lost after the utilisation of some inputs. Late emergers very often remain undersized because of a low relative growth rate and make an insignificant contribution to the ultimate produce.

By increasing the seed water content prior to sowing and sowing seeds more shallowly, the time from sowing to field emergence can be reduced and this can curtail the period in which the sown seeds could be affected by an adverse change in the seedbed, from the favourable conditions in which they are sown. This can be brought about by the equilibriation of the seeds with a humid atmosphere or by soaking seeds to allow the preliminary process of germination to proceed, but not nough to permit radicle emergence.

Certain seed treatments involving the imbibition of physiologically active substances may be of special value in promoting germination and early seedling growth, especially as the beneficial effects persist for some time even when the treated seeds are dried back to facilitate sowing. More intimate contact between the seeds and soil water is achieved by coating with a mucilaginous material with an inherently low affinity for water but pronounced hydraulic conductivity (Elliot, 1967).

When the sowing season approaches towards the rainy (humid/wet) period, seeds may be sown on dry soil but when the sowing season approaches towards the rainless (dry) period, seeds should be sown on semi-dry (optimum moisture for seed germination of the crop concerned) or wet soil conditions.

Slight compacting of the seedbed improves seed-soil contact, availability of moisture and imbibition and hastens germination, but may restrict subsequent root and shoot elongation and decrease field emergence.

c) *Soil air*: The seeds respire during and after imbibition. Soil microbes and other soil-inhabiting living entities compete with seeds for required ventilation in soil. The quantity and freshness of decomposable organic matter, the water content of the soil, degree of soil capping and the prevailing soil temperature determine the ambient soil respiration rate.

If structural collapse is complete, diffusion within the seedbed may be within the liquid phase and seedlings may subsequently have great difficulty in emerging through the resultant cap or crust when the soil surface dries before field emergence.

When seeds are sown in wet soil or the soil becomes saturated immediately after sowing, seeds suffer from oxygen deficiency. Rapid and efficient drainage to the desirable extent can save the life of the seeds in the soil.

d) *Soil temperature*: The temperature of the soil is determined by the balance between incoming and outgoing radiation at the soil surface and by the way that this energy is redistributed throughout the soil as heat. By day heat is gained, by night it is lost. There is also an annual fluctuation of soil heat. In winter the net daily gain is negative, while in summer it is positive. Where cold soil is a problem, soil temperature can be increased by the use of asphalt, plastic and other organic mulches and where very high soil temperature is a problem the use of organic mulches and irrigation (which can reduce soil temperature by around 10°C) can reduce the adverse effect of high soil temperature. Both the thermal conductivity and the volumetric heat capacity of soil increases with the water content. Thus though a wet soil may transmit heat more readily, more heat is required to raise its temperature (Currie, 1973). The surface of the standing water in the field may have 5 to 7°C higher temperature than the ambient atmospheric temperature towards the higher range of temperature of the atmosphere.

The draining of the field or maintaining a slow flow of water reduces the hot water injury to emerging seedlings.

e) *Light*: Some seeds especially those that contain the phytochrome system, germinate in response to light stimulus. The phytochrome system acts as an environment-detecting device as an informer to seeds to break the dormancy and thus causes germination under favourable conditions. In the field, light does not penetrate more than 1 mm except down drying cracks, therefore, exposing seeds to bright sunshine for at least half an hour before sowing stimulates seed germination along with the other positive effects of solar radiation on seeds. Positive photoblastic seeds should be exposed to light after imbibition but before sowing.

f) *Chemicals*: When the seeds are sown in the field, endogenous and exogenous inhibitory substances and other phytotoxins diffuse away from the seeds or detoxify for successful germination and establishment of crop plants.

g) *Predatory animals and birds*: Several frugivorous and

granivorous animals and birds feed on many seeds after they are sown in the field. Sufficient good seeds should be sown to ensure a desirable crop stand. If required, excess seedlidgs are removed leaving the strongest ones but if gap filling is intended, seedling vigour, age and stage of crop plants in a field will differ, which will affect in the scheduling of field operations and harvesting. The harvestable parts of crop plants in a field may ripen at different times. Gap filling with seeds staggers the period of germination and emergence. Thus ripening periods which extend over time affect the harvesting which is scheduled once for most crops, and this seriously impairs the quality of produce. There is an old axiom regarding seed rate "one for the rook, one for the crow, one to rot and one to grow" which is modifiable according to the growers' own experience and judgement, with possible measures to save the loss of seeds in these ways.

It is not always desirable to use a higher seed rate than the optimum for various reasons. The seed required in kg/ha can be calculated with the formula

$$\frac{100\ T}{P \times R} \times \frac{100}{pp \times g}$$ where, T=test weight (grammes)

P= spacing between plants within the row (cm)

R=spacing between rows (cm)

pp=purity per cent

g=germination per cent

or

$$\frac{P \times T}{pp \times g}$$ where, P=number of plants per sq m

T=test weight (grammes)

pp=purity per cent

g=germination per cent.

Other than the skill of the farmer, field factors, for instance climate, soil, pests including weeds, irrigation and method of sowing are important for crop establishment.

5.2 SEEDING IN NURSERY BED TO RAISE SEEDLINGS FOR TRANSPLANTING

A nursery bed is a consolidated area where seeds are sown densely to germinate and emerge; buds, roots or other propagules are allowed to sprout and/or produce roots and are nourished for a short period to enable them to grow into seedlings, saplings, stecklings, transplants or other planting materials until they are ready for transplanting in a manifold field area with adequate spacing as crop plants. A nursery bed may be dry or wet (with sand or soil) and hydroponic (soil-less or *dapog*) and for special purposes humid chambers, phytotron and glasshouses are used. A nursery bed may be raised, flat or flat with furrows. Seeds are also sown in earthen pots or vessels, polythene tubes and paper pots.

Plants which are able to endure the shock of uprooting and transplanting and have the capacity to produce newer roots or root branches rapidly, where there is less tendency to suberize or cutinize the root wounds formed on uprooting, or plants which have permeable cutin and pectinaceous layers of good water absorbing capacity in the shoot portion, can be used for transplanting. Those plants which have no such behaviour, may also be used for transplanting by lifting them with a lump of soil in the root-zone. Individual seedlings are also raised in earthen pots, polythene tubes and paper pots so that they can be transported and transplanted to any distant place without much injury to their root system.

5.2.1 Planting and Transplanting Materials

Seeds and other propagules are used for planting. Some vegetative materials can be planted directly in the field under good conditions without nourishment in the nursery. They may be tubers (potato), rhizomes (ginger), fingers (turmeric), crown (turmeric), bunches (onion), cloves (garlic), bulbil (agave), cormel (taro), vine (sweet potato), suckers (napier), setts (sugar cane), tillers or clones (rice), root cuttings (palwal).

Non-clonal (monophytes) forms of reproductive organs such as tubers, rhizomes, cormels and fingers may be used as they are or after cutting into pieces containing viable buds (eyes). In such organs the apical (rose) end contains more buds that sprout earlier than the buds located at the basal (heel) end.

The vegetative propagules that are planted directly on the field should be of good health, vigour, age, stage of growth and of unit structure and should have a desirable number of readily sprouting buds. Cut pieces of such organs should be disinfected properly and the healing of cut wounds should be allowed before putting them in the soil. Again, after cutting them into pieces of desirable size and weight they should not be allowed to dry and deform. Sometimes such organs are incubated for sprouting. Incubation may be before or after making them into pieces as individual units of reproduction. In general, cut pieces are more susceptible to adverse soil conditions than whole ones.

Transplanting is the removal of an actively growing plant from one place and planting it in another for further growth and production. This operation may be performed when the seedlings acquire some ability to endure such shocks and when environmental conditions are favourable for easy establishment in the new location.

In general, transplanting is done with seedlings (rice, tobacco, tomato, brinjal), saplings (subabool), stecklings (sugar beet), transplants (seedlings or saplings that are grown an additional year or two in the nursery to make them more stout and strong) or even wildings (naturally sown plants dug up for transplanting). Sometimes in rice, clones or tillers that form after transplanting are separated and used for second-time transplanting (double transplanting).

5.2.2 Preparation of Nursery Bed

A. *Site:* The location of the nursery bed should be in accordance with the type of crop and season for raising seedlings. Some important points are to be considered in selecting a site for the nursery bed as it is imperative to rotate the site every year if not in every crop season.

1. It must get bright sunshine throughout the day but be protected from the passage of high speed winds.

2. It should be slightly elevated above the surrounding areas or in such a topographical situation that it is not flooded,

moistened, leached or eroded easily. Extra water should be drained easily, quickly and uniformly.

3. It should be well protected from free grazing, trespassing and trampling. It should be sufficiently far from highways and railroads from where dust and dirt comes out through saltation and deposits, soils and discolours the seedlings.

4. It should be away from perennial grass, brush, shrub, and alternate hosts which may provide feeding, breeding and hiding sites for insect-pests, pathogens and parasites including rodents and vermin.

5. It should be close to the source of irrigation water and away from compost pits, quagmires and standing crops of similar types.

6. It should have a desirable type of soil—preferably light soil and its depth, reaction and water holding capacity should be favourable to the type of seedlings to be raised. In general, the soil should be free from problematic, perennial and objectionable weeds; soil-borne pests, pathogens and parasites; moderate to shallow in depth, fertile, rich in organic matter, moderate to high, in potential for receiving, retaining and releasing moisture.

7. It should not be used for growing crops or seedlings of the same family or even strains of sirailai agro-botanical and requiring agro-ecological conditions or those which act as collateral hosts.

B. *Time*: To obtain seedlings uniform in vigour, well developed, of good health, with strong and closer root: shoot (top) ratio as well as of the appropriate age, seeds or other propagules are to be planted sufficiently ahead of transplanting. During the nursery duration weather conditions such as light, temperature, rainfall and relative humidity should be favourable to the type of crop seedlings to be raised. In general, seedlings should bear four to five leaves during transplanting.

C. *Area:* The area of the nursery bed differs with crop variety, type of nursery, degree of care and management and the rate of recovery of seedlings from the nursery. The area of the nursery bed for one hectare of transplanting and the seeding rate in the nursery bed of different crops are as follows:

Crops	Area of nursery bed	Per cent of transplanted area	Seeding rate g/sq m
Rice			
(a) Dry bed	1000 sq m	10.3	50-60
(b) Wet bed	1250-1500 sq m	12.5-15	50-60
(c) Dapog	75-80 sq m	0.75-0.80	1000
Tobacco	50 sq m	0.5	0.5-0.6
Onion	100 sq m	10	1.0-1.25
Tomato, brinjal, chillies	150 sq m	1.5	2.5-3.0
Cabbage, cauliflower	500 sq m	5.0	1.0
Knolkhol	500 sq m	5.0	2.0-2.5

Seed requirement per hectare of transplanted area

$$= \frac{P \times T \times (100 + G)}{P \times g \times R} \quad \text{where} \quad P = \text{Number of plants per}$$

$T =$ Test weight in gram-mes.
$G =$ per cent gap-filling
$p =$ purity per cent
$g =$ germination per cent
$R =$ recovery per cent of seedlings from nursery bed.

5.2.3   Advantages of Raising Seedlings in the Nursery Bed and Transplanting

a) It is easy, convenient, and consumes less time and input to nourish the young and tender seedlings in a small but compact area of nursery bed;

b) Management decisions regarding water, nutrients, soil, crops, weeds, pests and diseases can be made easily, adequately and exactly as and when they are found necessary;

c) Seedlings grown in the nursery bed and then tranplant-ed provide:

1. economy of rare and costly seeds;

2. maintenance of desirable plant density with pure, true to the type, healthy, uniform in vigour, strong and stocky seedlings irrespective of size and germinability of seeds;

3. elimination of weak, thin, poor in vigour, diseased seedlings, rogues and seed contaminated weeds from desirable seedlings;

4. availability of enough time for ripening after maturity of the previous or standing crop in the field and land preparation for the succeeding crop in intensive cropping programmes; the opportunity of using seedlings of the same age for gap filling;

5. the opportunity to escape from unfavourable soil and atmospheric conditions during germination, emergence and establishment and the earlier and critical part of field duration of the direct seeded crop; the opportunity to mitigate the delayed onset of monsoon or other factors responsible for delayed planting of crops;

6. growing seedlings for quick establishment and close canopy development for the greater utilisation of natural resources such as light, carbon dioxide, water and land,

7. the opportunity to use favourable weather and soil conditions;

8. higher use-efficiency of residual plant nutrients, moisture and cultural practices along with the basal application of manures and fertilisers;

9. field duration of shorter length and thus reduced risks;

10. more uniformity and maturity, higher yield of produce with precise quality.

5.2.4 Disadvantages of Nursery Bed and Transplanting

a) Total duration of crop may be higher;
b) It increases the labour and power requirements in a peak period;
c) It increases the cost of land preparation, uprooting and transplanting of seedlings.

5.2.5 Preparation of Dry Bed (Upland) Nursery

The soil tilth of the beds is brought to the desirable level by repeated tillage operations after cleaning weeds and crop residues including roots. A liberal quantity (at least 2 kg/m²) of well decomposed organic manure (farmyard manure, leaf mould or compost that is free from viable weed seeds) and requisite quantities of fertilisers are to be mixed with the soil. If required the soil may be disinfected by the application of fumigants or BHC ten per cent dust at 2 g per m² and/or Aldrin five per cent dust at 2 g per m² or by rabbing (burning dry farm refuse on the soil).

Beds of 1.2 m width and any convenient length along the slope, if any, leaving a 40 to 50 cm path in between the beds are to be constructed. Beds may be raised (10 to 15 cm from the ground level) with the soil from the adjoining path which may be used as a drainage channel and walking space. Raised beds are prepared for raising seedlings which cannot tolerate waterlogging, for instance tobacco, cabbage, cauliflower and chillies.

Flat beds of convenient length and 1.2 m width separated by ridges are prepared for raising seedlings which need flood irrigation and standing water during uprooting and washing of the adhered soil in the roots, for instance rice, onion and *ragi*.

Flat beds separated by furrows are prepared for raising seedlings during the summer months, when flat beds can supply moisture to seedlings for a more prolonged period than raised beds, where there is an unequal distribution of moisture in the bed—seedlings nearer the path suffer from water stress because of the rapid loss of soil moisture from the sides and edges as compared with the centre of the bed. Furrows at 1.2 to 1.5 m apart, 15 to 20 cm wide at the top and 5 to 10 cm deep serve as the drainage channel and path.

Before seeding the beds should be perfectly levelled. The centre of the bed may be kept slightly elevated over the edges to drain rapidly and uniformly particularly in nursery beds prepared during the monsoon.

The small, irregularly shaped and sticky seeds may be pelleted or mixed with sand or ash to singulate them to make a steady flow of seeds in the seeding tube. Treated seeds may be sown by the broadcast method or in uni- or bi-directional rows after preparing furrows of 2.0 to 2.5 cm deep and width and 7.5 to 10 cm apart. The seeds should be covered with loose earth, leaf mould, well decomposed compost or ash and compressed lightly. It is better to sow dry seeds. Even if seeds need wet treatment they should be dried considerably to avoid clumping during seeding and desiccating in the dry soil environment. For the seeds which take a longer time to emerge the bed may be covered with a thin layer of straw and watered by rose-can over the straw to prevent the surface soil from drying or forming a crust. When the seedlings begin to emerge through the straw, these straws should be removed with great care.

Seeding in the bed should be such that no clumping or overcrowding of seedlings occurs at any stage of the nursery duration. Closer seeding makes the seedling leggy and slender with larger aerial parts and ill developed underground portions. Overcrowding of the seedlings in the nursery bed creates a microclimate congenial to disease and pest infestations and their rapid spread. Very often poor recovery per cent of seedlings (the number of good seedlings suitable for transplanting after uprooting per hundred seedlings raised in the nursery bed) is associated with closer spaced seedlings. Such seedlings are difficult to handle and result in poor survival after transplanting.

The wider the spacing the easier it is to obtain short, thick, uniform and disease-free seedlings which root quickly and once they root, have a profuse growth with broader and thicker leaves, stronger branches and tillers.

The right kind of fertiliser is also necessary for raising short and thick seedlings. The essential point to be observed in fertilising seedlings is to increase the application of potassium and phosphate compared with that of nitrogen. Though Indian soil is said to be rich in potassium for the present level of productivity. This level is very insufficient for a higher level of productivity, higher level of nitrogen and phosphate application and intensive cropping systems. Besides these, variable responses have been observed in different crops and seedlings with potassium application in different soils.

Seedlings which have absorbed potassium and phosphate in abundance are found to be strong and resistant to diseases, insect-pests and unfavourable conditions in the field in which they grow after transplanting.

The rate and proportion of nitrogen, phosphate and potassium varies with the chemical capacity of the soil, type of crop and nursery, season and agro-climatic situation of the prefecture. For example, the optimum proportion of N, P and K to be applied to the nursery as proposed by Matsuki (1956) is N: 10, P: 13-15 and K: 15-17. Matsushima (1967) from his experience thinks that the rates of N: 8 to 14 g, P: 8 to 17 g and K: 10 to 18 g per m^2 are the optimum rates which should be proportionately increased from the plains (warm regions) to the mountainous areas (cool regions) and also from clayish to sandy soils. By far the most important element in fertiliser application is nitrogen, Matsuki (1956) suggested that the dosage of N per m^2 should be 7 to 9 g in warmer regions, 10 to 13 g in intermediate regions, rising to 14 to 16 g in cooler regions since N is not released effectively in cooler regions.

Except in cool regions, it is probably safe to limit the rate of N to such an extent that on transplanting the seedlings are slightly pale coloured because such seedlings are stiff and highly resistant to rooting damage caused by transplanting. If the yellowish seedlings (having a high content of starch and a low content of N, and consequently a high value for their C:N ratio) are top-dressed with 6 g of ammonium sulphate per m^2 five days prior to transplanting and not before, the rice seedlings will root quickly in the field after transplanting.

In West Bengal the recommended doses of N, P and K are— 5 g each of P and K at sowing and seven to ten days prior to up-rooting of seedlings, 5 g N per m^2 as top-dressing for upland rice nursery bed.

For tobacco nursery N at 2.5 to 5 g per m^2 as basal and 2.5 g per m^2 as top-dressing in light soils and P at 7.5 to 8.0 g per m^2 as basal are recommended.

In soils having a deficient supply of micro and trace elements specific nutrients are to be supplied in adequate quantities. In alkali soils ferrous sulphate at 2 to 4 g per m^2, zinc sulphate at 2.5 to 7.5 g per m^2 and for highly leached acid soils borax at 0.1 to 0.2 g per m^2 should be applied.

The best method of basal fertiliser application in the nursery is mixing the fertilisers with the soil before constructing the beds. Broadcasting or spraying of solutions of fertilisers are also practised for top-dressing.

The secret of raising healthy seedlings is to make the plant develop with a closer root: shoot ratio. For this purpose a controlled supply of water is essential. The shallower the irrigation water in the nursery, the easier it is to obtain short and thick seedlings. A controlled water supply also hardens the seedlings against stresses and on transplanting in a favourable environment they exert their potential to utilise the ambient environment to the highest possible extent. For these an upland (dry bed) nursery is desirable and seedlings raised in an upland nursery are capable of producing high yield.

5.2.6 Preparation of Wet Bed Nursery

A wet bed nursery is prepared for wet land rice crops. When a dry bed is not practicable farmers switch over to this type of nursery. For raising seedlings of rice during winter months for *boro* crop, wet bed is practised to ameliorate the low soil temperature. Pre-germinated seeds are broadcasted uniformly on a raised or flat bed prepared by puddling (ploughing and harrowing the soft saturated soil) in a field free from fresh organic matter which may be the cause of destruction of tender seedlings due to the increased activity of soil microbes or earthworms. Seeds are allowed to set at the surface on which a thin film of muddy water settles. It is necessary to regulate the field soil in saturated conditions till coleoptiles turn green. During warm months and in light soils where deep cracks do not develop on drying, nursery beds are allowed to dry to a considerable extent. In winter months and in clay soils the water level of the wet bed are raised gradually and maintained at 2 to 3 cm.

The manuring and seeding rates are similar to those of a dry bed nursery. However, it is desirable to apply 5 g more phosphate per m² and top dressing of 2.5 to 5 g nitrogen per m² each at the three weeks stage and one week before the uprooting of seedlings raised for *boro* crop.

To raise the soil water temperature OED is used. The chemical diffuses easily on the water surface and it serves to raise the water temperature by preventing water loss by evaporation. Ashes either alone or in combination with well fermented compost of farmyard manure are used by the farmers. A thin layer of such mulch materials reduce evaporation loss and prevent crust formation and cracking of soil surface on slight drying, as well as raise the soil temperature and mineral supply.

Seedlings raised in this method become ready for transplanting within three weeks. A greater quantity of seeds are needed to sow in such a nursery to obtain the required number of seedlings for transplanting. A considerable part of the seeds decay if the level of the bed is not uniform or the seeds are placed below the soil surface which is under the water surface. A greater number of seedlings tear out or get damaged during uprooting particularly when seedlings are raised under continuous moist soil conditions and cloudy weather.

5.2.7 Preparation of Dapog (Soil-less) Nursery

This method consists of growing rice seedlings on a concrete floor or a raised bed covered with polythene sheets having no substratum except water. Since the food reserve in the seed is the only source of nutrition, seedlings after two weeks of age, start deteriorating. During the winter months when seedling growth is slow it is not desirable to raise seedlings by the dapog method particularly where hand transplanting is practised. In areas where the intensity and beating force of rainfall is high and hand transplanting is practised, seedlings raised by this method have been found to be unsuitable because they become small and delicate.

The beds of 1.2 to 1.5 m wide and of a convenient length are covered with polythene sheets or banana leaves removing the mid-rib. The sides are raised 5 to 7.5 cm from the ground to impound water. The bed is levelled uniformly and a spillway with a plugging arrangement is provided to drain water from time to time. The sprouted seeds are broadcast evenly. Gently pressing either by palm or flat wooden board should be done twice in a day for the first three to six days to set the heaved seeds at the bed surface.

Sprinkling of water two to three times a day or a continuous and slow flow of a thin film of water is maintained to keep the root-zone moist. At least once in a day the water film should be drained for half an hour and recharged thereafter with fresh water. The water level in the bed may be raised gradually to 2 cm.

By this method the seedlings are raised faster and are ready for transplanting within nine to fourteen days after the seeds are sown during the warm-wet period. The seedlings need not be pulled for uprooting. According to convenience the bed should be divided and rolled like a mat with roots outward. In no way should such seedlings be left in the nursery after 15 days of sowing. No fertiliser is applied to dapog beds.

The advantages of this method are labour saving because the bed is easily made, the short period for raising seedlings and the relative ease of transport of seedlings, because a mat of seedlings can be rolled; seedlings raised in wet or dry beds have to be pulled, bundled and tied for transport (De Datta, 1981).

5.2.8 Care and Management of Seedlings in the Nursery

In an upland nursery if there is moderate rainfall on the uncovered beds immediately after sowing there is every possibility of a crust formation of varied thickness. If there is no subsequent rain such a crust prevents the smooth and uniform emergence of seedlings. Under such conditions the crust is broken by raking or by applying water by a spray. If there is a possibility of rainfall or cyclone it is better to protect seedlings by providing an overhead cover of straw or polythene sheet. Seedlings of the broad-leaved group, for instance tobacco, cabbage and cauliflower are more susceptible to such weather conditions, which cause washing of the soil from the root-zone, making the seedlings weaker in anchorage and prone to be lodged, resulting in the twisting and bending of the stem.

Seedlings of some crop plants such as cabbage and cauliflower suffer from scorching sunshine in the early stages of their growth. The overhead cover should be so provided that seedlings get sunlight in the forenoon and afternoon for the first few days until they become hardy enough to tolerate full sunshine.

Watering by light sprinkling during the early and late portions of the day keep the soil and seedling moist and the humid microclimate is favourable for quick growth. Gradually the intensity and interval between sprinkling should be reduced.

Beds should be kept clean from weeds, insect-pests and diseases. Regular vigilance helps to indicate any abnormality or lack in the seedlings. Adequate measures are to be taken immediately to correct any deficiencies or supply any requirements. The application of pesticides and nutrients in the nursery should be done with great care and with the exact dose, time and method suitable for the crop concerned.

5.2.9 Pulling or Uprooting of Seedlings

A few hours before pulling up seedlings the bed should be watered well to moisten the root-zone and to soften the soil sufficiently. Seedlings belonging to monocot groups such as rice and onion are pulled and washed to remove soil from the roots (naked) while seedlings belonging to dicot groups such as tobacco, tomato, brinjal, and cabbage are pulled up with a lump of soil in the roots.

Individual seedlings are uprooted with sufficient care so that they do not get damaged except for the tearing of some roots. If seedlings are uprooted in phases, it is better to reduce the density on the principle of age and vigour by uprooting vigorous seedlings first leaving smaller ones to grow further. If seedlings are uniform in vigour uprooting may be done from one side.

Sometimes seedlings are uprooted and transplanted temporarily in vacant portions of the bed or elsewhere to harden them.

Seedlings after uprooting should not be allowed to stale. If staling is unavoidable keep them in a moist, cool shady place with frequent watering under loose conditions or with a lump of moist soil in the roots or in humid chambers. The seedlings should be kept erect otherwise they will bend upward by their negative geotrophic movement.

The handling of seedlings before transplanting should be with great care. Any injury to vital parts leads to the death of the seedling. Injury to vital parts may invite disease infestation. During transport seedlings should not be allowed to wilt or pile up which may cause suffocation and damage.

5.2.10 Transplanting

Seedlings are transplanted in a well prepared field for raising crop plants. The best time for transplanting is the late hours of the day when the temperature cools down and humidity rises but transplanting can be done at any time of the day if it is cloudy or raining or drizzling or about to rain.

Seedlings with more shoot growth should not be transplanted in dry-hot or chilly months. They should be trimmed to reduce transpirational surface. Overgrown seedlings (with larger and thinner leaves or weak stems) droop or lodge a few hours after transplanting. Such seedlings should be trimmed off by a sharp, clean knife with minimum injury.

Though early seedling age provides for quicker establishment yet the three-to-four-functional-leaf-stage is found to be a more appropriate stage for most of the crops for uprooting, transporting, transplanting and establishment. Such a stage coincides with a plant age of three to ten weeks.

Single seedlings may be transplanted in each station at predetermined spacings or at the rate of two to three seedlings in each hill.

Seedlings should be placed in such a way that the shoot remains upward and the root downward. Roots and the subterranean shoot portion should not be crooked or tufted as this may delay establishment or cause death. They should be well spread having enough contact with moist soil. The soil around the root may be compressed lightly to provide more root-soil contact and to remove any air gap and consequently the accumulation of more water in the rhizosphere. Otherwise, it may induce the withering or decaying of roots. Sufficient watering in non-plastic soils immediately after transplanting serves the same purpose along with supplying moisture to the root-zone.

Seedlings may be treated by dipping the root or immersing it in a solution or suspension for a few minutes for priming prior to transplanting. Such treatments may be to induce establishment and growth or to induce tolerance against insect-pest, and diseases as well as adverse weather and soil conditions.

During transplanting in wet lands the temperature of the standing water should be within the bearable limits (cardinal points) of the seedlings. The depth of water should be as low as possible. Deep water (more than 10 cm) may cause the floating or immersing of seedlings which seriously affects plant population. If the water temperature exceeds 40°C for half an hour or more it is better to drain it out. Deep water not only affects the establishment of seedlings but also impairs tillering.

The depth of transplanting should be shallow for rice and onion while it should be moderately deep for other seedlings. The colar zone (soil-air interface) of seedlings should be placed just below the soil surface.

Before transplanting wider spaced crops, fertiliser mixture may be placed 5 to 10 cm below the pocket for transplanting. Well decomposed organic matter may be used to fill the pockets either before or after transplanting.

The progress of transplanting should be with the backward movement of the transplanter. Transplanting becomes easy and quick if the transplanter moves to the leeward direction of the prevailing wind.

Straight row transplanting ensures a uniform plant population per unit area and subsequent field operations become easy and smooth. In closer spaced crops, for instance rice and onion a skip row after ten to fifteen rows provides a better micro-environment for crop growth.

The use of five to ten per cent more seedlings at regular intervals as buffer stock helps to fill up gaps if any, with seedlings of the same variety, age and at the same physiological stage.

CHAPTER 6

Soil and Soil Management

6.1 CONCEPTS OF SOIL

Soil may be defined as a thin layer of the earth's crust which serves as a natural medium for the growth of plants. It is the unconsolidated mineral matter that has been subjected to, and influenced by, genetic and environmental factors—parent material, climate, organism and topography—all acting over a period of time. Some soil scientists extend soil to include subaqueous materials which support plant and animal life. Soil is dynamic, three-dimensional (with length, breadth and depth) piece of landscape with a three-phase (solid, liquid and gaseous) system. To a farmer, the soil is a habitat for plants, 15 to 20 cm in depth from the surface and provides mechanical anchorage and supplies nutrients and moisture to the crop plants grown over it. However, the concept of soil may vary according to its user, for instance, to a road engineer or potter who may need a specific type of earth which may or may not be suitable for raising crop plants.

Soils of one location may vary from that of others depending on the genesis and environmental factors. Soil contains four major components, mineral material, organic matter, water and air, the proportions of which vary with respect to time, site and depth.

6.2 SOIL FORMATION

Soils are formed by the disintegration and decomposition of rocks due to weathering and the action of soil organisms and also the interaction of various chemical substances present in the soil. Although soils are normally formed from underlying rocks in a particular region (e.g. sedentary or residual soils), these may be transported to long distances (e.g. transported soils) by water (alluvial, marine, lacustrine soils), glacier (glacial soils), wind (aeoline soils) and gravity (colluvial soils) and deposited or silted to form soils.

Soil differs from the parent material in the morphological, physical, chemical and biological properties and with respect to workability.

Rocks are the chief sources for the parent materials over which soils are developed. There are three main kinds of rocks—igneous, sedimentary and metamorphic. Granites (acidic) and basalts (basic) are igneous rocks. Conglomerate, sandstone, shale and limestone are common sedimentary rocks. Gneiss, quartzite, marble and slate are common metamorphic rocks.

Weathering creates the parent material over which soil formation takes place. Later, weathering, soil formation and development proceed simultaneously. Temperature, water, wind, plants and animals are the principal agents of physical weathering whereas, solution, hydration, hydrolysis, carbonation, oxidation and reduction are the principal means of chemical weathering.

The mineral weathering combined with the associated physical and chemical phenomena constitute the processes of soil formation which include the addition of organic and mineral materials, the loss of these materials from the soil, the translocation of materials from one point to another within the soil column and the transformation of mineral and organic substances within the soil. The kind of intensity of weathering and the process of soil formation indicate the degree of soil development. Though soil differences in the initial period arise from parent materials, yet they are profoundly modified by other factors of soil formation, for instance, climate, organism, topography and time.

6.3 SOIL PROFILE

A vertical section of soil in the field extending up to the depth of the parent material shows the presence of more or less distinct horizontal layers. Such a section is called a profile and individual layers are regarded as horizons. A complete soil profile is a vertical exposure of a superficial portion of the earth's crust that includes all the layers that have been pedo-genically altered during the period of soil formation and also deeper layers that influenced pedogenesis (Buol et al. 1980). These horizons above the patent material are collectively referred to as the solum. A study of the soil profile is important both from the standpoint of soil formation and soil development (pedology) and crop husbandry (edaphology).

The horizons in the soil profile which vary in thickness may be distinguished from them morphological characteristics which include colour, texture and structure. Generally, the profile consists of three mineral horizons, 'A', 'B' and 'C'. The surface of some soils in forested areas has an organic horizon ('0') above the mineral soil. This horizon is denoted as '01' where the original forms of the plant and animal residues are recognised by the naked eye and '02' where the original plant and animal forms cannot be so distinguished. Such '0' horizon is absent in arable lands.

The 'A' horizon may consist of sub-horizons, Al, A2 and A3. Al is the topmost mineral horizon, containing a strong admixture of humified organic matter which tends to impart darker colour than that of the lower horizons. A2 is the penultimate horizon wrfn ifflotaKva. tluviation of clay, iron and aluminum oxides and a corresponding concentration of resistant minerals such as quartz in the sand and quartz sizes. This layer is generally lighter in colour than Al. A3 is the transition layer between 'A' and 'B' with properties more like those of Al or A2 than of the underlying 'B'. Sometimes this layer remains absent.

The 'B' horizon lies below the 'A' horizon and consists of sub-horizons namely, Bl, B2 and B3. The 'B' horizons may be incorporated at least in part in the plough layer or they may be considerably below the plough layer in the soils with deep 'A' horizon.

Bl 'is a transition layer between 'A' and 'B' with properties more like 'B' than 'A'. Sometimes this layer remains absent. B2 is the zone of maximum accumulation (illuvial) of clays, iron, aluminum oxides and also calcium carbonate, calcium sulphate and other salts in arid zones. Organic matter content is generally higher than that of A2. B3 is the transition layer between

'B' and 'C' with properties more like those of 'B' than those of 'C' below.

The 'C' horizon excludes the bedrock from which 'A' and 'B' horizons (solum) are presumed to have been formed. This horizon by itself is relatively less affected by pedogenic processes. Its upper layers may in time become a part of the solum as weathering and erosion continue.

Not all the profiles show the sequence of horizons mentioned above. Profiles developed *in situ* under intense pedogenic processes over a sufficient period of time show the presence of the horizons 'A', 'B' and 'C'. Since the formation of soil and the development of profiles are dependent on the genetic and environmental factors which vary considerably within and between regions, the variation in horizonation are frequent and common. The soils developed in a recent flood plain may have 'AC' profile without any A2 whereas those in the red and lateritic soils area may have A1, B2 and 'C'.

When a virgin soil is put under cultivation, the upper horizons become the furrow slice. Cultivation, of course, destroys the original layered condition of this portion of the profile and the furrow slice becomes more or less homogenous. Sometimes in lighter soils, distinct layers of sand, silt and clay form within the plough layer due to puddling.

Many times, specially on cultivated land, serious erosion produces a truncated profile. As the surface soil is swept away, the plough-line is gradually lowered in order to maintain a sufficiently thick furrow slice. Hence, the furrow, in many cases, is almost entirely within the 'B' zone and the 'C' horizon is correspondingly near the surface (Brady, 1974).

A study of the soil profile reveals the surface and subsurface characteristics and qualities, such as depth, texture, structure and water balance which directly affect plant growth. Hence, the study of the soil profile helps to classify soils and to understand the soil-moisture-plant relationship.

6.4 PHYSICAL PROPERTIES OF SOILS

Soil is a habitat for plant growth bears certain physical, chemical and biological properties which determine the degree of workability, suitability to the specific crop varieties, physical and chemical capacities as well as productivity. The physical properties of a soil largely determine the manner in which it can be used. The physical capacities of a soil are influenced by the size, proportion, arrangement and mineral composition of the soil particles.

The solid phase of the soil is composed of mineral and organic materials. A soil consisting predominantly of, and having its properties determined predominantly by, mineral matter is a mineral soil. Usually such soil contains up to 20 per cent organic matter but may contain an organic surface layer up to 30 cm thick. Most of the arable soils are mineral soils. In contrast, soils containing organic matter in sufficient quantities (20 per cent or more by weight) to dominate the soil characteristics are the organic soils. Soils in swamps, bogs and marshes commonly contain 80 to 90 per cent organic matter. These organic or muck or peat soils when drained and cleared are most productive.

The mineral particles of a soil vary in size, shape and function. They can be separated and grouped according to size (each group is called 'separate') to determine particle size distribution in a given soil.

6.4.1 Soil Separates

Some soils contain stone, gravel, chert and cobbles which are coarser and range in size from 2 mm upward. These separates may he more or less rounded, irregularly angular or even flat in shape.

Coarse sand grains are 0.2 to 2.0 mm in diameter. They may be rounded and quite irregular in shape. Fine sands are 0.02 to 0.2 mm in diameter. When not coated with clay and silt, sand particles are no?sticky even when wet. They do not possess the capacity to be molded (plasticity) as does clay. Their water-holding capacity is low. They facilitate drainage and encourage aeration.

Silt particles are 0.002 to 0.02 mm in diameter. They are irregularly fragmented, diverse in shape and seldom smooth or flat. The silt separates possess some plasticity, cohesion and adsorption, but to a much lesser degree than the clay separates.

Clay particles are the smallest separates, less than 0.002 mm in diameter. They commonly are mica-like in shape. A grain of fine colloidal clay has about 10,000 times as much surface area as the same weight of medium sized sand. Since the absorption of water, nutrients, gas and the attraction of particles for each other are all surface phenomena, the significance of the very high specific surface for clay is obvious.

6.4.2 Mineral Composition in the Soil Separates

The coarsest sand particles often are fragments of rocks as well as minerals. Quartz commonly dominates the finer grades of sand as well as the silt separate. In addition, variable quantities of other primary minerals usually occur, such as the various feldspars and micas. Gibbsite, hematite and limonite minerals also are found, usually as coatings on the sand grains. Since sand and silt are dominantly quartz (SiO_2) these two fractions are generally quite inactive chemically.

Some clay particles, especially those in the coarser clay fractions, are composed of minerals such as quartz and the hydrous oxides of iron and aluminum. Another is the complex aluminosilicates. Three main clay mineral types are kaolinite, illite and montmorillonite.

A unit kaolinite crystal lattice consists of one sheet of silica and one sheet of alumina (1 : 1 layer silicate). The two sheets are held together by mutually shared oxygen atoms. These units, in turn, are tenaciously held together through oxygen-hydroxyl linkage and consequently no expansion occurs when the mineral is wetted. It has a low specific surface, and a low cation-exchange capacity (CEC). Similarly, plasticity, cohesion, shrinkage and the swelling properties of kaolinite are low.

A unit montmorillonite crystal lattice consists of two sheets of silica and one sheet of alumina (2 : 1-layer silicate) held together by mutually shared oxygen atoms. Such units are held together through weak oxygen-oxygen linkages. There is some isomorphic substitution of iron or magnesium in the alumina sheet. Montmorillonite crystal can expand, and owing to this expansion, cations, and water molecules are able to move in between the crystal units. Montmorillonite has a high specific surface and CEC. Similarly, plasticity, cohesion, shrinkage and swelling properties are high.

Illite has the same general structural organisation as montmorillonite, except in respect of the linkages between the crystal units. Here, about 15 per cent of silica in the silica sheet is replaced by aluminum and potassium atoms and they supply the additional connecting linkages between the crystal units, due to which illite shows a lower expansion capacity. It has proper-ties intermediate between kaolinite and montmorillonite.

6.4.3 Soil Colloids

The most active portions of the soils are those which are in the colloidal state (a two-phase system in which fine clay and humus are dispersed in water). The mineral colloid is present as the clay of various kinds, Whereas, organic colloid is present as humus (containing fulvic acid, humic acid and humin). Clay particles less than one micron in diameter possess colloidal properties. The most distinctive colloidal properties are (a) the large specific surface or interface, and (b) the capacity to hold solids, gases, salts and ions.

The colloidal material may occur as a thin gelatinous film around coarser particles, or it may occupy a considerable part of the space between the larger particles, thus serving as a binding material.

Soil colloids exhibit plasticity, cohesion, adhesion, swelling, shrinkage, dispersion, flocculation, and CEC.

6.4.4 Textnral Classes

The proportion of particles of different size groups in a soil, generally remains unchanged. According to the relative proportions of the various soil separates in a soil material (textural class) it is classed as clay, sandy clay, silty clay, clay loam, sand, Jay loam, silty clay loam, loam, sandy loam, silty loam, sand, loamy sand and silt.

The sand group includes all soils Of which sand separates make up 70 per cent or more of the material by weight. The clay group must carry at least 35 per cent of the clay separate and in most cases not less than 40 per cent. The silt group must dominate with silt separates. The loam group contains many subdivisions. An ideal loam is a mixture of sand, silt and clay particles and which exhibits light and heavy properties in about equal proportions.

Sandy soils are light as they are easily tilled and cultivated. On the other hand, silty and clayey soils are heavy as they have high plasticity and stickiness.

The textural classes have specific bearing on some factors affecting plant growth, such as (a) the movement and availability of water, (b) aeration, (c) workability, and (d) the chemical capacity.

6.4.5 Soil Colour

Colour gives a ready clue to soil conditions and important chemical, physical and biological properties of the soils in that area. It is due either to mineral matter or organic matter and mostly to both. Because soil colour is moisture dependent, both moist and dry soil colours are to be recorded.

Red, yellow and brown colours are mostly related to different degrees of oxidation, hydration and diffusion of iron oxides in the soils. Dark colours of soils are associated with one or a combination of several factors including impeded drainage conditions, content and state of decomposition of organic matter and the presence of titeniferous magnetite. Upon the removal of free iron under reducing conditions, the soil appears grey or bluish grey in colour. In some instances, relict colours i.e., those inherited from the initial materials, persist in the soil.

Colour is composed of three measurable variables; hue, value and chroma. Hue is the dominant spectral colour and is related to the wavelength of light. Value is a measure of degree of darkness and lightness of the colour and is related to the total amount of light reflected. Chroma is a measure of the purity or strength of spectral colour.

Mottling is described in terms of three characteristics: contrast, abundance and the size of area of each colour. Abundance of mottles is subdivided into three classes: few, common and many as they occupy less than two per cent, two to 20 per cent and more than 20 per cent of exposed surface respectively. Size refers to the approximate diameters of individual mottles. There are three relative size classes, namely, fine (less than 5 mm), medium (5 to 15 mm), and coarse (greater than 15 mm) in diameter along with the greatest dimension. Contrast may be described as faint, distinct or prominent depending on the degree of difference.

6.4.6 Density

Soils with larger particles are usually heavier in weight per unit volume than those with smaller particles. The bulk density is the weight per unit volume of dry soil as a whole. This volume includes both solids and pores (voids). Therefore, the bulk density or the apparent density is lower than the true density (particle density). The relationship between true density (T) and the apparent density (A) and the pore space (P) is

$$P\% = \frac{(T-A)}{T} \times 100$$

In most mineral soils the true density varies within narrow limits of about 2.5 (in very loose soils) to 2.7 (in compact soils) and the apparent density between 1.0 (in clay soils) to 1.8 (in sandy soils). Bulk density is influenced by texture, organic matter content, and degree of compactness of the soil.

6.4.7 Pore Space

The pore space of a soil is the portion occupied by air and water and it is largely determined by the structure. The size, shape and continuity of the pores determine to a large extent the movement of air and water in the soil. Pores may be macro (non-capillary) and micro (capillary). The macropores allow the ready movement of air and percolating water whereas, in the micropores air movement is greatly impeded and water movement is restricted primarily to slow capillary movement. The shape of the pores and the degree of interconnection influences infiltration and permeability.

Sands have low pore space (about 30 per cent) with few macropores whereas, clays may have as much as 50 per cent to 60 per cent with many micropores. Good loams may have pore space of 40 per cent to 50 per cent. The macropores as in sands are more conducive to good drainage and aeration. In soils with high organic matter content, the pore space is high.

6.4.8 Plasticity and Cohesion

Plasticity is the property that enables a moist soil to change shape on the application of force and retain this shape even when the force is withdrawn. According to the degree of plasticity (pliability and the capacity to be molded) soils may be classified into non-plastic, slightly plastic, plastic and very plastic. Sandy soils may be considered to be non-plastic whereas, clayey soils to be plastic.

Cohesion is the tendency of the particles to stick to one another. For wet soils, the degree of stickiness (quality of adhesion to other objects) may be non-sticky, slightly sticky, sticky and very sticky.

Moist consistency soils are grouped into loose, very friable, friable, firm, very firm and extremely firm. Dry consistency soils are classified into loose, soft, slightly hard, hard, very hard and extremely hard.

The cementation of some soil horizons appears to be independent of the soil moisture level and the degree of cementation is grouped into weak, strong and indurated.

Plastic soils are cohesive. Plasticity and cohesion reflect the consistency and workability of the soil.

6.4.9 Soil Temperature and Heat

Soil temperature refers to the mean annual soil temperature within the main root-zone (5 to 100 cm) depth, plus the mean seasonal fluctuations and the average warm or cold seasonal temperature variation with depth in the root-zone. The temperature of the soil is influenced by its colour, composition, slope and its direction, degree of shading of the surface and water content. Generally sandy soils absorb heat during the day and lose it during the night more quickly than the clay soils, because the latter retain more water and the specific heat of water is four to five times more than that of soil particles.

The soil temperature categories are: Pergelic—mean annual soil temperature is less than 0°C, and Cryic—mean annual soil temperature is in between 0°C and 8°C with summer temperature less than 15°C, and Iso—(used as a prefix to other temperature descriptions) when the average soil temperature for the three warmest months differs by less than 5°C from the average temperature of the three coldest months—as it is in tropical latitudes.

The temperature of the surface layers vary more or less according to the air temperature and therefore, exhibit a greater fluctuation than the subsoil. On an average, the surface 15 cm layer of soil is warmer than the air at every season of the year, while the subsoil is warmer in autumn and winter but cooler in the spring and summer because of its protected position and the lag in conduction.

Soil temperature is one of the important factors that control the moisture regime, microbiological activity and all the processes involved in the growth of plants. The thermic capacity of a soil determines the seed germination, root growth, nitrification, absorption and transport of water and nutrients, decomposition of organic matter and also other physical, chemical and biological activities in the soil.

Soil temperature can be modified to some extent by the use of mulches, irrigation and bulky organic manure.

6.4.10 Soil Air

Soil air is the gaseous phase of the soil, being that volume not occupied by solids or liquids. The pore space not filled by water, is occupied by air. Under moist field conditions macropores generally constitute the air space. Ordinarily, the occupation of nearly one-third of the pore space in the soil by air and two-thirds of it by water constitutes the most favourable condition for plant growth.

The consumption of oxygen and liberation of carbon dioxide by plant roots and the soil organisms tend to increase the difference in the composition of the soil air and the atmosphere above the soil surface. The rate of gaseous interchange tends to the normalization of the composition of soil air.

The soil air is composed largely of nitrogen and oxygen. With more moisture and impeded gaseous exchange, there is more carbon dioxide and a little oxygen. Under continuous submergence, organic matter decomposition leads to the production of methane, hydrogen sulphide etc. Compared to the atmosphere, soil air usually is much higher in water vapour, being essentially saturated except at or very near the surface of the soil.

The rate of aeration (the oxygen diffusion rate) depends largely on the volume and continuity of pores within the soil. Drainage and tillage induce soil aeration. Aeration has a contributory role on root respiration, nutrient absorption, nitrogen fixation, organic matter decomposition, and soil moisture status.

6.4.11 Soil Water

Water is an essential part of plant food. It serves as a solvent and carrier of plant nutrients. It maintains cell turgidity and regulates temperature. The water content regulates most of the physical, chemical and biological properties of the soil as well as plant growth. Soil without water is barren. Soil moisture is one of the most important ingredients of the soil. It is also one of its most dynamic properties. Soil serves as the storage reservoir for water. Only the water stored in the root-zone of a crop can be utilised by it for its transpiration and the building up of plant tissues. Optimum soil moisture determines the workability of the soil. Moist soil escapes wind erosion whereas excess water leads to loss of soil and plant nutrients.

Water is held in the soil in the following forms:

a) Hygroscopic water: It occurs as a thin film (4 to 5 millimicron) and is held tenaciously with a tension of 31 atmospheres or more. It is not available to plants.

b) Capillary water: It forms a continuous film around soil particles (outside the film of hygroscopic water) and in the micropore spaces. It is held by surface tension. Capillary water is held at a tension ranging from 1/3 to 31 atmosphere.

c) Gravitational water: Water that moves freely in response to gravity and drains out of the soil. It saturates the soil and is held at a tension below 1/3 atmosphere.

The maximum amount of capillary water remaining in the soil after the removal of gravitational water is called its field capacity. Capillary water held at tensions greater than 15 atmosphere is not available to plants. At this point of soil moisture, the plant wilts permanently and, hence, the percentage moisture at 15 atmosphere is called its wilting point or wilting percentage. Comparatively easy movement does not occur until the water film thickens and pressures near 1/3 atmosphere are reached. As a result of its energy relations, the capillary water is the only fluid water bearing solutes, that remains in the soil for any length of time, if drainage is satisfactory. Thus, it functions physically and chemically as the soil solution.

Capillary water is capable of movement upwards, downwards or laterally, the movement taking place from the thicker part of the film to the thinner part. Similarly, as the capillary water zone moves further from the water-table below, the rate of movement becomes less and the suction needed to draw up water increases. Soil water stored in the upper 80 cm of soil can be as high as 250 mm just after surface drainage, so that the crop becomes less dependent on daily rainfall or irrigation or ground water contribution.

The principal factors influencing the amount of capillary water in soils are the structure, texture, and organic matter. The finer the texture, the greater is its capillary capacity. Granular soil structure produces higher capillary capacity. Presence of organic matter increases the capillary capacity.

Capillary water is said to be usable by plants and as such is available water. The rest of the forms are regarded as unavailable water.

6.4.12 Soil Structure

Structure refers to the aggregation of individual soil particles into larger units with planes of weakness between them. Individual aggregates are known as peds or secondary units. Soils that do not have aggregates with naturally preserved boundaries (peds) are considered to be structureless. Two forms of a structureless condition are recognized, that is, single grain (particles are easily distinguishable) or massive (individual particles adhere closely to each other but the mass lacks planes of weakness).

Three features of structure are usually described in each

horizon: grade, class, and type.

The grade or strength of structure is moisture dependent. The grades are weak, moderate, strong and very strong depending on the stability of aggregates when disturbed.

The class of soil structure refers to the size of the peds. Five size classes are recognized in each of the primary types. They are very fine, fine, medium, coarse and very coarse.

The type of soil structure refers to the shape and arrangement of peds. They are platy, prismatic, columnar, angular blocky, subangular blocky, granular and crumb.

Soil conditions and characteristics such as water movement, heat transfer, aeration, bulk density, porosity, ease of tillage, resistance to erosion, root penetration and productivity are influenced by structure. The important physical changes imposed by the farmer while tilling, draining, liming, and manuring his land and by the activities of soil organisms are structural rather than textural. From the farmers' point of view, crumb and granular structure (spheroidal) is considered favourable to plant growth.

The structure and physical qualities of soils can be unproved by adopting a suitable system of soil management, including legumes in the rotation system, green manuring and regularly supplying the soil with organic manure.

6.5   CHEMICAL PROPERTIES OF SOILS

6.5.1   Minerals

The principal minerals found in the soil are feldspars (anhydrous alumino silicates of K, Na and Ca), quartz (SiO_2), mica (alumino silicate of K, Fe, Mg and Na), limestone [$CaCO_2$, $CaMg(CO_3)_2$] horneblende and augite [$Ca_2Al_2Mg_2$ Fe_2 Si_2 O_{23} $(OH)_2$ and Ca_2 (Al, Fe)$_4$ (Mg, Fe), Si_2 O_{23}], olivine and serpen-tine (olive green ferro-magnesium silicate and hydrated silicate of Mg), clays (hydrated alumino silicates) and others (such as tourmaline, rutile titanium oxide, zirconzirconium silicate).

6.5.2   Inorganic Components

Besides the chemical constituents of the mineral matter, the soil contains small amounts of a large number of other mineral elements, e.g. P, B, Mn, Cu, S, Zn and Co.

Thus the soil supplies all the following essential mineral elements required by the plants.

1. Macros: P, K, Ca,' Mg, and S.
2. Micros: F, Mn, Zn, Cu, Mo, B and Cl.

Plants obtain C, H and O from carbon dioxide and water. Soil is also the source of N for plants though the ultimate source of N is the atmosphere from where N fixation takes place physico-chemically and biologically.

6.5.3 Ioa Exchange

This is a chemical process involving the reversible exchange of ions (cations and anions) in solution and ions bound to an insoluble substance. Ion exchange is the most important of all the processes occurring in a soil. Soil colloids are the seats of ion exchange. Chemical and physical processes connected with ion exchange include the weathering of minerals, nutrient absorption by plants, leaching of soluble salts and swelling and shrinkage of clay.

The capacity of soils to adsorb and exchange cations (posi-tively charged ions, e.g. Ca^{++}, H^+, Mg^{++}, K^+ etc.) and anions (negatively charged ions, e.g. NO_3^-, Cl^-, SO_4^{--} etc.) varies greatly with the nature and amount of clay and the organic matter.

Cation-exchange capacity (CEC) is the sum total exchangeable cation adsorbed by a soil at pH 7 expressed in milli-equivalent per 100 g of soil.

The CEC values of different clay minerals are: Kaolinite–3 to 10, iilite-10 to 30 and montmorillonite-80 to 150 m e/10 g. Organic colloids may have CEC of 200 m e/100 g. or more.

6.5.4 Base Saturation

The portion of CEC accounted for by the basic ions (Ca, Mg, K, Na) and expressed as a percentage of the CEC is the base saturation percentage. A soil saturated with Ca and Mg is considered normal and fertile. If a soil has more than 15 par cent exchangeable sodium, it is considered to be an alkali soil. On the contrary, if the soils are base-unsaturated i.e., the proportion of exchangeable hydrogen is more, the soil tends to be acidic.

6.5.5 Soil pH

Soil pH is the negative logarithm of the hydrogen-ion activity of a soil. The degree of acidity (or alkalinity) of a soil as determined by means of a glass, quinhydrone or other suitable electrode or indicator at a specified moisture content or soil to water ratio is expressed in terms of the pH scale, from 0 to 14. The pH values for most agricultural soils range from 5 to 8.5. On the basis of pH measurements (with soil: water at 1:2 proportion) the soils are classed as: extremely acid (4.5 or below), very strongly acid (4.6 to 5.0), strongly acid (5.1 to 5.5), medium acid (5.6 to 6.0), slightly acid (6.1 to 6.5), nearly neutral (6.6. to 7.5), slightly alkali (7.6 to 8.0), moderate alkali (8.1 to 8.5), strongly alkali (8.6 to 9.0) and extremely alkali (9.0 or more).

The soil pH has strong correlations with soil nutrients, soil organisms and thus has a relation with crop plants.

6.6 ORGANIC MATTER

Organic matter plays a vital role in the productivity and conditioning of soils. It serves as a source of food for soil bacteria and fungi which are responsible for converting complex organic materials into simple substances readily used by the plants. The intermediate products of decomposition of fresh organic matter help to improve the physical condition of the soil. The organic matter content of the soil improves the biological and chemical capacities of the soil. Growth promoting substances and enzymes induce crop growth and yield and improve the quality of the produce.

The organic matter in the soil consists largely of plant remains, the residues of soil micro-organisms feeding on them and several products of their decomposition. The plant remains may occur in a recognisable form, but more commonly they are found as a fairly stable, dark, amorphous complex colloidal substance called humus. Humus exhibits exchange properties.

The content of organic matter varies with the kind of the soil, vegetation, climate, cultivation, manurial and rotation practices, and biological activities. The vegetation determines the quality and quantity of organic material added each year, whereas the climate, particularly the temperature and moisture conditions, determines the rate of decomposition. Organic matter content in Indian soils is low because of the high rate of decomposition under tropical and subtropical climate. Except in a few localised areas in the hilly and high-altitude regions, the organic matter in most of the cultivated soils rarely exceeds one per cent.

The organic combination of nitrogen, with organic matter in the soil constitutes the major store-house from which nitrogen is slowly made available to the crops. The ratios among nitrogen, carbon and organic matter in the cultivated soils are fairly constant. In humus the carbon: nitrogen ratio is 10:1. In most of the Indian soils the carbon: nitrogen ratios fluctuate widely from 5 to 25 with the average value around 14.

6.7 SOIL ORGANISMS

A variety of organisms inhabit the soil. They decompose organic matter, fix atmospheric nitrogen, cause denitrification and plant disease.

Cultivated soils harbour bacteria, actinomycetes, fungi, algae, protozoa, nematodes, worms, insects and rodents. Their population, proportion and distribution are determined by food supply, moisture, temperature, physical conditions, reaction, and degree of disturbance (physically, chemically and biologi- cally) of the soil. Specific groups of organisms are responsible for specific activities in the soil. Such activities may be benefi- cial or harmful to the crop or its yield potential.

6.8 LAND CAPABILITY CLASSES

Land is that part of the surface of the lithosphere which is not usually covered with water. Land is a broader term than soil. In addition to soil, its attributes include other physical conditions such as mineral deposits and water supply; location in relation to centres of commerce, population and other lands, the size of the individual tracts or holdings; the existing plant cover, and works of improvement. Land capability refers to the suitability of land for use without damage or capability of land for intensive use and the treatments required for sustained use.

The capability classes are grouped under arable and non-arable lands.

6.8.1 Arable Lands

Class 1: These soils have only a few limitations that restrict their use. The soils are nearly level, deep, well drained, with good water holding capacity. They are productive and suitable for intensive cropping. Such soils need ordinary crop management practices such as the use of fertiliser, lime, crop residues and green manure to maintain their productivity.

Class 2: These soils have some limitations, for instance gentle slopes, moderate erosion hazards, inadequate soil depth less than ideal soil structure and workability, slight to moderate alkali or saline conditions and somewhat restricted drainage that reduces the choice of crops and requires moderate conservation practices to prevent deterioration. They are capable of sustaining less intensive cropping systems. The management practices that may be required for such soils include terracing, strip cropping, contour tillage, rotation involving grasses and legumes and grassed waterways. These are in addition to the practices required for the soils in class 1.

Class 3: The soils in this class have severe limitations such as moderately deep slopes, high erosion hazards, very slow water permeability, shallow depth and restricted root-zone, low water-holding capacity, low fertility, moderate alkalinity or salinity and unstable soil structure. All this reduces the choice of crops and requires special conservation practices when used for raising field crops. In addition to management practices required for soils in class 2, special management practices including tile drainage may be needed.

Class 4: Soils in this class can be used for cultivation, but there are very severe limitations on the choice of crops. Also, very careful management may be required. The alternate uses of these soils are more limited than the previous classes. Close growing crops must be grown extensively, and row crops cannot be grown safely in most cases. The choice of crops may be limited by excess moisture as well as by erosion hazards.

The most limiting factors in these soils may be one or more of the following: severe erosion susceptibility, steep slopes, severe past erosion, shallow soils, low water-holding capacity, poor drainage and severe alkalinity or salinity. Soil conservation practices must be applied more frequently.

6.8.2 Non-arable Lands

Class V: The soils in this class have little or no erosion hazard, but have other severe limitations, for instance, wetness or overflow, as in the case of bottom lands which prevent the normal tillage for common cultivated crops; a growing season too short for crop plants, stony or rocky soils and ponded areas where drainage is not feasible. These restrictions limit their use for pasture or adaptable tree species.

Class VI: The soils in this class have extreme limitations of depth, slope and erosion hazards that make them unsuitable for the cultivation of normal crops. Their use is restricted to pasture and silviculture.

Class VII: The soils in this class have more limitations than those of class VI and their use is restricted only to grazing.

Class VIII: The soils under this class have very severe limitations that preclude them from being used for agriculture or silviculture. Their use is restricted to recreation, wild life, water supply or aesthetic purposes. The soils or land forms included in this class are sandy beaches, river washes and rock outcrops.

6.9 MODERN CLASSIFICATION SYSTEMS OF SOILS

The modern classification system (the U.S. comprehensive soil classification system based on the 7th approximation) is completely new in design and nomenclature. This system contains six categories. From the highest to lowest levels of generalisation these are order, suborder, great group, subgroup, family and series in the present form of soil taxonomy. Each

Orders	Simplified key	Great groups
1. Histols (organic soils)	More than 30 per cent organic matter to a depth of 40 cm.	Bog soils.
2. Spodosols (with subsoil accumulations of sesquioxide and humus)	Other soils with a Spodic horizon within 2 m.	Ground water Podzols, Podzols, Brown Podzolic.
3. Oxisols (Sesquioxide rich highly weathered soils of the Intertropical regions)	Other soils with an Oxic horizon within 2 m and no Argillic horizon.	Laterite soils, Latosols.
4. Vertisols (Shrinking and swelling dark clay soils)	Other soils with more than 30 per cent clay in all horizons; some cracks when dry at 50 cm.	Grumusols.
5. Aridisols (Soils of arid regions)	Other soils that are dry more than 50 per cent of the year and no mollic epipedon.	Desert, Reddish Desert, Serozem, Solanchak, some Brown and Reddish Brown soils, associated Solonetz.
6. Ultisols (Low base status forest soils)	Other soils that have an Argillic horizon but a base saturation at pH 8.2 less than 35 per cent at a depth of 1.8 m.	Red-Yellow Podzolic, Reddish-Brown Lateritic, associated Planosols, and some Half Bogs.
7. Mollisols (Grassland soils of Steepes and prairies)	Other soils that have a Mollic epipedon.	Chestnut, Chernozem, Brunizem, Rendzina, some Brown, Brown Forest, associated Humic Gley, and Solonetz.

8.	Alfisols (High base status forest soils)	Other soils that have an Argillic horizon.	Grey-Brown Podzolic, Grey Wooded, Noncalcic Brown, Degraded Chernozem, associated Planosols and Half-Bogs.
9.	Inceptisols (Embryonic soils with few diagnostic features)	Other soils that have an Umbric, Mollic, or Plaggen epipedon or a Cambric horizon.	Ando, Sol Brun Acide, some Brown Forest, Low Humic Gley, Humic Gley.
10.	Entisols (Recently formed soils)	Other soils.	Azonal soils, some Low Humic Gley.

category has a variable number of taxa.

The simplified key to soil orders and their relations to great groups of earlier (1938) classification are:

6.10 SOIL GROUPS OF INDIA

The soils of India are derived from a wide variety of minerals. Thus they differ physically, chemically and biologically. Their distribution does not follow any regular pattern. The soils of India are broadly divided into five major groups. Soils with traditional nomenclature and their corresponding orders under soil taxonomy are: alluvial soil (Entisol, Inceptisol, Alfisol), black soil (Vertisol, Inceptisol, Entisol), red soil (Alfisol, Inceptisol, Ultisol), laterite soil (Alfisol, Ultisol, Oxisol), and desert soil (Entisol, Aridisol). The other recognised group of soils includes forest and hill soils, saline and alkali soils, peat and marshy soils.

6.10.1 Alluvial Soils

The alluvial soils include the deltaic alluvium, calcareous alluvial soils, coastal alluvium and coastal sands. Alluvial soils are formed by transportation in streams and rivers and are deposited in flood plains or along coastal belts. They are generally deep soils. The fresh alluvium shows little or no horizonation, while older alluvium shows distinct profile development. Such soils occur in the basins of the Indus, Ganges, Brahmaputra, Godavari, Krishna and their tributaries. Geologically the alluvium is divided into *khadar* or newer alluvium which is sandy, generally light coloured and less *kankary* in composition, and *bhangar*, or older alluvium of a more clayey composition, generally dark and full of *kankar*.

Alluvial soils of Assam are acidic. Those on the old alluvium and hills are more acidic than the new alluvial soils along the river banks. The latter are often neutral or alkaline. The soils of the Brahmaputra Valley are sandy; their available and total potash contents are fairly good; their P_2O_5 content is good and the organic matter and nitrogen content are fairly moderate. The soils of the Surma Valley are fine in texture.

In W. Bengal, the portions of Murshidabad, Bankura, the

whole of Burdwan and the western half of Midnapore are composed mainly of old alluvium. Some of the deposits may be different from one another in texture, colour, profile, chemical composition and mechanical and other physical properties.

The alluvial soils of Bihar may be divided into two main divisions, depending on the distinctness of their characters: (a) the alluvium north of the Ganges, and (b) the alluvium south of the Ganges.

The northern alluvium comprises the area between the Himalayas in the north and the Ganges in the south. The soil is alluvial, with a calcareous belt in the form of a triangle in the west, and broken inundated areas in the middle. These areas remain flooded for different periods in the year. The soils are sandy loam to clayey loam and neutral to alkaline. Their CaO content ranges from 0.5 to 20 per cent. They are rich in total and available K_2O but deficient in P_2O_5.

The southern alluvium comprises the area between the Ganges in the north and the hilly region in the south. The soils vary in colour and texture from light greyish loams to heavy black clays. The middle of the area has a depressed feature which floods during the monsoon. The area has been subdivided into: high land soils, soils liable to inundation, saline soils, and *diara* land soils.

The soils of Uttar Pradesh are divided into four classes: (a) the alluvium of the west and northwest, lighter in texture, (b) the alluvium in the centre, with a texture intermediate between light and heavy, and (c) the alluvium in the northeast developed on a calcareous parent material.

The soils contain varying amounts of $CaCO_3$ and soluble salts and are neutral to alkaline. The lime content usually increases at lower depths. They are generally poor in P_2O_5, N and organic matter.

On the coast of Orissa, there are stretches of sand and sandhills, alternating with deltaic swamps. Behind this coastal belt is an area of cultivated alluvial and lateritic formations. Soils are both sandy and of finer texture. There is sufficient K_2O but not enough P_2O_5.

The alluvial soils of Tamil Nadu are found in the deltaic areas and along the coast. A section of its profile reveals a deposit of alternate layers of sand and silt. The composition of the strata varies with the nature of the silt brought by the rivers, which in turn, varies with the catchment areas and the tracts through which they flow. The stratified deposition is confined to areas very near the river courses   But away from the rivers, the soils are heavy throughout, the texture ranging from clay loam to heavy clay through silty clay. In such cases, sandy layers occur at very low depths.

In the Gujarat state, the alluvial soils are confined to the northern Gujarat tract, Ahmedabad and Kaira districts and are locally known as *goradu*. The *gorat* soil of Baroda corresponds to the older alluvium, consisting of brown clay with *kankar*. The soils are secondary deposition, fairly deep, poor in organic matter and N, but fairly rich in P_2O_5 and K_2O.

The light sandy, red and yellow soils found in the Mahanadi basin (Madhya Pradesh), including the Balaghat and three districts of Durg, Raipur and Bilaspur, are of alluvial origin.

The soils of the Punjab and Haryana plains belong to the same class of alluvial soil that is typical of the Indo-Gangetic plains. The majority of the soils are loams or sandy loams consisting of a soil crust of varying depth. Soluble salts are present in considerable amounts. The lower layer contains *kankar* nodules. Owing to the presence of sodium in the clay complex, the soils are generally alkaline. They are adequately supplied with P_2O_5 and K_2O, but are deficient in organic matter and N.

In Kerala, there are two types of alluvial soils on river banks, coastal alluvium and alluvium. In central Kerala, the width of the coastal alluvial tracts increases, whereas in the north and south, they are comparatively narrow. The alluvial soils of Kuttanad form a low-lying area, believed to be once a part of the sea and later filled up by the silt carried down by the Pampa and other rivers. The coastal alluvium is sandy, having a low water-holding capacity and a low nutrient status. The alluviums on the banks of the rivers are fertile.

6.10.2 Black Soils

The typical soil derived from the Deccan trap is the *regur* or black cotton soil. These soils vary in depth from shallow to deep. It is common in Maharashtra, western parts of Madhya

Pradesh, parts of Andhra Pradesh, parts of Gujarat, and some parts of Tamil Nadu.

Many black soil areas have a high degree of fertility, but some, especially in the uplands, are rather poor. They are somewhat sandy on the slopes and uplands and are moderately productive. In the broken country, between the hills and the plains, they are darker, deeper and richer and are constantly enriched by the additions washed down from the hills.

Black soils are highly argillaceous, very fine-grained and dark and contain a high proportion of calcium and magnesium carbonates. They are very tenacious of moisture and exceedingly sticky, when wet. Owing to considerable contraction on drying, large and deep cracks are formed. These soils contain abundant iron and fairly high quantities of lime, magnesia and alumina. Potash has a wide range. They are poor in P_2O_5, N and organic matter. In all *regur* areas there is generally a layer rich in *kankar* nodules formed by the segregation of $CaCO_3$ at some depth below the surface and above the weathered rock. The soils are generally rich in the montmorillonitic and beidellitic group of clay minerals.

In Maharashtra, the soils derived from the Deccan trap occupy quite a large area. On the uplands and slopes, soils are light-coloured, thin and poor. On the low lands and in the valleys, deep and relatively clayey black soils are found. Along the Ghats, the soils are very coarse and gravelly. In the valleys of the Tapti, Narmada, Godavari and Krishna rivers, heavy black soil is often 6 m deep. The subsoil contains a good amount of lime. Outside the Deccan trap area, the black cotton soil predominates in Surat and Broach districts. Degraded solonized black soils occur in areas in the canal zones of the Deccan in Maharashtra.

In Tamil Nadu, the black soils are either deep or shallow and may or may not contain gypsum in their profiles. Soils are fine-textured, have high pH (8.5 to 9.0) and are rich in lime (five to seven per cent). They have low permeability and high values of hygroscopic coefficient, pore space, maximum water-holding capacity and true specific gravity. Black soils have generally a high base status and a high CEC (40 to 60 m e/100 g). The analysis of clay fractions shows that the iron content varies from 3 to 3.5, CaO and MgO contents are high. The soils are formed from a variety of rocks which include traps, granites and gneisses.

In Madhya Pradesh, two distinct kinds of black soils are found: (1) deep heavy black soils covering the Narmada Valley and (2) shallow black soils in other areas. The cotton-growing areas are mainly covered by the deep heavy black soils, but there are also soils of lighter texture. The clay content usually varies from 35 per cent to 50 per cent and the SiO_2/R_2O_3 ratio from three to three-and-a-half. The organic matter content is low.

The black soils of Karnataka are fine-textured with varying salt concentration. The soils are generally rich in lime and magnesia, the SiO_2/R_2O_3 ratio of clay fraction is 3.6.

6.10.3 Red Soils

The red soils comprise vast areas of Tamil Nadu, Karnataka, Goa, Daman and Diu, southeastern Maharashtra and eastern Andhra Pradesh, Madhya Pradesh, Orissa and Chhotanagpur. In the north, the red soil area extends into and includes the greater part of the Santhal Parganas in Bihar, the Birbhum district of West Bengal, the Mirzapur, Jhansi and Hamirpur districts of Uttar Pradesh.

The ancient crystalline and metamorphic rocks on mateoric weathering have given rise to the red soils. The colour of the soil is due to the wide diffusion of iron rather than to a high proportion of it. The soils grade from poor thin gravelly and light-coloured varieties of the plains and valleys. They are generally poor in N, P_2O_5 and humus. These soils are poorer in lime, potash, iron oxide and phosphorus than the *regur* soils. Many of the so-called red soils of southern India are not red. On the other hand, some red soils are of lateritic origin and of a quite different nature.

The clay fraction of red soils is rich in Kaolinitic type of minerals. Red soils have also been found under forest vegetation. Red and yellow soils are also seen side by side.

Morphologically, the red soils can be divided into two broad sub-groups: (1) red loams, characterised by argillaceous soils with a cloddy structure and the presence of only a little concretionary material, and (2) red earths where the topsoil is loose and friable and rich in secondary concretions.

Red soils in Tamil Nadu occupy the largest area. They are all *in situ* formations from the underlying rock under the influence of climatic conditions. The rocks are micaceous or red granites; the latter are acidic. The soils are rather shallow, open in texture, have a pH ranging from 6.6 to 8, they have a low base status and their exchange capacity is low. They are also deficient in organic matter and poor in plant nutrients. An anylysis of their clay fractions reveals SiO_2/R_2O_3 ratios of two-and-a-half to three.

The predominant soil in the eastern tract of Karnataka is the red soil overlying the granite from which it is derived. Especially in the districts of Bangalore, Kolar, Mysore, Tumkur and Mandya, this is the chief type, of varying depth. There are shades of red which pass on to yellows. Loamy-red soils are predominant in the plantation districts of Shimoga, Hassan and Kadur. They are rich in the total and available K_2O and contain sufficient amounts of the total P_2O_5 and their lime content varies from 0 1 to 0.8 per cent. N is below 0.1 per cent Fe and Al are high (30 to 40 per cent).

A broad strip of the area running between the eastern and western parts of Coorg is red loam, easily drained, and with a fairly dense growth of trees.

The acid soils towards the south of Bihar, those of Ranchi, Hazaribagh, Santhal Parganas, Manbhum and Singbhum, are red soils. The pH of the soils varies from 5 to 6.8. Another distinctive feature is the higher percentage of acid soluble Fe_2O_3 than that of Al_2O_3. Available K is quite sufficient, but P_2O_5 is low.

In West Bengal, the red soils (sometimes mis-represented as laterites) are the transported soils from the hills of the Chhotanagpur plateau.

A typical profile of red soil at the Chandkhuri Farm, Raipur, Madhya Pradesh, reveals that the percentage of concentrations increases down the profile. The total exchangeable cations are about 20 me/100 g. Thus SiO_2/R_2O_3 ratio of the clay fraction varies between two and three and the C/N ratio is approximately ten.

A part of the Jhansi district in Uttar Pradesh comprises red soils. There are two types, locally known as *parwa* and *rakkar*.

The *parwa* is a brownish-grey soil varying from good loam to sandy or clay loam. The *rakkar* is the true red soil which is not generally useful for cultivation. The soils of Banaras and Mirzapur developed on the Vindhyan parent materials, have also been classified as tropical and subtropical red loams.

In the Telengana division of Andhra Pradesh, where the predominating geological formation is granite and a gneissic complex, both red and black soils predominate. The red soils are sandy loams located at higher levels. Such soils are utilized for the cultivation of *kharif* crops.

6.10.4 Laterites and Lateritic Soils

Laterite is a formation peculiar to India and some other tropical countries, with an intermittently moist climate. It is a compact to vesicular rock composed essentially of a mixture of the hydrated oxides of aluminum and iron with small amounts of manganese oxide, titania and other minerals. It is derived from the atmospheric weathering of several types of rocks. Under the monsoon conditions of alternating wet and dry seasons, the siliceous matter of the rocks is leached away almost completely during weathering.

Laterites are specially well developed on the summits of the hills of Karnataka, Kerala, Madhya Pradesh, the Eastern Ghat regions of Orissa, Maharashtra, West Bengal, Tamil Nadu and Assam. All lateritic soils are very poor in lime and magnesia and are deficient in N. Occasionally, P_2O_5 as iron phosphate may be high but K_2O is deficient. There is occasionally a higher content of humus.

In Tamil Nadu, there are both high-level and low-level laterites. They are found all along the West Coast and also in some parts of the Eastern Ghats.

At lower elevations paddy is grown, whereas at higher elevations plantation crops are grown. The soils are rich in nutrients and contain 10 to 20 per cent organic matter. The pH is generally low particularly of the soils under tea (pH 3.5 to 4), and the higher the elevation, the more acidic the soils are.

In the laterite soils of Ratnagiri (Maharashtra), the coarse material is found in large quantities. These soils are rich in plant food constituents, except lime.

In Kerala, both high-level and low-level laterites occur. The soils are generally poor in N, P, K and organic matter, the pH ranging from 4.5 to 6. The SiO_2/R_2O_3 ratio gradually increases with the decreasing elevation (from 1.32 to 2.08).

The laterite soils in Karnataka occur in the western parts in the districts of North Kanara and South Kanara, Shimoga, Hassan, Kadur and Mysore. All the soils are comparable with the laterites and similar formations in Malabar, and Nilgiri districts. These soils are very low in bases, because of the severe leaching and erosion.

In West Bengal, the area between the Damodar and the Bhagirathi is interspersed with some basaltic and granitic hills, with laterite capping. The SiO_2/Al_2O_3 ratio of the clay fraction is quite high. The percentages of K_2O, P_2O_5 and N are low due to leaching and washing out. In Bihar, the laterite occurs principally as a cap on the higher plateaus, but is also found in fair thickness in some valleys.

The laterites of Orissa are found largely capping the hills and plateaus occasionally in considerable thickness. Large areas in Khurda are occupied by laterites; those of Balasore are gravelly and appear to be detrital.

6.10.5 Desert Soils

A large part of the arid region, belonging to western Rajasthan, Haryana and Punjab, lying between the Indus river and the Aravalli range is affected by desert canditions of geologically recent origin. This part is covered under a mantle of blown sand which, combined with the arid climate, results in poor soil development. The desert sand is composed of quartz but feldspar and hornblende grains also occur with a fair proportion of calcareous grains. The sands are partly derived from the disintegration of the subjacent rocks, but are largely blown in from the coastal regions and the Indus valley. Some of these soils contain high percentages of soluble salts, possess high pH, have a low loss on ignition, a varying percentage of $CaCO_3$, and are poor in organic matter.

The Rajasthan desert is a vast sandy plain, including isolated hills or rock outcrops at places. Though, on the whole the tract is sandy, the soil improves in fertility from west and northwest to east and northeast. In many parts, the soils are saline or alkaline, with favourable physical conditions and pH (Murthy, 1980).

6.11 PROBLEM SOILS

The problem soils are those which, owing to land or soil characteristics, cannot be economically used for the cultivation of crops without adopting proper reclamation measures. Highly eroded soils, ravine lands and soils on steeply sloping lands constitute one set of problem soils. The shallow soil depth, deep gullies and steep and complex slopes are some of the problems which require to be tackled in such areas.

Acid, saline and alkali soils constitute another set of problem soils, in the case of which acidity, soluble salts and exchangeable sodium limit the scope of cultivation.

6.11.1 Acid Soils

Although soils with pH below 7 are considered to be acidic, from the practical standpoint, those with pH 5.5 and which respond to liming may be designated as acid soils. The per cent base saturation and the pH are used as criteria to distinguish acid soils from non-acid ones.

Acid soils occur widely in the Himalayan region, the great eastern plains of extra-peninsular India, the peripheral peninsula and the coastal plains, including the Gangetic delta. They are found to occur on different geological formations under varying physiographical, climatic and vegetational environment. In all these regions the rainfall appears to have a dominating influence on the formation of acid soils.

Soil acidity affects availability of plant nutrients, microbial activities, soil structure and overall plant growth and production. Correction of soil pH by liming is of prime importance. The pH measurements of the soils are widely used for estimating the lime requirement. The lime requirement as determined in the laboratory is multiplied by a 'liming factor' of 1.5 to 2 to achieve the desirable results.

The liming material (limestone) should be finely ground and evenly broadcast and worked into the soil several weeks before sowing a crop to allow time for completing the reaction. Keeping the land moist hastens the exchange process. Although liming once in five years may serve the purpose, the frequency of liming should be determined by making periodic pH measurement.

Many crops are sensitive to the pH range of soils. For example, maize, wheat, barley, sugar-cane and berseem have an optimum pH range of 6.0 to 7.5 while rice and tea have 4.0 to 6. Thus in estimating the lime requirements for a particular crop the optimum pH range of that crop must be considered.

6.11.2 Saline and Alkali Soils

Three classes of saline and alkali soils are recognised. They are briefly described below:

Saline Soils

The soils containing toxic concentrations of soluble salts in the root-zone are called saline soils. Electrical conductivity (EC) in the saturation extract of such soils is greater than 4.0 mmhos/cm at 25° C. Exchangeable sodium percentage (ESP) is less than 15 and the pH is less than 8.5. The soluble salts mainly consist of chlorides and sulphates of sodium, calcium and magnesium. Because of the white encrustation due to salts, the saline soil is also called white alkali.

Non-saline Alkali or Sodic Soils

These soils do not contain any large amount of neutral salts and as such EC is less than 4 mmhos/cm at 25°C. The detrimental effect of alkali soil on plants is largely due to toxicity of a high amount of exchangeable sodium and pH. They have an ESP of more than 15 and a pH greater than 8.5. Such soils have low infiltration rate and the physical condition is unfavourable. Because of high alkalinity, resulting from Na_2CO_3, the surface soil is discoloured and black and hence called black alkali.

Saline-alkali Soils

This group of soils is both saline and alkali. They have appreciable amount of soluble salts as indicated by the EC values of more than 4 mmhos/cm at 25°C. The ESP is greater than 15 and the pH is likely to be less than 8.5.

The soil salinity or alkalinity or both have many adverse effects. They may be:

1. causing low yields of crops or crop failure in extreme cases;
2. limiting the choice of crops as some crops are sensitive to salinity or alkalinity or to both;
3. rendering the quality of produce poor as, at times, the fodder grown on alkali soils may contain a high amount of molybdenum and a low amount of zinc, causing nutritional imbalance and diseases among livestock;
4. causing excessive run off and floods owing to poor infiltration, resulting in damage to crops in the adjoining areas.

In arid and semi-arid areas, salts formed during the weathering of soil minerals are not fully leached. During the periods of higher-than-average rainfall, the soluble salts are leached or washed out from the more permeable high-lying areas to the low-lying areas, where, if the drainage is restricted, salts accumulate on the surface as water evaporates. The excessive irrigation of the uplands containing salts results in the accumulation of salts in the valleys. In areas having a salt layer at lower depths in the profile; faulty water management or even seasonal irrigation may favour the upward movement of the salts. Salinisation is also caused by irrigation with saline water. In all these cases, restricted drainage is usually the main reason. A rise in the water-table within 2 m of the surface due to irrigation. the obstruction of natural drainage by roads or canals and the siltation of natural drainage may also cause soil salinity. In the coastal areas, the ingress of sea water induces salinity in the soil. When sodium ions predominate in the soil solution, and carbonates are present, alkali soils are formed.

Different types of saline and alkali soils may occur singly or in association, depending upon their mode of formation and physiographic position. Also, the degree of salinity or alkalinity may vary. Thus the salt affected soils differ a great deal in their physico-chemical characteristics and, as such, methods of their reclamation also differ.

If the problem is only of salinity, the salts need to be leached below the root-zone and not allowed to come up. In practice, this might be difficult to accomplish, especially in deep and fine-textured soils containing more salts in the lower layers. Under these conditions, a provision of some kind of subsurface drains becomes important. If the soil contains a sandy layer at a lower depth, the leaching of the salts below this layer will check the rise of salts.

The reclamation of alkali soils needs the addition of a soil amendment, containing soluble calcium salts. The commonly used amendment is gypsum. In the course of reclamation, sodium on the exchange complex is replaced by calcium. The sodium salts, thus formed, are leached down.

The number and frequency of leaching, the quantity of gypsum to be added and the techniques involved vary from region to region, depending upon the clay mineralogy of the soils, the intensity of the problem, the subsequent use of the soils, the availability and quantity of irrigation water and the economics of these operations. The operations may also include other simple agronomic techniques to reclaim soils and to know the crops and varieties that may be grown and other management practices that maybe followed on such soils (Murthy, 1980).

6.12 SOIL MANAGEMENT

Successful farming concerns the appropriate management of soil, plants and the environment in such a way that a maximum return can be obtained not only in a season or year but also over centuries. Thus the physical, chemical and biological properties of soils and their modifying factors regulate the present and future state of soils, the source of infinite varieties of life. The most important consideration in soil management is the correct application of the relationships among the soil, the environment and the crops to be grown. Although the problems of soil management vary according to the soils and their situation in the land, the climatic conditions, biotic influences and crops to be grown, yet there are fundamental factors which govern the choice of a suitable soil-management practice.

Good soil tilth is the first feature of good soil management. It means a suitable physical condition of the soil and implies, in addition, a satisfactory regulation of soil moisture and air. The maintenance of soil organic matter which encourages granulation is an important consideration of good tilth. Tillage operations and timings should be so adjusted as to cause the minimum destruction of soil aggregates. Good tilth minimises erosion hazards.

The choice and sequence of adaptable crops or crop rotation are other very important considerations. These are related to climate, particularly rainfall and its pattern (unimodal, bimodal or pseudomodal) of distribution and the characteristics of the soil profile, including drainage and extent and duration of available soil moisture. A proper sequence of crop varieties greatly influences soil conditions. It is more realistic to evolve cropping patterns and soil management practices according to the land-capability maintaining the aim of raising lower capabilities to higher and more productive ones. The cropping pattern and the management practices should be based on the principles of soil and water conservation and efficient nutrient and moisture utilisation. In irrigated areas, special management practices become necessary to avoid salinity, alkalinity, waterlogging, leaching and the loss of plant nutrients. In rain fed areas special management practices include improving soil conditions to receive, retain and release more soil moisture, harvesting water to use as life-saving irrigation or extending the cropping season when there is insufficient rainfall for raising crops, protecting the soil from degradation both in cropped and bare fields. Land shaping, levelling, mulching and the use of wind brakes and vegetative cover are the other aspects. The productive capacity of the soil should never be allowed to diminish, but rather should be improved and maintained by providing adequate organic manures and plant nutrients through fertilisers and by including legumes in the rotation and the use of biofertilisers. Similarly, the provision of irrigation facilities in semi-arid and arid areas, the adoption of different remedial measures against excessive salinity and alkalinity or acidity in humid areas, the use of specific soil amendments to correct imbalances of plant nutrients and the application of trace elements where they are deficient, are some of the other improvement measures needed in different and special circumstances.

Economic plant-protection measures against pests, pathogens and parasites including weeds should form part of the management practices in the cropping system. This object may be achieved by regulating the cultural practices or by using recommended practices regarding the application of pesticides including herbicides. The use of these inputs should never be to replace or supplement the deficiencies of management practices.

The economics of selected management practices is of vital importance. Emphasis should be laid on maximising sustained income, rather than yields for the time being. Unless the recommended practices are economically profitable and within the limits of the present state of adoption they are not likely to be adopted, however sound they may be on scientific considerations. Therefore, a package of practices for the integrated land use plan, including all the above points and yet economically profitable, is necessary (Murthy, 1980).

CHAPTER 7

Tillage

Tillage is the tilling of land for the cultivation of crop plants, i.e. the working of the surface soil for bringing about conditions favourable for raising crop plants. Work involves a transference of force from one body or system to another. Force requires some energy or power. Therefore, tillage operations require power that may come from manual, animal, mechanical and other sources as well as tools, implements and equipment as aids to disturb and disrupt the normal state of the soil to a desirable extent. Tillage is a laborious and expensive cultural practice.

Tillage operations in various forms have been practised from the very inception of growing crop plants. To prepare a virgin or fallow land and use it for growing crops, tillage in any form is an indispensable practice even today. Tillage is one of the forms of management practices of soil, water, nutrient, crop and pest. Tillage helps to replace natural vegetation with useful crops and is necessary to provide a favourable edaphic environment for the establishment, growth and yield of crop plants.

The physical condition of the soil brought out by tillage, that influences crop emergence, establishment, growth and development is termed *tilth*. It is a loose, friable, airy, powdery, granular and crumbly structure of the soil with optimum moisture content suitable for working and germination or sprouting seeds and propagules.

7.1 OBJECTIVES OF TILLAGE

The objectives of tillage operations may be grouped into physical, chemical and biological benefits derived from the soil. These are applicable to both cropped and uncropped arable lands.

7.1.1 Physical

a) to cut, heave and shatter the dense soil to a desirable depth and to break the clods and crusts to a desirable extent for a suitable seed-bed or planting field; to bury in or cover and pack the seeds or plant propagules with soil;

b) to improve the capacity of soil in receiving rain or irrigation water, retaining and releasing moisture for crop plants or to increase percolation or drying of excess soil water;

c) to redistribute the soil constituents particularly soil particles, organic matter, micro-organisms, moisture and air;

d) to incorporate crop residues, green manure and other organic manures, fertilisers, soil amendments and conditioners and soil applied agro-chemicals and to remove stiff roots, stubbles, root stocks and stumps;

e) to maintain the proper structural condition of the soil;

f) to prepare the land surface suitable for easy, early and uniform movement of irrigation and drainage water;

g) to increase soil aeration particularly in non-capillary spaces;

h) to reduce soil erosion, degradation and depletion;

i) to modify (slightly) the thermic capacity of the soil;

j) to prepare soil mulch or to preserve the particle mulch to conserve soil moisture and to make an impervious soil layer in wet land paddies to check loss of water through percolation.

7.1.2 Chemical

a) to accelerate the weathering of the soil;

b) to expedite the reclamation of problem soils;

c) to improve the availability of plant nutrients by enhancing the decomposition of organic matter, mineralisation etc.;

d) to expel toxic gases from the prolonged reduced soil conditions and to detoxify soil from any erratic use of agro-chemicals.

7.1.3 Biological

a) to control weeds, soil-borne insect-pests, pathogens, larger soil animals including rodents and vermins. Tillage may induce the germination of weed seeds which can then be destroyed by subsequent tillage or chemical treatment (cf. activise dormant weed seeds to put in perilous conditions) to impoverish the population of weed seeds in the soil. Tillage may also prevent weed seed germination by the burial effects on seeds of deep ploughing. Tillage physically alters the weed relationship with the soil. It may destroy weed seedlings by uprooting, smothering, desiccating, decomposing or merely weakening,weed plants through dislodging, damaging, disorienting, depleting food reserves, root pruning or other injury. Tillage may bring out roots, stolons and other such fleshy and regenerative organs of weeds to the surface where they dry out.

   Tillage reduces or eliminates weed competition for moisture, nutrients, light, carbon dioxide in the microenvironment and thereby improves crop growth. It would have been a sorry thing for agriculture if there had been no weeds. They make us stir the soil, and stirring the soil is the foundation of food farming (Bailey, 1950).

   Tillage, particularly ploughing exposes the lower soil to weather and places the surface soil underneath; thus soil inhabiting organisms are destroyed by heat, desiccation, predatory animals and birds, pathogenic and parasitical organisms or by direct injury and the burial effect. Other pests and parasites are also controlled by breaking the green-bridge by tillage;

b) to improve distribution, nutrition and growth of roots by reducing penetration resistance of the soil, promoting root respiration and providing accessibility of roots to the moist zone in the soil;

c) to provide an optimum habitat that encourages early and uniform emergence and establishment of seedlings;

d) to provide better anchorage to crop plants and greater

space for the underground development of storage roots and stems;

e) to prune older and inactive roots, to earth up and to suppress late tillers, shoots, suckers and stolons;

f) to encourage growth and activity of soil-inhabiting beneficial flora and fauna including symbiotic bacteria;

g) to provide conditions suitable for necessary field operations, for instance planting and harvesting quickly, smoothly and uniformly;

h) to provide a better total underground environment for a prolonged period of growth and yield of crop plants and better care and management of the soil even when there is no crop in the field.

7.2 DISADVANTAGES OF TILLAGE

Though tillage is an indispensable operation in crop production, it has some drawbacks. As field soil is a habitat for several organisms including crop plants, it is exposed to weather. Therefore, soil environment is greatly modified by the ever dynamic atmospheric environment. In general, a very short time is available to prepare all the fields at their optimum conditions for seasonal cropping. If the fields are not prepared within the required period, the soil may either lose or gain moisture and become unworkable or the sowing season is delayed considerably. This condition is particularly true for rain fed farming or areas under a monsoon climate with poor soil conditions. Tillage operations scheduled at beyond optimum conditions cause the physical condition of the soil to deteriorate.

Repeated tillage operations or tillages over longer periods have detrimental effects on surface soil granules. Tillage hastens the oxidation of organic matter from the soil. Tillage operations with heavy equipment tend to break down the stable soil aggregates and form a hard pan immediately below the tilled layer (Brady, 1974). All this reduces infiltration and induces run off, soil erosion and other negative effects.

Among draft animals, cattle bullocks are generally lighter and place their hind feet in the footprints of their fore feet whereas, buffaloes do not; thus buffaloes trample and compact the soil more than cattle bullocks. Light soils require less draught power for tillage. Therefore, buffaloes are generally unfit for tillage operations in light and shallow soils whereas, cattle bullocks are unfit for heavy soils but where power requirement is higher two or more pairs of cattle bullocks may be used replacing one or two pairs of buffaloes.

In non-plastic or less plastic soils (which when moist cannot be rolled into a long cylinder and even if they can be rolled, they will break on bending, cf. light soils) more than adequate tillage operations render the soil more droughty and lacking fertility. In plastic soils (cf. heavy soils) if a tillage operation is scheduled at beyond the optimum moisture conditions, negative results occur. If such soils are ploughed when too wet, the soils get puddled and become practically impervious to air and water and also structureless; on drying the soils become hard, dense and unworkable until remoistened. On the other hand, if heavy soils are ploughed when too dry, great clods are turned up which are difficult to work into a good seed-bed. It may require more draught power.

Frequent tillage after ploughing pulverizes the soil into dust, breaks down the soil aggregates and enhances wind and water erosion. Repeated cultivation even under optimum soil moisture conditions impairs the tilth more than ploughing. The soil surface is puddled forming a reduced condition in the root zone due to the beating action of rain drops on powdery structureless conditions. On drying it forms a layer of crust which causes mechanical resistance to the emergence of crop plants, restricts gaseous exchange and thus impairs crop stand and growth. It also creates more intimate soil-seed contact for weed seeds to germinate since most weed seeds are quite small.

Intertillage after a certain growth of crop plants may cause damage to them both above and below the ground. This is most deleterious when the crop is in the crook or spike stage.

Tillage operations immediately after the harvest of crop plants very often incorporate weed seeds in soil. Deep ploughing increases the volume of soil containing weed seeds and other propagules and preserves a source of infestation for a longer time. Tillage helps to spread the colonies of different soil-borne pests, pathogens and parasites. Weeds that propagate vegetatively (particularly rhizomatous and stoloniferous perennials) multiply and spread profusely from the fragments of stem and root stocks by the severing, shearing and tearing action of tillage implements. Tillage operations may cause shifting of the weed flora to difficult-to-control species. Tillage has limited effectiveness against weeds within crop rows. Tillage requires a post-operation warm-dry period to destroy weeds. Thus tillage as an independent factor seldom controls weeds.

The development of herbicides has made it possible to study the effects of tillage as an agricultural practice independent of weed control. In general, it has been found that tillage is of little benefit, other than for weed control, under most soil conditions (NAS, 1971).

7.3 TYPES OF TILLAGE

Tillage operations may be grouped into on-season and off-season tillage.

7.3.1 On-Season Tillage

Tillage operations that are done for raising crops in the same season or at the onset of the crop season are known as on-season tillages. They may be preparatory tillage and inter-tillage.

A. Preparatory tillage: This refers to tillage operations that are done to prepare the field for raising crops. It consists of deep opening and loosening of the soil to bring about a desirable tilth as well as to incorporate or uproot weeds and crop stubbles when the soil is in a workable condition. Preparatory tillage is divided into primary and secondary tillage operations.

1) *Primary tillage*: In preparatory tillage, the first cutting and inverting of the soil that is done after the harvest of the crop (turn around period) or untilled fallow or to bring virgin or new land under cultivation (as it is in shifting cultivation) is known as primary tillage. This primary tillage may be done once or twice a year in normal and settled agriculture or once in four to five years in dryland agriculture. The depth of primary tillage may range from 10 to 30 cm. Primary tillage includes ploughing in which the soil is cut, lifted, shattered, twisted, inverted and sheared for further preparation. Ploughing continues to be an important factor in the structural

management of land and the maintenance of crop yields.

2) *Secondary tillage*: This refers to those tillage operations that are done after primary tillage to bring a good soil tilth. In this operation soil is not inverted but instead is stirred and conditioned by breaking clods and crusts, uprooting and removing weeds, stubble, and root stocks, incorporating manure and fertilisers, closing cracks and crevices that form on drying and accelerate drying of deeper soil layers, levelling, preparing ridges, furrows, and irrigation and drainage channels in the fields, conserving moisture, increasing infiltration and aeration and checking erosion. Post-ploughing but pre-seeding tillage operations with their stirring effect on the soil, destroy weed seedlings and prevent weed-seed germination.

Secondary tillage includes cultivating, harrowing, pulverizing, grazing, raking, hoeing, levelling and ridging.

Only secondary tillages may be sufficient as preparatory tillage, in cropping under aerobic (upland) soil conditions and also when the preceding crop is harvested by tillage operations as well as in other similar situations. In ratoon cropping, *paira* cropping, till planting and also in some high intensity cropping systems no preparatory tillage is essential provided the soil is in good tilth.

B. Inlertillage: This refers to tillage operations done in the field after sowing or planting and prior to the harvesting of crop plants, i.e. during field duration. This is also known as intercultivation or post-seeding/planting cultivation. It includes cultivating, harrowing, hoeing, weeding, earthing up, ridging and furrowing. Intertillage is shallower in nature and allows the crop to emerge, keeps the field free from weeds for a significant period of time by killing germinated but pre-eraerged or emerging weeds. Post-planting broadcast tillage which provides the minimum disturbance to crop plants in rows is more advantageous to destroy. weeds that come out from different depths of the soil and have different root systems. It is advantageous to the thining of extra plants.

Intertillage helps to incorporate top dressed manures and fertilisers, to earth up and to prune roots. In wet land direct-seeded and transplanted paddies it induces puddling and bushening of rice plants by inducing tillering.

7.3.2 Off-season Tillage

Tillage operations done not for immediately raising crop plants but for conditioning the soil suitably for the forthcoming main-season crop are said to be off-season tillages. These are scheduled during uncropped seasons for levelling to the desirable grade, leaching to remove salts, lowering of a seasonal water table and reducing the population of harmful flora and fauna in the soil. Off-season tillages may be post-harvest tillage, summer tillage, winter tillage and fallow tillage.

A. Post-harvest tillage: Immediately after the harvest of the crop from the field, the soil is tilled if it is in a favourable condition for such an operation. Ploughing becomes indispensable after a wet land crop (rice), long duration crop (sugar-cane, napier), sod or turf crop, green mulch crop or green manure crop. If the preceding crop is grown in aerobic soil conditions, harrowing serves the purpose.

Removing stiff stubble, turning under trashes and weeds, improving the physical condition of the soil, mixing the stratified layers of soil constituents, breaking up clay pans and other soil barriers and making the soil receptive to rain water are the major objectives of such tillage operations. Post-harvest tillage prevents seed setting of late maturing annuals and eliminates the seedlings of annuals and biennials that come up at the later stage of crop growth. Shallow post-harvest tillage which causes a minimum disturbance of crop plant residues, may place newly matured weed seeds in a more favourable environment for over-ripening that has variable effects on survival. It also provides a better environment for the germination of volunteer crop plants from shattered crop seeds and thus exhausts the population of such seeds which may cause an admixture in the forthcoming crop seasons.

B. Summer tillage: In tropical zones the warm humid period is preceded by hot summers. There may be intermittent rains with variable intensity which induce weed seed germination even though conditions are not favourable for raising any crop. Tillage operations done during this period are known as summer tillage. This is done mostly as pre-preparatory tillage for main season crops. It is generally of the soil stirring type to destroy weeds and other soil-borne pests, pathogens and parasites by baking under the sun, receiving and retaining rain water and checking water erosion. Summer tillage affects soil aggregates and soil organic matter and induces wind erosion.

C. Winter tillage: In temperate zones or in areas with severe winters that are unfit for raising crops or in fields where cropping is not possible during winter months for various reasons, (e.g. lacking irrigation facilities) but the soil is in good condition for tillage, ploughing and harrowing are practised. In low-land areas or in areas where the water-table is high and drainage is impeded during the post-monsoon period but the soil moisture is in the optimum condition for tillage during the cool-dry period, winter tillage is practised. This is mostly as pre-preparatory tillage for raising spring planted crops. It is practised mainly to destroy weeds, improve the physical condition of the soil and incorporate crop residues. In some areas autumn ploughing is practised after harvesting short duration *kharif* rice.

D. Fallow tillage: Fallow refers to the leaving of arable land uncropped for a season or seasons for various reasons which may be intentional or need-based or due to unfavourable weather and/or soil conditions. Tilled fallow represents an extreme condition of soil disturbance in that the soil is tilled frequently to eliminate all vegetation. Intentionally leaving the land fallow is done mostly for the eradication of pernicious, problematic, poisonous, perennial weeds, for exhausting the population reserve of weed and volunteer crop plant seeds, the destruction of soil-borne ubiquitous pests, pathogens and parasites of monophagous in particular and polyphagous in general or for the improvement of edaphic conditions of the field by changing the agro-ecological situation.

Regarding the control of perennial weeds, repeatedly (16 tillage operations at intervals of 12 days) tilled fallow continues to exhaust food reserves in the roots or underground parts particularly when they are not in a dormant state. Any serious lapses in the tillage sequence permit storage of food reserves and prolong the time required for eradication (NAS, 1971). The beginning of fallow tillage should be when food reserves in the underground portions of the weed plants are at the lowest level.

If the soil moisture is inadequate during the season, the survival of annual weed seeds or disease spores in a dormant condition is enhanced under this tilled fallow system and tillage intended to promote their germination and destruction during fallow tillage is futile and is wasteful of soil moisture. Soils dry rapidly as the depth of tillage increases; the shallow and stirring type of tillage is less wasteful of rainfall than deeper tillage. Fallow tilled soil is prone to be eroded both by wind and water, degraded and depleted.

7.3.3 Special Purpose Tillages

Tillage operations intended to serve special purposes are said to be special purpose tillages. They may be subsoiling, levelling, blind tillage, clean tillage, mulch tillage, contour tillage, wet tillage and zero tillage systems.

A. Subsoiling: To break the hard pan beneath the plough layer special tillage operations (chiselling) are performed to reduce compaction so that a greater volume of soil may be obtained for the cultivation of crops, excess water may percolate downward to recharge the permanent water-table and reduce run off and soil erosion. Roots of crop plants can penetrate deeper to extract moisture from the parched water-table.

Sub-soiling is essential once in four-to-five years where heavy machinery is used for field operations such as tillage, seeding, harvesting and transporting and where there is a colossal loss of topsoil due to carelessness.

B. Levelling: Arable fields require a uniform distribution of water and plant nutrients for uniform crop growth. This is achieved when the fields are kept fairly levelled. In levelled fields soil erosion is restricted and other management practices become easy and uniform. Erosion, siltation and the wrong operation of tillage implements disturb the levelled condition of fields. Some special tillage operations are needed for shaping, scraping and levelling the fields as well as the holding as a whole at regular intervals and as a routine measure for land improvement. This levelling is very important for large fields having many crop-beds, irrigation and drainage channels, or ridges and furrows in different directions. It necessitates the shifting of a layer of soil from one side to the other or vice versa. Levellers, scrapers and rollers are used for such operations.

C. Blind tillage: Blind tillage refers to the tillage of the soil after seeding or planting the crop, either at the pro-emergence stage of crop plants or while they are in the early stages of growth so that crop plants (e.g. cereals, tuber crops etc.) do not get damaged but broad-leaved weeds and extra plants are uprooted. The germination of weeds should have begun but emergence is not necessary before tillage starts. This tillage is mostly of the soil stirring type and the depth of it is restricted to above the seeding zone of the crop. This is practised mainly when crop emergence is delayed because of low temperatures or the intrinsic nature of the seeds, but soil conditions are favourable for profuse weed infestation. Tillage operations are done blindly assuming that there is no standing crop in the field. Tillage implements should be light, shallow and speedy and therefore secondary tillage implements are used for this purpose. Tillage operations may be along or across the crop tows.

D. Clean tillage: This refers to the working of the soil of the entire field in such a way that no living plant (weed, crop plant or crop residues) is left undisturbed. It is adapted to destroy weeds and crop plants that may regenerate or volunteer as weeds or may trap insect-pests, pathogens and parasites as direct or alternate hosts or provide feeding, breeding, heeding and hiding sites for them by maintaining a green-bridge for subsequent infestation of the succeeding crop or crops. It is practised between two cropping periods with repeated tillage operations. Such practice enhances the loss of soil, water and soil structure. Soil turning implements are used for this purpose.

E. Mulch tillage: It refers to tilling the soil in such a way that the mulch materials (particle mulches), surface clods or crop residues are least disturbed. It is also known as stubble-mulch tillage or subsurface tillage according to the nature and form of tillage. In areas and seasons with deficient and precarious rainfall and high speed desiccating winds, stubble-mulch tillage is beneficial for various reasons. Tillage implements are light and subsurface soil stirring types.

F. Contour tillage: This refers to the tilling of the land along contours or lines of uniform elevation in order to reduce soil erosion. Tillage operations, such as ploughing, disking, harrowing, cultivating, and others, can be done on the contour but it is usually more difficult because there are no permanent markings to follow (USDA, 1964). It is practised in gently to moderately sloping land (five to twelve per cent slope), subject to wind or water erosion or both. Contour tillage in such lands is helpful for prevention of run off and erosion.

G. *Wet tillage*: This refers to tillage operations done when the soil is in a saturated (anaerobic) condition. Ploughing and harrowing the soft saturated soil (puddling) are practised for preparing an impervious soil layer to reduce loss of water by percolation from wet land paddies and wet nurseries, for the incorporation of green manure and for weed control and the reclamation of salt affected soils. In areas and seasons which have a prolonged humid period (in which total precipitation exceeds potential evapo-transpiration) or in low lying areas with impeded drainage or widespread surface flood irrigation from canals for a prolonged period and beyond the control by the farmer locally (at the field level), wet tillage is the only means of land preparation for seasonal cropping with semi-aquatic crop plants such as rice.

Wet tillage destroys the soil structure, induces stratified layering of soil particles of different densities (mass per unit volume), controls soil erosion by permitting a layer of standing water over it, prolongs the saturated condition of the soil, reduces loss of plant nutrients by leaching, prevents cracking of the soil and thus prevents the tearing of crop roots and their injury, converts soil pH towards neutrality, reduces weed emergence and growth, facilitates uniform levelling of the field, hastens transplanting operations easily and smoothly as well as the establishment of seedlings and modifies the agro-ecological system thus producing bio-chemical and bio-physical reactions, (De, 1981).

H. *Zero tillage system*: This refers to the growing of a crop with the least possible soil disturbance, which involves controlling unwanted vegetation by other than mechanical means or by the minimum use of tillage equipments. This is also known as minimum or reduced tillage, as seeding requires some tillage operations. Thus zero tillage is a tillage system in which mechanical soil manipulation is reduced to traffic and seed-bed preparation only (Baeumer and Bakermans, 1973). Even for wide spaced crops seed-bed preparation can be dispensed with and instead root-beds or pits, holes or stations may be prepared for seeding. On soils of good tilth, there, is usually no advantage in cultivation if weeds can be controlled in some other manner. Where weeds are to be destroyed with a minimum of soil upheaval, blade or sweep implements are used.

Weed seeds that remain close to the soil surface do not remain dormant as they are exposed to fluctuations in soil temperatures, soil moisture and the action of micro-organisms. Properly planned tillage frequently promotes the emergence of one or more 'crops' or flushes of weeds before the crop is planted (stale seed-bed technique). Plough-plant procedure, wheel track planting, strip planting, seeding from aeroplanes or by hand or by spreaders on the moist soil surface or immediately preceding rain, or splashsprinkler irrigation are the different methods of sowing under minimum tillage practices.

The zero tillage system was first used successfully in 1950 in pasture renovation in USA with mulch farming practices and the tilt-plant system for row crops in order to provide year round protection of the soil from erosion and to minimise planting cost. This system is found to be an alternative to the conventional tillage system where soils are subject to wind and water erosion, the timing of tillage operations is too difficult, the requirements and costs of energy and labour are too high and performance is insufficient.

A non-tilled site is always covered by some plant or animal residues. A recently tilled field has a rugged surface but zero tilled soil surface is relatively smooth and even. Undisturbed soil appears more dense and firm and bears the characteristic structure, texture and other physical, chemical and biological capacities of typical soils rich in organic matter. The population of both micro- and macro-organisms increase and beneficial organisms exceed and dominate over harmful ones. The activity of earthworms increases resulting in biological turning and opening of the soil, forming tunnels and mittens in or on the soil. Undisturbed decayed roots also provide channels. Some soil inhabiting or migratory animals such as moles, voles, mice, eels, pisces, crustaceans, reptiles and mites burrow through the soil, turn it and make larger openings in search of food or shelter or both. All these influence the rate of water infiltration and the mixing of soil constituents and protect the soil from erosion, degradation and depletion but their activities are restricted in dry and cool soils.

With zero tillage, plant residues remain on the soil surface. This provides a clothing that adequately protects the soil from erosion without presenting problems with seeding, weed control, soil fertility, infiltration of rain water and aeration. Instead, it improves soil capacity and workability. In undisturbed soils, the concentration of organic matter is highest near the soil surface and declines steadily with depth.

A minimum volume of ten per cent pore (air) space is necessary for adequate gas exchange. Impeded aeration is recorded in heavy soils during humid periods or where and when drainage is delayed or restricted or there is higher compaction due to heavy traffic particularly on moist soil. Raising crops requiring well drained, well aerated soil conditions on such soils and seasons under zero tillage is a futile attempt. Though the total pore-space in zero tilled soil is lower in surface area, in the deeper soil layer, the differences tended to diminish.

Zero tilled surface soil holds more available water immediately after rain whereas, the reverse is true in the subsurface layer. Zero tilled soil restricts the infiltration rate particularly with low rainfall. Most of the rainfall is soaked up by the organic matter on the surface and a little infiltrates downwards at a slow rate. Under normal situations soil water tension remains high on the surface of the untilled soil and decreases rapidly in the subsoil.

Organic mulches physically absorb raindrop impact energy, thus the slaking and sealing of the soil surface is prevented or at least retarded. Therefore, zero tillage reduces run off, prevents soil erosion and crust formation and thus protects both soil and water for better use by crop plants.

The minimum tillage procedures reduce labour but crowd work into a shorter time period immediately prior to planting. The fact that tilled soil dries more rapidly than untilled soil often affects the choice of method. Since timeliness is critical to farm operations, this is the significant disadvantage of minimum tillage (NAS, 1971). In plough-plant procedure farming, the soil over the crop row creates a depression. Heavy rainfall may erode herbicides applied to the soil from the sides of the depression and concentrate them in the crop row. Thus the expected results may be negated. Land improvement, reclamation of problem soils, cultivation of tuber, root and rhizone crops that bear an economic yield in subterranean parts are seriously affected in the zero tillage system of farming.

7.4 TILTH

Tilth, which is defined as the physical condition of the soil in its relation to plant growth, not only depends on granulation and its stability but also on other factors such as, moisture content, degree of aeration, rate of water infiltration, drainage and capillary-water capacity. Soil tilth is not static but changes rapidly and markedly. For instance, a slight change in the moisture content may alter the workability of the soil. Mechanical forces may alter the structure, the total porosity and bulk density of the tilled layer. One of the main objectives of tillage is the encouragement and maintenance of good tilth. Administering tillage in improper soil conditions may seriously impair tilth directly or set the stage for later deterioration. However, soils with a mellow tilth can be cultivated at both higher and lower moisture contents than soils with a raw tilth.

In spite of its dynamic nature, measurements of tilth can be made by aggregate analysis, determination of porosity, density, water-holding capacity, saturated hydraulic conductivity and force of penetration to study the suitability of soil management practices and crop cultivation. From the field point of view the following properties of a good tilth are of great importance:

a)   There should be a continuous system of wide pores from the soil surface down to the water table or the land drains through which surplus water can move rapidly and which will allow rapid diffusion of CO_2 from the subsoil to the atmosphere.

b)   These pores should be sufficiently stable to last several years before being filled up.

c)   There should be volumes in between these pores that hold as much water as possible against drainage, but which are readily accessible to the plant roots.

d)   The surface soil should be crumbly, and the crumbs

should be large enough not to blow away, but small enough to allow good germination of the seed, and sufficiently unsticky when moist for them to keep their individuality when tractors or implements move over them (Russel, 1961).

CHAPTER 8

Types of Tillage Operations

Tillage operations are broadly grouped into ploughing, harrowing and cultivating.

1. PLOUGHING

After the harvest of the crop the field soil is undercut, inverted and exposed to weather by the use of ploughs. Ploughs may cut the soil forming a V shaped furrow (as in country ploughs) or U or L shaped furrows (as in mould board and disc ploughs). The furrow slice may be laid on one or both sides of the furrow forming either a flat or ridge shape at the crest. A country plough forms a notch whereas mould board and disc ploughs form a flat type furrow sole. Between two V shaped furrows a portion of the soil is left uncut which does not take place with the U or L shaped furrows. Shifting and turning of the furrow slice are complete with U and L shaped forrows but not with V shaped furrows; thus the latter leaves a corrugated surface on the dead furrows. Mould board and disc ploughs need more draft power than that of indigenous ploughs. Country ploughs made of wood or iron are simple in design and light and may be used for other types of tillage, for instance harrowing, cultivating, sowing and covering seeds, whereas improved ploughs are stronger and heavier and designed to meet specialised operations and thus cannot be used for all purposes except ploughing, ridging and furrowing. To achieve the complete severing and inverting of the soil up to the desirable depth more than one ploughing is needed with a country plough while one operation is enough with mould board or disc ploughs. The progress of ploughing with country ploughs is gradually deeper and closer whereas, it is gradually shallower with mould board and improved ploughs if they are needed at all for further use.

Different types of ploughs are in use. They have been grouped in various ways:

A. 1) Rolling type: disc plough with smooth or serrated edge of the discs which may be arranged single-way or double-way.

2) sliding type: mould board plough with single or double mould board, Bose plough and may be lister or buster type.

B. 1) Subsurface plough: shovel, chisel plough.

2) Surface plough—a) soil stirring type, e.g. country plough

b) soil turning type, e.g. mould board plough, disc plough

C. 1) Animal drawn plough—a) walking type: country plough, Bose plough

b) riding type: gang plough

2) Tractor drawn plough—a) trailing type: disc plough

b) mounted type: mould board plough

2. HARROWING

The soil needs to be worked after ploughing to bring it into a desirable tilth condition for sowing or planting. For this purpose soil is stirred, clods are crushed, stubbles are removed or incorporated, weeds are destroyed and land surface is levelled and compacted. All these operations are brought about by harrowing. Harrowing is also done during uncropped seasons for levelling, controlling weeds, receiving and conserving moisture.

Different types of harrows are used in different agro-climatic zones to meet different tillage requirements. They may be spike tooth, spring tooth, disc, knife, blade-harrows, ladder or planking type harrow, rake, clod crusher, pulveriser, roller or leveller.

Harrows may be pull or push types and rolling, sliding and grazing types. Harrowing is done mainly for the final preparation of the land.

3. CULTIVATING

The working of soil after seeding, planting or transplanting becomes essential for various purposes such as loosening surface crust due to the post-planting rains or irrigation, weeding thinning, hoeing, earthing up, mixing fertilisers and agro-chemicals and to bring a good physical condition of soil for the underground growth of the crop plants. All these operations are brought about by special types of tillage implements. Large implements have a provision for changing the spacing between the tynes or blades for adjusting them with crop rows. Cultivators work mostly by a sweeping action in or on the soil surface.

Different types of cultivators are of

A. Shovel type—e.g. shovel;

B. Tyne type—e.g. duck foot tyne;

C. Blade type—e.g. disc cultivator;

D. Hoe type—1) hand hoe—e.g. *khurpi*, spudder, trowel, bull tongue.

    2) spade—e.g. hand spade, phowrah, mattock, delver, pick axe, crow bar, single hoe, double hoe.

    3) rotary hoe e.g. wheel hoe, star pulveriser

    4) weeder—e.g. finger weeder, rod weeder, sweeps.

Cultivators may be push or pull types and sliding, scraping, rolling and digging types.

8.1 METHODS OF PLOUGHING

Ploughing may be of the pull type with straight and forward movement or rotary (or rotovation) type where soil is cut and inverted by the rotary motion of the discs of the plough. Soil may be ploughed forming beds or furrows leaving a smooth, rugged or ridged surface.

Ploughing types may be:
side to centre, centre to side and ploughing in lands.

8.1.1 Side to Centre Ploughing

In fields of irregular or circular configuration ploughing may be started from the side and gradually proceed towards the centre. This type of ploughing is particularly important where the field surface is shaped like a pan, i.e. the outer sides are elevated and the centre is depressed or one side is elevated and arched. This type of ploughing helps to shift the soil towards the centre or depressed site and gradually the field becomes levelled. This type of ploughing is also known as the splitting type in which the diameter of the unploughed area is shortened gradually by the inward spiral movement of the plough. In this type the last run of the plough leaves an open furrow in the centre.

8.1.2 Centre to Side Ploughing

This is the opposite of side to centre ploughing. This is important where the top of the field is of the turtle back type in which the centre is more elevated than the outer sides. This type of movement of the plough helps to level the field gradually. This type of ploughing is also known as the gathering type as each run of the plough gathers some more ploughed land. As the movement of the plough is spiral and outward the last run is at the boundary line.

8.1 3 Ploughing in Lands

In this method a thoroughly levelled and regular shaped (rectangular or square) field is divided into some small and workable units, either along the length or breadth and each one is called land. These lands should be such that turning becomes easy and does not cause repeated ploughing or heaping of earth in the head lands (turning sites of the field).

In such a type of ploughing prior to the last run of the plough in the land an extension (gathering) is made to mark the second land and before the last run of the plough in the second land, an extension is made to mark the third land.

In ploughing with larger machinery or two or more pairs of bullocks one after another, a larger land is marked at first and after ploughing about half of the first land, an extension is made for the second land. The second land is then ploughed by splitting and onward movement and the rest of the first land is ploughed during the return of the plough. At the last run of the first land leaving an open furrow an extension (gathering) is made to the third land and thereafter the third land is ploughed by onward movement and the second land is ploughed during the return of the plough. In this way the entire field is ploughed down. Each land is ploughed two ways during the onward and return type of ploughing in contrast to the centre-to-side or side-to-centre ploughing where there is one-way ploughing. There remains an open furrow in the middle of each land.

In moderately irregular shaped land all these three types of ploughing may be needed to cover the whole area.

In two-way ploughing head lands are heavily trampled by traffic and therefore should be ploughed separately and along the boundary.

Except with horses as the source of draft power left hand turning is always beneficial. During tillage operations in any field one-way turning has many more advantages than two-way or irregular turning.

The movement of the plough may be straight and turned or circular or the gathering and splitting type. If a second ploughing is needed it should be across the first ploughing. The areas prone to be eroded or in contour ploughing areas where second ploughing should not or cannot be across the first ploughing, zig-zag ploughing may be adopted.

8.2 PREREQUISITES FOR PLOUGHING

8.2.1 Soil with Optimum Moisture

For dry or wet ploughing the soil should not stick to the plough, the bullock's feet, the operator's feet or the wheels of the tractor. The soil should not be dry and hard enough to resist penetration of the share to the desirable depth. It should not take more draft power than usual. Optimum soil moisture for ploughing decreases penetration resistance and increases the shattering action of the furrow slices whereas moist soil has a low load bearing capacity, increases rolling resistance and destroys the structure of the soil to an unworkable level until it is made thoroughly wet. Optimum moisture of the soil should be uniform throughout the field to be ploughed.

8.2.2 Field Should be Free from Scraps

Field soil should be free from stiff stubbles, strong stumps, hard sods, turfs, shrub, brush and creepers as well as from stones, rubble and boulders which may hinder the movement of, or cause damage or even breakage of the plough.

8.2.3 Suitable Type of Plough and Draft Power

For ploughing a clean and level field under optimum moisture conditions, the country plough or mould board plough may be used. For early ploughing when soil is nearing the optimum moisture level, a country plough is better as it does not disturb the bottom-soil and exposes the soil for drying. In fields that are heavily infested with weeds, trash and particles of soft stubble, the disc plough is better. Under wet conditions puddling is better with a country plough but for incorporating weeds and green manure crops a Bose plough is better after planking.

The plough should be in good condition, sharp and well adjusted.

Draft power should be such that it is not bothered by a little more resistance than usual in certain weak spots. Draft power should have uniform and rapid movement and should trample less during traffic.

8.2.4 Clear Weather Conditions

Weather conditions during ploughing should be such that they will not hinder the workability of the soil, draft power or the operator. High speed winds or high intensity rain during ploughing may induce loss of soil by erosion.

8.2.5 The Field Should be of Optimum Size and Shape

Fields that are either too large or too small create difficulties in uniform ploughing. The field should be of a convenient size for draft power and capable of being ploughed in a day's work. Fields that are too small require a short run and frequent turns which waste time and energy.

The field should be as regular in shape as possible. Odd areas require repeated turns and runs and thus raise the cost of ploughing which is a substantial part of the overall cost of cultivation.

8.3 QUALITIES OF A GOOD PLOUGHING OPERATION

a) No uncut soil should be left;
b) Trashes, weeds and other crop residues should be buried in;
c) No large clod difficult to pulverise should be there;
d) There should be straight and uniform furrows or beds;
e) Furrow tops should be slightly ridged or rugged;
f) Depth of ploughing should be uniform;
g) Dead furrows should be clean and should expose soil inhabiting insect-pests and disease germs to weather;
h) Turning areas or head lands should not be left uncut or with heaps of earth;
i) Boundary bunds should be kept away from direct attachment with furrow slices;
j) Ploughed land should have a characteristic look, receptive to rain or irrigation water, a drainage facility without loss of soil and a good tilth, good granulation, low bulk density, well aerated, high pore space, soft and mellow with optimum moisture for subsequent tillage operations and seed germination.

8.4 FACTORS INFLUENCING THE PREPARATORY TILLAGE

Several factors are responsible for the type of preparatory tillage selected. The most important ones are: (1) type of farming, (2) cropping system, (3) crop variety, (4) systems of culture, (5) climate and season, (6) soil type, (7) extent of infestation of weeds and soil-borne pests, pathogens and parasites, (8) tillage implements and draft power, (9) economic condition of the farmer, (10) knowledge and experience of the farmer.

8.4.1 Type of Farming

Preparatory tillage for normal arable farming is quite different from that for dry farming where ploughing is omitted for four to five years and instead harrowing is adopted. A specialised farm differs from a diversified and integrated farm with respect to preparatory tillage for raising crops. Land preparation in a vegetable farm is distinctly different from a fodder, ley or pasture farm. A nursery or seed multiplication/production farm needs thorough land preparation by more repeated tillage than that of a commercial farm. In a mechanised farm preparatory tillage differs from a chemicalised farm. In an irrigated farm preparatory tillage is quite thorough compared with that of a rain fed farm.

8.4.2 Cropping System

The cropping sequence determines the degree and extent of preparatory tillage. When wheat or potato is grown during the *rabi* season after wet land rice it requires more ploughing and harrowing than when the same *rabi* crops are grown after maize or groundnut which grows on well drained aerobic soil. Crops grown after a long duration crop that leaves stiff stubbles, for instance, sugar-cane or napier need more tillage operations for the removal of the stubbles. Immediately after the harvest of tuber or root crops such as potato, sweet potato and sugar-beet no preparatory tillage is required except for the sowing or planting of a subsequent crop such as black gram, green gram, sesame and summer maize. For *paira* cropping after wet land *aus* rice no tillage even for seeding is required. In monocropped fields preparatory tillage is needed for sole crops whereas for intercrops or companion crops it may be need-based.

8.4.3 Crop Variety

Each crop and its varieties have a specific root system (shallow, medium, deep or very deep; fibrous or tap rooted;

spreading or tufted; with or without storage organs) and underground stem portions (rhizomes, stolons, suckers, tubers, crown) that require adequate space for geotropic and diageotropic growth, anchorage and storage organs as well as spread and coverage over a greater volume of soil for functioning well.

The size of the crop seed is important with respect to preparatory tillage. Smaller seeds require a finer seed bed for more intimate soil-seed contact and larger seeds require a coarser seed bed with sufficient airspace. Hardy crops such as sorghum, cow-pea and safflower do not require finer tilth whereas, delicate crops such as tobacco, chillies, cabbage and cauliflower, require finer tilth and a shallower cover over the seeds.

Crops requiring earthing up, for instance potato, groundnut, maize and sugar-cane need deep tillage so that such crops get sufficient space below the furrow from where soil is being shifted to the base of the plant. Crops grown under residual soil moisture should be tilled deep so that the roots can reach the deeper moist zone of the soil to survive. On the contrary, modern dwarf varieties are not only dwarf above ground but below ground too. Deep preparatory tillage which involves an expense of energy, effort and time can be replaced by shallower tillage. Good water and nutrient management restrict the penetration of roots to the deeper layer of soil and deep preparatory tillage can be dispensed with under such situations.

8.4.4 Systems of Culture

When rice is grown as upland direct seeded, preparatory tillage distinctly differs from direct seeded puddled rice or a wet land transplanted crop. Large power units can be employed for preparatory tillage of direct seeded upland rice and pre-monsoon rains can be utilised considerably for initial crop growth. Soil structure is more favourable for stand establishment and root development of the succeeding non-rice crop. On the other hand puddling improves weed control, eases transplanting, reduces draft requirement, reduces percolation loss of water and therefore, conserves rainfall as standing water during crop growth, reduces soil erosion and loss of nutrients, thus improving soil fertility and fertiliser management.

A drilled crop requires better land preparation than the crop raised by sowing broadcast or behind the plough.

8.4.5 Climate and Season

Preparatory tillage in arid and semi-arid zones is quite different from humid and sub-humid zones. In tropical and subtropical climates there is no possibility of heaving and thawing whereas there is a strong beating action of rain drops and hailstones which increase the density of the soil. Such soils require thorough preparatory tillage.

In plains, three crop seasons have distinct characteristics which influence preparatory tillage. During *kharif*, *rabi* and summer warm-wet, cool-dry and warm-dry conditions prevail. As the monsoon approaches the soil gains moisture and becomes prone to be eroded by water and as the monsoon retreats, the soil loses moisture and becomes hard. During the summer tilled soil is subject to wind erosion. Therefore preparatory tillage should be scheduled in time considering the prevailing weather conditions characteristic of the season.

8.4.6 Soil Type

Under aerobic soil conditions non-plastic soil can be prepared well with fewer tillage operations whereas, plastic soil requires careful handling. Light soils lose moisture rapidly, therefore preparatory tillage must be scheduled early. Heavy soils lose moisture slowly but once dry become difficult to till. Shallow soils require early and rapid preparation compared to deep soils.

Puddling becomes easy and smooth in clay soils and can be retained a few days prior to transplanting, whereas puddling in sandy soils forms stratified layers of soil particles, in the form of sand, silt and clay from bottom to top, which on settling result in difficulty in transplanting. Transplanting should be prior to the settling of such soils.

8.4.7 Extent of Infestation of Weeds and Soil-borne Pests, Pathogens and Parasites

Good land preparation reduces the infestation of weeds and soil-borne pests, pathogens and parasites. If the field or the locality is susceptible to infestation of a praticular individual or group of pests including weeds in an epidemic or endemic form, stale seed-bed technique, trap cropping, flooding, drying, *rabbing* (heat treatment applied to the soil by burning refuse placed over it), disinfecting/sterilising and fallowing are the measures taken during or prior to preparatory tillage. Special tillage operations may be needed to eradicate perennial and objectionable weeds.

8.4.8 Tillage Implements and Draft Power

With weak, worn, infirm and shaky tillage implements and incapable draft power preparatory tillage will be of a poor type. Improved implements and equipment and adequate draft power help to prepare the field to the desirable extent.

8.4.9 Economic Condition of the Farmer

Even if the farmer does not own appropriate implements and draft power but is economically sound, he can hire them for his use provided they are available for hiring.

8.4.10 Knowledge and Experience of the Farmer

The farmer should acquire, adapt and adopt appropriate knowledge and skill regarding management practices of newer crop varieties including preparatory tillage for the same. An experienced farmer generally prepares his land to the desirable extent considering the crop variety and its system of culture. Sometimes the unskilled assistant of the farmer prepares land improperly. Erratic weather may cause the soil to be soggy or desiccated and unworkable while the sowing time is delayed. In such a situation the novice farmer prepares his land hurriedly but improperly whereas an experienced farmer prepares the land well as he stresses the quality factor along with other factors of preparatory tillage.

8.5 PREPARATORY TILLAGE FOR SEED-BED

Well begun is half done. To have a good beginning land should be prepared up to the desirable extent. In the seed-bed crop seeds are sown directly. Therefore, the seed-bed should be such that all stages of crop growth and development starting from germination and emergence to maturity are sustained by the soil without great deterioration and depletion. Poor management should never be compensated by other means.

The seed-bed should be free from large clods, crop residues and established weeds. It should be properly levelled and of a desirable physical state with the correct moisture content for good growth and yield of crop. Adequate irrigation and drainage arrangements are to be provided during preparatory tillage. The soil should be supplied with a basal dose of manures and fertilisers.

Stiff stubbles and other readily decomposable organic matter should be removed otherwise they may invite infestations of pests and pathogens. More than adequate tillage should be discouraged as it results in poor granulation and more erosion. The depth of the tillage should be need-based. The building up and consolidation of tilth ensures an adequate moisture and air supply. Rough seed-beds are favoured for winter and large seeded crops. The smaller the seed the firmer the seed-bed and shallower the seed should be sown.

Seed-beds for summer crops are much finer than for winter crops in order to conserve soil moisture particularly on light soils.

It is a good practice to work from the soil surface downwards to avoid bringing unweathered cloddy soil to the surface. If the surface soil becomes compacted after ploughing due to heavy rain, lifting and stirring the surface soil with light implements speeds up the drying. To conserve soil moisture for a few days more for preparatory tillage, breaking and sealing of pores and levelling the field to reduce exposed area are done by light harrowing and planking. These operations destroy emerging weeds and allow a fresh flush of weeds from the lower layer of soil.

In the fields having turf or trash or heavy infestation of weeds inverting and compacting the soil is necessary to decompose them. Alternate harrowing and ploughing or rolling or rototilling are the effective methods for producing a fine, firm and level seed-bed by breaking clods, removing weeds and stubbles and mixing applied manures and fertilisers.

The seed-bed should be prepared quickly with the minimum loss of soil moisture and structure and if required several tillage operations at one pass should be done. A loose granular seed-bed is the objective of dryland tillage whereas, a puddled seed-bed is the objective of wet land tillage.

8.6 PREPARATORY TILLAGE FOR NURSERY BED

A nursery for arable crops is generally for a shorter duration. This does not mean careless land preparation. As the tender seedlings are raised densely, proportionate root and shoot growth of seedlings of uniform size and vigour are essential.

In general, the nursery area is a fraction of the transplanted area and it is easy to prepare land to the desirable extent.

Repeated ploughing, harrowing and planking bring the soil under good tilth. After the removal of stubbles, trashes, weeds including their roots and fragments, the soil is tilled when it is at optimum moisture. If the area is heavily infested with weeds scrapping of the soil surface may be done.

Nursery beds are tilled shallowly so that roots do not penetrate too deep. During uprooting, deeply penetrated roots tear out and such roots cause the tearing or breakage of tender shoots resulting in poor recovery of quality seedlings per unit area of bed. Very shallow soil also causes poor root growth and anchorage resulting in lodging and deformed seedlings.

Nursery beds should be supplied with a sufficient quantity of well decomposed organic manure that is free from weed seeds, virulent germs and toxins. A basal dose of manures and fertilizers should be mixed well with soil during preparatory tillage.

The nursery bed may be dry or wet, raised, flat, turtle back shaped or flat furrowed. It should be provided with irrigation and drainage facilities.

8.7 PREPARATORY TILLAGE FOR TRANSPLANTING FIELDS

The transplanting field may be wet or dry. For the establishment of uprooted grown up seedlings a well prepared land with sufficient moisture in the root-zone is of primary consideration for rapid root initiation and growth.

Seedlings of rice, *ragi* and onion are transplanted in wet fields but tobacco, brinjal, tomato, cabbage and cauliflower are transplanted in well drained aerobic fields as they cannot tolerate waterlogged condition. For wet land rice, the field is puddled by ploughing and harrowing the soft saturated soil. If weeds and green manures are to be incorporated they are first dislodged under the wet soil, ploughed down and allowed a few days to decompose. Puddling is done thereafter. For *ragi* and onion, land is prepared under aerobic soil conditions but they are transplanted under wet conditions or under moist soil conditions and then irrigated. For other transplanted crops the field is prepared like the seed-bed. The field is also shaped with ridges and furrows for crops which cannot tolerate a waterlogged condition.

Incubated and sprouted seed tubers of potato, setts of sugar-cane, stem cuttings of sweet potato, napier, para grass etc. need preparatory tillage similar to the seed-bed but they need watering immediately prior to or after planting. Row crops are also planted on ridged or listed fields. Such ridges and furrows should be made just before planting.

8.8 INTERCULTIVATION

Intercultivation should be scheduled between pre-emergence or post-establishment of the crop until it closes in canopy or prior to the beginning of reproductive growth. It should not be too frequent and the depth should be shallow. It should be of the soil stirring type and the entire surface of the field should be worked. During such operations care must be taken not to disturb the crop too much. Intercultivation is to be done with light tools and implements, quick in operation and effective. Work should start prior to attaining the optimum moisture condition of the soil. The soil surface should be left rugged.

It should be scheduled after rain or irrigation. Top-dressed fertilisers and agri-chemicals may be incorporated during such operations.

Intercultivation in any form in wet soil results in puddling and expelling of toxic gas occluded in the soil.

In some crops during intercultivation irrigation and drainage channels are prepared and earthing up is done.

8.9 EARTHING UP

Earthing up consists of shifting the soil from the central portion of the space between rows towards the base of the plant (hill, clump, station) so as to cover and close up the spreading tillers. As a result the initiation of new (late) tillers in rice and sugar-cane or rhizomes in taro, turmeric and ginger are restricted whereas, the pegs of groundnut, the stolons and the tubers of potato and the stilt roots of maize are covered with earth which protects them from solanization, provides better anchorage to the plant, keeps the field weed free and serves to make drains as well as watering channels.

Earthing up may be done both under wet and dry conditions of soil. For dry soil it must be with optimum moisture to stabilize the earth in position without compaction. In wet soil the shifted soil should not slip down for which stable slices should be lifted. A great lump of soil should not be lifted as it may affect the crop particularly when the water supply is restricted as earthing up exposes a greater soil surface which may accelerate the loss of moisture.

CHAPTER 9

Nutrient Management

INTRODUCTION

Plants require food for growth and development. At the beginning of the new phase of the life process from a seed or other propagule, plants are nourished by the food reserved in such structures or organs. This period is very short and is when the plant must get established. Green plants are autotrophic or self nourishing, that is, they are able to manufacture carbohydrates from raw or inorganic materials and thus nourish themselves. Non-green plants on the other hand, are heterotrophic. Such plants cannot prepare carbohydrate and nourish themselves. They are either parasites when they depend on other living plants or animals or saprophytes when they depend on the organic materials present in the soil or in the dead bodies of plants and animals.

9.1 ESSENTIAL ELEMENTS FOR PLANTS

Plant-food is composed of certain chemical elements. Most of the plants require eighteen essential elements. These are C, H, O, N, P, K, Ca, Mg, S, Fe, Mn, B, Co, Cu, Zn, Na, Mo and Cl (*C. HOPKiNS Ca Fe Mighty good, Many CurB Zones Commonly Close Naturally*).

Arnon (1954) has laid down the criteria for the essentiality of elements in plant nutrients. They are:

(a) the plant must be unable to grow normally or complete its life cycle in the absence of the element;

(b) the element is specific and cannot be replaced by another; and,

(c) the element plays a direct role in metabolism.

Recent investigations show that point (b) Cannot be accepted absolutely as Mo may be substituted by V, similarly, Cl by Br,

K by Rb, Ca by Sr in some instances.

The elements such as Na, Co, V, Si, I, Se, Ga and Al are found to be required by a particular group of plants or plant species.

There are several other elements such as Rb, Sr, Ni, Cr and As which at very low concentrations and often under specific conditions have been shown to stimulate the growth of certain plants or to have other beneficial effects. These elements are often referred to as 'beneficial elements', 'potential micro-nutrients' or 'micro-elements' (Agarwala and Sharma, 1976).

9.1.1 Sources of Plant Nutrients

The growing plants have different sources from which they get necessary nutrients. Basically these sources are the air (C and O), water (H), and soil (remaining other elements except N, a part of the total requirement of which is supplied from atmospheric N by symbiotic bacteria, and azofication and also from rain and snow).

9.1.2 Macro and Micronutrient Elements

Depending upon the quantitative requirement, the essential elements are grouped into macronutrient and micronutrient elements. Elements which are required by plants in considerable concentrations (greater than 1 ppm) are called macronutrients e.g. C (45 per cent), H (6 per cent), O (43 per cent) which are used by plants from air and water constitute 94 per cent, N (1 to 3 per cent), P (0.005 to 1 per cent), S (0.05 to 1.5 per cent), K (0.3 to 6 per cent), Ca (0.1 to 4 per cent), Mg (0.004 to 1 per cent) and the elements which are required by plants in minute quantities (0.01 ppm to 1000 ppm) are termed micronutrient or trace elements, e.g. Mn, Cu, Zn, Mo, B, Cl, Co, Na and Fe.

Among macronutrients plants utilise greater quantities of N, P and K from the soil and these need to be replaced in greater quantities. Hence, they are regarded as major nutrients and the other macronutrients, e.g. Ca, Mg and S that are utilised by plants from the soil are considered secondary nutrients.

9.1.3 Role of Essential Elements

The essential elements play important but different roles in plants. The main functions of essential elements are:
a) as structural components of cell constituents and its metabolically active compounds;
b) in the maintenance of cellular organisation;
c) in energy transformation; and
d) in enzyme action.

Individual elements have a number of functions at the different stages of the growth and development of plants. Individual plant parts such as roots, stems, leaves, flowers, fruits and seeds differ with respect to the composition and concentration of nutrients. Individual plant species also differ with the distribution of the elements within the plant. The nature and extent of supply or the availability of nutrients also determines the composition and concentration of elements in plants.

9.2 FUNCTIONS OF ESSENTIAL ELEMENTS

9.2.1 Nitrogen

It is a structural component of protein molecules, amino acids, nucleic acids, nucleotides, enzymes, alkaloids, vitamins, chlorophyll and other constituents. It is involved in photosynthesis, respiration and protein synthesis. It imparts the dark green colour of the leaves, promotes vegetative growth, improves the quality of produce including fodder, leafy vegetables and food crops.

When N supply is limited leaf enlargement and thickness, tillering, branching, internode elongation, flowering, fruit setting and other such developments are adversely affected. The flower bud often turns pale and sheds prematurely. The maturation is hastened but the size and quality of the fruits are poor. Root growth is severely restricted. Roots become finer and thinner and reddish brown. The growth of the plant is stunted with restricted foliage forming an acute angle with the stem axis.

9.2.2 Phosphorus

It is a structural component of the membrane system of the cell, the chloroplasts and the mitochondria. It is a constituent of ADP, ATP, nucleic acid, phospholipids, the co-enzyme NAD, NADP. It is involved in photosynthesis and dark reaction, glycolysis, respiration and fatty acid synthesis and also N metabolism. It stimulates early root development and growth and thus helps in plant establishment. It enhances the development of reproductive parts, thus bringing about maturity. It stimulates blooming and fruit setting and seed formation. It improves the quality of the produce. In legumes it induces rhizobial activity, nodule formation and thus N fixation.

With a poor supply of P, the growth of shoots and roots is restricted. P deficient plants are thin and spindly. Foliage is sparse and restricted, the development of lateral buds is suppressed, leaves are narrow and make an acute angle with the stem. The foliage turns bluish-green, with purple or red anthocyanin pigmentation which is more conspicuous during cold weather. Leaves may shed prematurely and flowering and fruiting may be delayed considerably.

9.2.3 Potassium

K does not constitute any of the important constituents such as protein, fat, carbohydrate or chlorophyll but K stress is known to affect diverse metabolic processes. K plays an important role in the maintenance of cellular organisation by regulating the permeability of cellular membranes and keeping the protoplasm in a proper degree of hydration by stabilising the emulsions of highly colloidal particles. It acts as a chemical traffic policeman, root booster, stalk strengthener, food former, sugar and starch transporter, protein builder, breathing regulator, water stretcher and as a disease retarder but it is not effective without its co-nutrients such as N and P (Raju, 1978). K is an enzyme activator. It increases the plumpness and boldness of seeds. It induces winter hardiness. Legumes, cereals and root and starch crops are very much susceptible to K deficiency.

K deficient plants show a reduced rate of photosynthesis, a mottled chlorosis. Under conditions of acute deficiency the interveinal areas become very pale and chlorotic and the leaf tip, and margins become scorched and necrotic. The older leaves show the deficiency symptoms earlier. The root system develops poorly. Roots become thin and brown. Plants become stunted in growth with shortening of internodes and bushy in appearance.

9.2.4 Calcium

Ca is involved in the maintenance of cellular organisation by regulating the permeability of cellular membranes and by the hydration of protoplasm by stabilising the emulsions of highly colloidal particles. Its hydration effect is, however, opposite to that of K which promotes hydration whereas Ca depresses it.

Ca is a structural component of chromosomes. It is an essential co-factor or an activator of a number of enzymes, for instance lipase and apyrase. It favours the assimilation of N in proteins. Ca helps to make leguminous plants susceptible to infection by *Rhizobium*. It induces stiffness of straw and promotes early root development and growth. Ca neutralises organic acids (such as citric, malic and oxalic acid) which may become poisonous to plants.

Ca does not move freely from the older to the younger parts of the plants. Thus the first symptoms of Ca deficiency appear as chlorosis of the young leaves followed by distortion of the tips of shoots. The terminal buds with young leaves become hooked in dicot plants while in monocots such as cereals the young leaves fail to unroll and appear thread like. Roots may become short, stubby and brown.

9.2.5 Magnesium

Mg is a constituent part of chromosomes, polyribosomes and chlorophyll. Mg plays a catalytic role as an activator of a number of enzymes that are responsible for carbohydrate metabolism, phosphate transfer, decarboxylation and organic acid metabolism.

Mg is a mobile element within the plant, thus symptoms of deficiency become apparent first in the basal leaves. Under acute deficiency, symptoms appear in the younger leaves. Fading of the normal green colour of leaves to interveinal chlorosis, marbling or chlorotic mottling in the reticulate-veined leaves take place under Mg deficiency. Chlorosis is often followed by anthocyanin pigmentation and necrotic spotting in acute deficiency.

9.2.6 Sulphur

S is a constituent of the amino acids, cystein, methionine and as such of proteins. Vitamins (thiamine and biotin), lipoic acid and acetyl co-enzyme A contain S. It is involved in the synthesis of glucosides in brassicaceous plants. S is involved in dark reaction and CO_2 metabolism. S is an important element for oil seeds, root, starch and pulse crops.

S deficiency appears as chlorosis of the younger leaves first. In cereals, chlorotic stripping appear between the veins. In *Brassica* species the lamina is restricted with cupping due to inward curling of the margins. The older foliage may develop orange or reddish tints and may shed prematurely. The stem and leaf petioles become brittle and may collapse.

9.2.7 Iron

Fe is a constituent of cytochromes, haem and non-haem enzymes, ferredoxin and haemoglobin. It has some role in chlorophyll synthesis, pigmentation and N fixation. Fe has a catalytic role in the activities of several enzymes.

Fe deficient plants show chlorosis first in the younger leaves. Under moderate deficiency parallel-veined leaves show chlorotic stripping and the reticulate-veined leaves show yellowish-green chlorotic mottling or marbling. Under severe deficiency leaves become dry and papery and may later turn brown and necrotic.

9.2.8 Zinc

Zn is involved in the biosynthesis of IAA thus resulting in flowering and fruiting. It participates in the metabolism of plants as an activator of several enzymes such as triphosphate dehydrogenase and tryptophan synthetase. Zn has a positive role in photosynthesis and N metabolism.

Zn deficiency appears as interveinal chlorosis first in the older leaves, starting at the tips and margins. There is a reduction in the size of young leaves. Bronzing and purple, violet, reddish brown or brown colouration develop on the foliage. Deficiency of this element causes a number of disorders such as the white bud of maize and the *khaira* disease of rice.

9.2.9 Molybdenum

Mo is a constituent part of the enzyme nitrate reductase. Activities of several enzymes are regulated by Mo. It is associated with the biochemical fixation of N, nitrate assimilation, protein synthesis, carbohydrate metabolism and inhibiting acid phosphatase.

In dicot plants deficiency symptoms of Mo appear as chlorotic mottling between the veins of old or middle leaves or all over the surface. *Brassica* species and legumes are very susceptible to Mo deficiency. Under severe stress mottled areas may become necrotic, leading to scorching and withering. The effects extend to the young leaves which fail to expand fully. Growing points become necrotic and further growth ceases. Deficiency of Mo causes 'whip tail' in cauliflower, downward 'cupping' in radish, and scald in beans. In lucerne the young leaves show downward curling of leaf margins, which later become scorched, distorted and withered.

Among cereals wheat shows deficiency symptoms of Mo at a later stage when the ears are about to emerge. The tips of the middle leaves lose their green colour and appear golden yellow. Discolouration and yellowing later spread over all the leaves which become dry and papery. In Mo deficient plants flower formation is inhibited and if flowers do form, they abscise before setting fruit.

9.2.10 Manganese

Mn is an essential factor in respiration, N metabolism and photosynthesis. It is known to activate a variety of enzymes concerned in hydrolysis, synthesis of chlorophyll, development of chloroplast and other functions to the plant system. The symptoms of deficiency may appear first on the young leaves of some species while on other species they may appear first on the older leaves. Mn deficiency is characterised by the appearance of chlorotic and necrotic spots in interveinal areas of the cereal leaves and fine chlorotic mottling in the leaves of dicot plants. The chlorotic areas become necrotic and turn red, brown and reddish brown. Mn deficiency may result in disorders such as the 'grey speck' of oats, 'speckled yellows' of sugar-beet, 'marsh spots of peas', and 'Pahala blight' of sugar-caue.

9.2.11   Copper

Cu is a constituent part of several enzymes participating in the cellular oxidation-reduction. Cu containing compounds are involved in the electron transport during photosynthesis. It is involved in carbon assimilation and other metabolic processes.

Cu deficiency is evident as chlorosis, withering and often distortion of the terminal leaves. In cereals symptoms appear as bleaching and withering of the apices of the younger leaves. The youngest leaves may be chlorotic, or majr fail to unroll and appear withered. In broad-leaved plants, younger leaves develop fine chlorotic mottling between the veins; mott areas often develop white necrotic patches particularly along the margin giving it a withered appearance. In extreme deficiency of Cu. cereals and legumes may show 'reclamation disease'.

9.2.12   Boron

It plays an important role in the development and differentiation of tissues, particularly the vascular elements, carbohydrate metabolism, translocation of photosynthates as sugar-borate complex and nucleic acid metabolism.

In B deficient plants there may be the death of stem and root tips and abscission of flowers and areas of the plant high in metabolic activity. B deficiency assumes symptoms of sugar deficiency. B has been implicated in cellular differentiation and development, in N metabolism, fertilisation, active salt absorption, hormone metabolism, water relations, fat metabolism, P metabolism and photosynthesis.

B is associated with the reproductive phase in plants and its deficiency is often found to be associated with sterility and mal-formation of reproductive organs. B inhibits the production of phenolic substances the accumulation of which is responsible for tissue necrosis.

In some crops deficiency symptoms of B appear first as the death of the growing points. In legumes the deficiency becomes apparent at the seedling stage with chlorotic mottling and downward rolling of margins of the leaves. In cabbage, the leaves may develop dull-green, water soaked and translucent areas. In many root crops there are discolouration and browning and curling of the tender terminal leaves which die and decay later. The development of roots is severely retarded and develop dark spots, wrinkles or cracks and often become rough and distorted. The core may become discoloured and disorganised. In most plants the external symptoms are accompanied with anatomical or histological abnormalities. The *Brassica* species and the root crops are the most sensitive to B deficiency.

B deficiency results in disorders such as 'heart-rot' of sugar-beet, 'browning or hollow stem' of cauliflower, and 'top sickness' of tobacco.

9.2.13 Chlorine

Cl has been found to be involved in the oxygen evolution in primary photosynthetic reactions and cyclic photophosphory-lation.

Cl deficient plants display a wilted appearance of the foliage and stuffy roots with profusely branched laterals. In some species leaves show chlorotic mottling, bronzing and tissue necrosis.

9.2.14 Sodium

In halophytes (salt tolerant plants) such as sugar-beet, Na plays an important role in the maintenance of osmotic relations of the cells. Na helps to maintain a high internal osmotic concentration and thus the plant may withstand desiccation when the external concentration of salts is high. It also influences water uptake.

In normal field crops no apparent deficiency symptomes

observed but the number and size of the younger leaves are found to be markedly reduced in the salt loving *Atriplex vesicaria.*

9.2.15 Cobalt

Co is required by rhizobia for the fixation of atmospheric N by both leguminous and non-leguminous symbionts. It is a structural component of vitamin B_{12} which is essential for the formation of leghaemoglobin. It may also play a catalytic role by being an activator of certain enzymes.

No apparent deficiency symptom of Co is noticeable in plants but reduced activity of biochemical N fixation has been observed.

When plant nutrient elements are short in supply plants show specific symptoms of their deficiency. The nutrients such as N, P, K, Mg, Zn, Mo and S show mobility within the plant and show deficiency symptoms from lower or older to upper or younger leaves. On the other hand Ca, Fe, Mn, Cu and B show deficiency symptoms from upper or younger to lower leaves.

Nutrient elements that are moving out of the leaves are N, K, P, S, Cl and under certain conditions Fe and Mg. Those remaining include Ca, B, Mn and Si (Biddulph, 1959).

Elements such as N and K are luxuriantly used by plants when their supply is abundant while elements such as Al, Ni, Pb, Fl, Ba and B become toxic to plants when their levels in the soil or soil-water increase beyond certain limits, although all essential elements in excess are poisonous or harmful to plants.

9.3 FORMS OF NUTRIENTS USED BY PLANTS

Plants absorb nutrient elements from the soil as:

N	—	NH_4^+, NO_3^{-*}	Ca	— Ca^{++}
P	—	HPO_4^{--}, $H_2PO_4^-$	Mg	— Mg^{++}
K	—	K^+	S	— SO_3^{--}, SO_4^{--}
Fe	—	Fe^{++}, Fe^{+++}	Zn	— Zn^{++}
Mn	—	Mn^{++}, Mn^{++++}	B	— BO_3^{---}
Cu	—	Cu^+, Cu^{++}	Cl	— Cl^-
Mo	—	MoO_4^{--}	H_2O	— H^+, OH^-

*Plants can absorb organic and molecular N in addition to the above forms.

For proper growth and development of field crops the nutrients should fulfil the following conditions:

a) They must be present in available form or in a form usable by plants;

b) The nutrient must be in a concentration optimum for plant growth;

c) There must be a proper balance among the concentrations of various soluble nutrients in the soil (Yawalkar and Agarwal, 1962);

d) The availability of the nutrients should be according to the requirements of the plant concerned at its different stages of growth and development.

9.4 ABSORPTION OF NUTRIENTS BY PLANTS

Nutrient solubility and availability depend on various edaphic, climatic and biotic factors. Nutrient uptake by plants requires an intimate root-soil association. It is accentuated by root exudates and microbial activity in the rhizosphere. Such biochemical phenomena greatly increase the rate and ease of transfer of nutrients from the soil to the plant (Brady, 1974). Besides root uptake of nutrition plants are capable of absorbing nutrients through their leaves. This method is known as foliar nutrition.

9.5 LOSSES OF PLANT NUTRIENTS FROM THE SOIL

Soil is the storehouse of plant nutrients. There are always in-and-out flows of plant nutrients from this storehouse. Plant nutrients are lost from the soil mainly by:

a) removal by crops and weeds;

b) leaching;

c) erosion or run off;

d) transformation to gaseous form.

A crop field producing a harvestable biomass of 000 kg on an average depletes 60 kg plant nutrients from the soil. The extent of loss of nutrients by other means depends on the management factors and other related conditions.

9.6 SOURCES OF PLANT NUTRIENTS

Enrichment of soil with nutrient elements takes place mainly from:

a)   green manures;
b)   organic manures (bulky and concentrated);
c)   commercial fertilisers (organic and inorganic);
d)   plant and animal residues;
e)   soil amendments, pesticides and agrichemicals other than fertilisers;
f)   rain-water.

9.7 SOIL FERTILITY AND PRODUCTIVITY

Soil is a 'substratum obliged to innumerable lives' and the 'source of infinite lives' therefore, maintaining soil in conditions suitable for crop plants and the farmer is of prime importance. It must be protected from impoverishment. Efforts should be made to upgrade the soil for higher fertility and productivity. *Soil fertility* refers to the ability of a soil to supply all the essential nutrients in an optimum amount and balance and in a form readily available to the plants concerned. *Soil productivity* refers to the capacity of a soil for producing a specified plant or sequence of plants under a specified system of management, i.e. the capacity of the soil to produce crops per unit area.

A soil may be fertile, i.e. having chemical capacity but may not be productive because of excessive acidity or alkalinity or the presence of toxic substances, poor physical properties or an excess or deficiency of water.

9.8 ORGANIC MANURES

Some of the organic wastes or by-products (excreta of animals and birds, litter, crop refuse, and other by-products) either decomposed or treated or fresh are used to enrich soil fertility. These are called manures. Manures may be bulky (nutrient contents are very low per unit volume) such as farmyard manures (FYM), and compost or concentrated (containing a higher per cent of nutrients) such as oilcakes, meals of blood, meat, bone, fish, horns and hooves.

The effects of organic manures are many. The most important ones are:

a) they enrich soil with all the essential elements required for plant growth,
b) they improve the physical properties of the soil;
c) they provide food for soil micro-organisms which are responsible for various activities in the soil;
d) they provide a buffering action in soil reactions;
e) they prevent loss of nutrients by leaching or erosion.

9.9 GREEN MANURES

This refers to the practice of incorporating plant materials while they are green for improving the soil. In this practice the decomposition of plant materials takes place in the soil. Thus in addition to the application of organic materials through decomposed organic manures, there is the production of enzymes, vitamins, hormones and antibiotics which induce crop growth. Green manuring practices may be green leaf manuring (turning under of green leaves and tender green twigs collected from shrubs and trees grown on bunds, waste lands, road sides and nearby forest areas; the most useful such plants are *Glyricidia maculata, Leucaena leucocephala* etc.) or green manuring *in situ* (turning into the soil of undecomposed green manure crops in the same field where the crop is grown; the most common plants are sunnhemp, *dhaincha*, pillipesara, cow-pea, green gram, cluster bean and berseem from legumes and sunflower, *Cannabis sativa, Vernonia cinerea, Echinochloa colonum* and *Sorghum vulgare* from non-legumes.

Different organic manures contain variable amounts of major nutrients:

Manures	Per cent of		
	N	P_2O_5	K_2O
1	*2*	*3*	*4*
FYM	0.5	0.2	0.5
Compost (from farm litter)	0.5	0.15	0.5
Castor cake	4.3	1.8	1.3
Cotton seed cake			
undecorticated	3.9	1.8	1.6
decorticated	6.4	2.9	2.2

(Contd.)

(Contd.)

1	2	3	4
Safflower cake			
undecorticated	4.9	1.4	1.2
decorticated	7.9	2.2	1.9
Mahua cake	2.5	0.8	1.8
Neem cake	5.2	1.0	1.4
Coconut cake	3.0	1.9	1.8
Groundnut cake	7.3	1.5	1.3
Linseed cake	4.9	1.4	1.3
Niger cake	4.7	1.8	1.3
Rapeseed cake	5.2	1.8	1.2
Sesame cake	6.2	2.0	1.2
Blood meal	11.0	1.5	—
Meat meal	10.5	2.5	—
Fish meal	4-10	3-9	0.3-1.5

The amount of nitrogen turned under from the parts above ground, of some of the green manure crops are:

Crop	Average green matter yield (kg/ha)	N in the green matter	Amount of N (kg/ha)
Sunhemp	19,504	0.43	83.9
Dhaincha	18,400	0.43	79.1
Black gram	11,040	0.41	45.3
Green gram	7,360	0.53	38.8
Cluster-bean	18,400	0.34	62.6
Cow-pea	13,800	0.49	30.4

Adapted from Singh and Lal, 1976.

9.10 COMMERCIAL FERTILISERS

Fertilisers are the organic or inorganic materials of natural or synthetic origin which are added to the soil to supply certain elements essential to the growth of plants. The term is now commonly restricted to commercial products.

Each fertiliser contains only one or two or a few essential elements in specific concentrations. Fertilisers can be used on the needs of individual nutrients in desirable quantities to supplement nutrients supplied from organic sources and those that are depleted in the soil. They can also be used to improve soil reactions as some of them are basic or acidic in residual effects.

Fertilisers may be straight (having a declarable content of one major nutrient only) such as ammonium sulphate and urea, binary (containing two major nutrients) such as potassium nitrate, ternary (containing three major nutrients) such as ammonium potassium phosphate, compound (having a declarable content of at least two of the major nutrients obtained chemically or by mixing or both) such as nitrophosphate, ammonium phosphate or mixed (individual or straight fertiliser materials blended together to permit application in the field in one operation; they supply two or three major nutrients in a definite proportion or grade, for instance, nitrophosphate with potash 15 : 15 : 15 of N, P and K. They are also referred to as complete fertilisers.

Fertiliser mixtures can be prepared at the farm or at home but care must be taken to avoid the uneven mixing of incompatible fertilisers, or the mixing which leads to a loss of some of the fertilising nutrients in the form of gas, converts soluble nutrients into insoluble ones or induces caking. The following combinations should be avoided in preparing the mixed fertilisers at home.

a) Ammonium sulphate, ammonium chloride, other ammoniacal fertilisers and nitrogenous organic manures with lime.
b) Sodium nitrate or potassium nitrate with superphosphate.
c) Nitrochalk with superphosphate or lime.
d) Ammonium sulphate-nitrate with lime.
e) Urea with superphosphate.
f) Superphosphate with lime or calcium carbonate or wood ashes (Patnaik, 1980).

Fertilisers may be high analysis (containing not less than 25 per cent of the major plant nutrients, for instance, urea) or low analysis (containing a low percentage of plant nutrients; usually less than 25 per cent, for instance single superphosphate and sodium nitrate).

Fertilisers are classified according to the nutrient contents present in them.

9.10.1 Nitrogenous Fertilisers

They are again classified according to the form of N present in them:

a) *Nitrate fertilisers*:

Sodium nitrate	16% N
Calcium nitrate	15.5% N

b) *Ammoniacal fertilisers*:

Ammonium sulphate	20% to 21%N
Ammonium phosphate	20% N, 20% P_2O_5
Ammonium chloride	24 to 26 % N
Ammonia anhydrous	82% N
Ammonia solution	20 to 25% N

c) *Nitrate and ammonia fertilisers*:

Ammonium nitrate	33 to 34% N
Calcium ammonium nitrate (CAN)	25% N
Ammonium sulphate nitrate (ASN)	26% N

d) *Amide fertilisers*:

Urea	46% N
Calcium cyanamide	21% N

Most of the field crops except rice, in the early stage of plant growth take up N in nitrate form. Nitrate fertilisers are suitable for top- and side-dressings. On dry soils these are superior to other forms of N fertilisers. In moist soil they leach rapidly.

Ammonia fertilisers are preferred by the rice crop in the early stage. For other crops these fertilisers need to be nitrified to nitrate N. These fertilisers are more resistant to loss by leaching because they are readily absorbed on the colloidal complex of the soil. Nitrate and ammoniacal fertilisers are readily soluble in water and suitable for use under a wide variety of soils and crops. The nitrate form is readily available to plants for rapid and early growth and the ammonia N resists leaching losses and can be utilised by the plant at a late stage. Such fertilisers are not well suited to wet land rice culture.

Amide fertilisers are readily soluble in water and are easily decomposed in the soil where they are quickly changed into ammoniacal and thence to nitrate form.

All nitrate fertilisers are basic and ammoniacal and nitrate and ammonia fertilisers are acidic in residual effects.

9.10.2 Phosphatic Fertilisers

The phosphate content in such fertilisers is expressed in terms of phosphorous pentoxide (P_2O_5) which is readily dissolved in water and produces salts of phosphoric acid (H_2PO_4, HPO_4). Phosphate fertilisers are classified according to solubility types:

a) *Water-soluble or monocalcium phosphate*:
   Single super phosphate   16% P_2O_5
   Triple super phosphate   46 to 48% P_3O_5
   Ammonium phosphate   20% or 53% P_2O_5

b) *Citric acid soluble or dicalcium phosphate*:
   Basic slag   14 to 18% P_3O_5
   Dicalcium phosphate   34 to 39% P_2O_5

c) *Insoluble or tricalcium phosphate*:
   Rock phosphate   20 to 40% P_2O_5
   Raw bone-meal   20 to 25% P_2O_5
   Steamed bone-meal   22% P_2O_5

Water soluble phosphates can be absorbed quickly by plants. They should be used on neutral to alkaline soils. They convert into unavailable Fe and Al phosphate in acid soils.

Citrate soluble phosphates are suitable for acid soils where they convert into water soluble phosphates and there are less chances of fixation (conversion of soluble P nutrient in the soil into less soluble and unavailable forms). They also improve acidic soils because of Ca content.

Tricalcium phosphates are soluble for strongly acidic or organic soils. Their availability is also increased if they are applied or incorporated with green manuring crops or composts.

9.10.3 Potassic Fertilisers

a) *Having K in the chloride form*:
   Muriate of potash   58 to 60% K_2O

b) *Having K in non-chloride form*:
   Potasssium sulphate   48% K_2O

Chloride form of K fertilisers are used extensively and in all crops except where no chlorine is desired in the fertiliser. Non-chloride form of K fertilisers are in demand by cultivators growing special crops such as potato, tobacco and tomato, in which the quality of the produce is as important as the quantity of yield.

9.10.4 Compound Fertilisers

Potassium nitrate	13.0% N and 44% K_2O
Nitrophosphate	12.9% N and 12.9% P_2O_5
Monoammonium phosphate	11% N and 48% P_2O_5
Diammonium phosphate (DAP)	21% N and 54% P_2O_5
Crude ammonium phosphate	16% N and 20% P_2O_4

Use of such fertilisers supplies N and P/K together. Ammonium phosphatic fertilisers have residual acidity thus they are well suited for use in calcareous and alkaline soils.

9.10.5 Fertilisers and Soil Amendments Containing Secondary and Micronutrients

In the manufacture of nitrogenous, phosphatic and potassic fertilisers Ca, Mg and S get mixed in as impurities or additives. Thus secondary nutrients are supplied to the soil through some of these fertilisers.

N, P and K fertilisers and soil amendments containing Ca, Mg and S nutrients

Fertilisers	Content per cent of		
	Ca	Mg	S
Calcium nitrate	19.5	1.5	—
Calcium ammonium nitrate	8.1	4.5	—
Calcium cyanamide	39.1	—	—
Ammonium sulphate	—	—	24.0
Superphosphate (ordinary)	19.5	0.3	11.6
Superphosphate (triple)	14.3	—	—
Bone-meal (raw)	22.4	—	—
Basic slag	—	3.4	—
Sulphate of potash	—	0.6	18.0
Soil amendments :			
Limestone	32.2	—	—
Gypsum	29.2	—	20.5

Micronutrient content of some important fertilisers and manures

Fertilisers/manures	Cu	Zn	Mn(ppm)	B	Mo
Ammonium sulphate	upto 0.5	0.33	70.0	6.0	0.1
Ammonium nitrate	—	4.5—6.00	5.0	—	—
Ammonium chloride	0.3—1.3	—	—	2.7	—
Urea	0.3—6.0	0.5	0.5	5.0	0.7—6.2
Calcium ammonium nitrate	upto 18.0	8.35	10.50	trace	—
Sodium nitrate	0.1	1.0	8.0	255	—
Ammonium liquor	10.0	—	—	—	—
Superphosphate	26.0	50.16	65.27	9.5	3.3
Superphosphate (triple)	2—12	53—100	175—245	529	9.1
Ammonium phosphate	3—4	about 80	115—220	—	2.2
Basic slag	9.2—56.4	4—59	689	33.4	10.0
Rock phosphate	5.6—9.5	24—137	130—320	19	5.6
Bone-meal	270	660	500	715	—
Muriate of potash	3.0	3.0	8.0	14.0	0.2
Potassium sulphate	5.6—10.4	2.0	2.2—13.0	4.0	0.2
FYM	10	43—247	201	17.4	0 13
Compost	300—600	9—40	1,240	5.8	0.10
Limestone	4—70	3.13	40—65	15.4	2.20

Adapted from J.S. Kanwar, 1976.

Soil amendments are the substances that are added to the soil for the purpose of improving its physical or chemical character, enhancing soil productivity or promoting the growth of crops but this does not include commercial fertilisers and organic manure. Different fertilisers are used to supply micronutrients. They may be used as direct fertiliser to the soil or as foliar spray or be mixed with other fertilisers or manures. The nutrient content of micronutrient fertilisers is as follows:

Fertilisers	Nutrient content
Ferrous sulphate	20% Fe + 18.8% S
Manganese sulphate	23% Mn + 12.4% S
Zinc sulphate	23-35% Zn + 17.8% S
Copper sulphate	25-35% Cu + 12.8% S
Sodium borate (borax)	10.6% B
Sodium molybdate	37-39% Mo

9.11 PLANT AND ANIMAL RESIDUES

The modern concept is that the periodic addition of large quantities of crop residues to the soil, supplied as a result of the increased production of plant material, brought about through adequate fertilisation with inorganic fertilisers, will permit nitrogen and organic matter to be maintained at high levels without using legumes or sod crops in the rotations.

Continued cultivation without the return of adequate crop residues will, on the other hand, lead to a decline in the humus content of soils. Evidently, if any significant increase in the supply of organic matter in the soil is to be effected, it must be provided by crop residues. For this purpose, high yields of crops must be produced. If livestock is not a part of the farming enterprise, stalk and straw should be incorporated into the soil (Singh and Lal, 1976).

Soil micro and macro flora and fauna add nutrients through their products and dead bodies. Some of them fix atmospheric N, for instance blue green algae, some help to solubilise fixed nutrients such as phosphobacterin and S bacteria, some of them such as earthworms turn and mix the soil and enrich the root-zone.

9.12 SOIL AMENDMENTS, PESTICIDES, AND AGRO-CHEMICALS OTHER THAN FERTILISERS

Soil amendments such as lime or gypsum supply only one or two elements but improve the soil reaction and physical properties which in turn improve the availability of a large number of essential elements and thus increase the yield of the crops. They also improve activities of soil organisms which enhance decomposition of organic matter, fixation of atmospheric nitrogen and mineralisation of nutrients.

Pesticides and other agro-chemicals supply only a limited amount of essential elements.

9.13 RAIN-WATER

Rain-water carries a considerable quantity of nitrate and ammoniacal N (upto 8.5 kg/ha/year) and S (upto 100 kg/ha/

year near cities and industrial sites). Although a portion of these is lost through leaching and run off a greater portion is utilised by the crop plants. Rain-water also supplies Mg. The annual addition in this way may be 1.1 to 11.2 kg/ha.

9.14 TIME AND METHOD OF NUTRIENT APPLICATION

Manures and fertilisers are limited in supply and are costlier inputs. The maximum benefits should be achieved from such inputs. For this they need to be applied at the proper time and site and by the correct method. If they are applied much earlier than usual they may be lost in various ways and thus use-efficiency may be less than expected. Similarly, if they are not applied at the proper place they may not be properly utilised by the plants and thus the necessary objectives may not be fulfilled. If they are applied later than usual or at wrong sites they may cause harmful effects on the crop. The time for the application of manures and fertilisers may be:

9.14.1 Basal Dressing

Application before the sowing of the crop. Such application may be:

a) *Before the preparatory tillage:* Bulky organic manures, green manures, soil amendments and conditioners (chemicals which are added to maintain the physical condition of the soil such as polyvinylites, cellulose gums and silicates) di- and tri-calcium P fertilisers are generally applied at this time as they require sufficient time to decompose or react or mineralise (conversion of an element from an immobilised form to an available form as a result of microbial decomposition) after thorough mixing with the soil.

b) *During preparatory tillage:* Concentrated manures, compound and mixed fertilisers and biofertilisers are applied during this time to mix them thoroughly with the soil.

c) *At sowing or planting:* Concentrated manures, straight fertilisers, fertiliser mixtures and the readiy soluble but highly mobile fertilisers, slow-release fertilisers (whose nutrients are present as a chemical compound or in a physical state such that their availability to plants is spread over a period of time for instance, urea super granules, S or lac or tar coated or neem blended urea), starter dose of N fertilisers to legume crops and fertilisers for specific nutrient deficient soils are applied during this time.

9.14.2 Top-Dressing

This is the application of manures or fertilisers to the established crop or crops, within the crop duration. Top-dressing may be done to the soil or to the foliage. Concentrated manures and readily soluble but highly mobile fertilisers are applied in splits in medium to long duration crops and in light soils. Slow release fertilisers are also applied as top-dressing. All types of manures and fertilisers may be applied as top-dressing in ratoon and multicut crops as a booster dose after the harvest of the previous cycle of crops but prior to regenerative growth.

9.15   METHODS OF APPLICATION OF MANURES AND FERTILISERS

Methods of application of manures and fertilisers depend on the form in which they are available.

9.15.1 Solid Form

A. *Broadcast application*: Scattering by hand or spreader to provide uniform distribution of the material over the entire area of the field is broadcast application. Well decomposed FYM, compost, oil-cake, bone-meal, urea, superphosphate and lime are applied by this method.

B. *Placement application*: This may be applied by the following methods:

1) *Plough-sole placement*: In dry soils where there is moisture only in the plough-sole layer and in problem soils where there is the problem of fixation, manures and fertilisers

are placed in the plough-sole after opening the furrow by the plough. Such furrows are covered immediately during the next run of the plough.

2) *Deep placement*: In wet land rice a reduced form of N fertiliser (ammonium sulphate) is placed deep in the reduced layer to avoid denitrification (the bio-chemical reduction of nitrate or nitrite to gaseous N in the soil either as molecular N or as an oxide of N usually carried out by denitrifying bacteria, which results in the escape of N to the atmosphere).

3) *Subsoil placement:* In strongly acidic soils P and K fertilisers are placed in deeper layers by heavy machinery to avoid fixation.

C. *Localised placement*: This method may be as follows:

1) *Contact placement or combine drilling*: Well decomposed manures, ashes and P and K fertilisers in small quantities are used along with seeds during sowing. Seed-cum-fertiliser drill is used for such placement. In this method care must be taken so that seeds ore not burned by the fertiliser.

2) *Band placement*: This is the placement of manures or fertilisers or both in bands on one side or both sides of the row at about 5 cm away from the seed or plant in any direction. These bands may be continuous or discontinuous. Such band placement is of two types: hill and row placement.

a) *Hill placement*: In wide spaced (both between rows and between plants) plants such as cotton, castor and cucurbits manures and fertilisers are placed on both sides of the plants only along or across the row but not along the entire row.

b) *Row placement*: In row crops with wide spaces between rows, for instance sugar-cane, potato, maize and tobacco, manures and fertilisers are placed on one or both sides of the row in continuous bands.

3) *Pocket placement*: In drylands and in wide spaced crops such as cotton, castor, cassava, cucurbits and chillies manures and fertilisers are placed deeper into the pocket (dibble) and seeds are sown in the same pocket about 5 cm above the fertiliser or manure or their mixture.

4) *Side dressing*: This is the method of application of manures and fertilisers along the side of a row or around a plant. This method encompasses hill and ring placement.

5) *Pellet application*: Sometimes N fertilisers are pelleted in various types of mudbolls and placed deep into the soft saturated soils of wet land rice.

9.15.2   Liquid Form

Liquid forms of manures (urine, sewage, shed washing) and fertiliser solutions (liquids or solutions) are applied to the soil by various methods. The important ones are:

a) *Starter solutions*: These are solutions of fertilisers prepared in low concentrations which are used for soaking seeds, dipping roots or plants after uprooting or are sprayed on the seedlings to strengthen the seeds or seedlings for early rooting, establishment and growth. In nutrient deficient areas starter solutions help the plants in many ways.

b) *Foliar application*: This is a method of feeding nutrients to the plants by applying fertilisers to the foliage usually in the form of spray. This method has rcme advantages over soil application of fertilisers, almost all the nutrient elements may be applied by this method.

c) *Direct application to the soil*: Liquid fertilisers such as anhydrous ammonia are applied directly to the soil with special injecting equipment. Liquid manures such as urine, sewage, water and shed washing are let into the field.

d) *Application through irrigation water*: Fertilisers containing soluble nutrients are dissolved and diluted with irrigation water and applied either through surface, subsurface or overhead irrigation. Liquid and slurry (received from cowdung gas plant) manures are also diluted with irrigation water and supplied through surface irrigation.

9.16   PRINCIPLES GOVERNING SELECTION OF PROPER TIME AND METHOD FOR APPLICATION OF MANURES AND FERTILISERS

The time and method for application of manures and fertilisers depends on various factors. The most important ones are:

a) The nature of the manures and fertilisers;
b) The soil type and soil-water balance;
c) The nature of the crop and the cropping system.

9.16.1 The Nature of Manures and Fertilisers

In general organic manures mineralise slowly but steadily; di- and-tricalcium phosphates and slow release N fertilisers become available slowly while other forms of fertilisers are readily available to the plants with brief residual effect. Bulky organic manures are applied much ahead of sowing so that they mix well, decompose and mineralise in the soil. Concentrated organic manures may be applied during sowing or planting as pocket or side dressing or top-dressing as side or band placement. N fertilisers are highly mobile whereas K and P fertilisers are slightly mobile and almost immobile respectively in the soil. Highly and slightly mobile nutrient containing fertilisers should be applied in splits with lower doses or as foliar spray. Some fertilisers have an antagonism effect, for instance Zn and P fertilisers as the increased amount of available P reduces the uptake of Zn; some fertilisers have a complementary ion effect, (the influence of one adsorbed ion on the release of another from the surface of a colloid) for instance, the high concentration of K in the soil reduces the uptake of Ca and Mg by plants. Therefore such nutrient containing fertilisers should not be used together or should not be used with proportions that induce such effects.

Some manures and fertilisers may be used directly on the standing crop such as urea and oil-cakes, while some require treatment before application, for instance sulphates containing micronutrient fertilisers should be mixed with half the amount of lime to avoid toxicity when they are used for the foliar feeding; some require incubation, for instance urea with moist soil or di- and-tricalcium phosphates with organic matter, which show better efficiency than direct application; some need to be fortified with other nutrients for instance, sodium molybdate + superphosphate gives a better result than sodium molybdate alone.

Green manuring should be done at least a week before sowing or planting crops so that they decompose well before seeding or planting. Biofertilisers such as *Azolla* sp. may be grown and incorporated before transplanting rice, or may be grown in a standing crop of wet land rice and incorporated thereafter. Blue green algae can be grown in moist but bare or cropped fields and incorporated before sowing or planting the crop. Bacterial fertilisers may be used as seed or soil inoculant before sowing for their early population and infection.

9.16.2 The Soil Type and Soil-water Balance

In light soils the loss of nutrients are greater and more rapid. Therefore all of bulky organic manures and slowly available fertilisers and a part of concentrated manures and readily soluble nutrients are to be applied immediately before or at the time of sowing or planting and the remaining parts are to be applied in splits as top-dressing and as side placement.

In moist soil, both bulky and concentrated manures can be applied by broadcasting or by localised placement. In drylands, manures and fertilisers should be placed below the seed as pocket or contact placement. All the manures and fertilisers need sufficient moisture for their solubility and release of nutrients. Therefore, they should be applied in the moist zone in dryland crops and soil application of top-dressed manures and fertilisers should be preceded or succeeded by rainfall or irrigation. In wet soil the loss of N is more. Slow release N fertilisers may be used safely immediately after the establishment of the crop. Ammoniacal fertilisers should be applied in the reduced zone and nitrate fertilisers should be applied in the oxidised zone in wet land rice as top-dressing in splits.

Some manures and fertilisers need incorporation (mechanical mixing of the materials into the soil), for instance all solid manures and N fertilisers in particular but incorporation of P and K fertilisers in acidic soils increases soil-fertiliser contact which induces fixation. Such fertilisers under such conditions need placement in the rhizosphere. Green manures should be thoroughly incorporated into moist or wet soils. Incorporation in dry soil delays decomposition.

9.16.3 The Nature of the Crop and Cropping System

Most of the crops require N throughout their growth period but the rate of uptake varies considerably. At the beginning of their growth the plants rate of uptake of N is slow and gradually increases to a maximum and then declines to a minimum or nil. This uptake pattern also varies with a crop and its management. It is better to apply N fertiliser in two, or more splits of the total quantity as per the duration and requirement of the crop, soil type and type of fertiliser. In a very short duration crop, such as leafy vegetables or seedlings in the nursery, the entire dose of N may be applied as basal dressing to the soil. Foliar feeding can supplement the basal application. In long duration crops such as sugar-cane, split application (dividing the total dose in two or more instalments) of manures and fertilisers may be adopted. In dryland crops N may be applied along with other manures and fertilisers as basal. Foliar feeding or root application may be adopted under special situations. In legume crops a starter dose of N and a full dose of P and K should be applied as basal to boost initial growth.

In a multiple cropping system, bulky organic manures and green manures, if possible, may be applied at least once in a year. In mono- or double-cropping systems each crop may be supplied with such manures during land preparation.

In irrigated crops phosphates may be applied to legumes in rotation and N to cereals or other crops. In rain fed or dryland cropping P fertilisers should be applied to the *kharif* crop while N and K to each crop of *kharif* and *rabi* if none of the crops are leguminous.

In crops in which the vegetative part is the economic yield, for instance sugar-cane, sugar-beet, potato and tobacco, N supply must be stopped much before maturity otherwise the quality and maturity of the crop may be affected. In seed crops, the last application of N may be during the seed development phase to improve the quality and germinability of seeds. The application of the last dose of N should be through the foliage is annual plants lose uptake capacity through their roots

In determinate grain crops such as rice, wheat and maize, he supply of N at the beginning of the reproductive phase (e.g. panicle initiation) as the last dose increases the number and weight of the grains. In indeterminate plants such as rape, mustard, sesame and cotton an application of N at the flowering stage and another at the late flowering stage increases the yield and the quality of the produce.

If there is luxuriant consumption of N, a growth modifier

such as cycocel may be applied to enhance reproductive growth and yield.

Plants require a greater amount of P at their early stages. The entire P dose may be applied as basal before or at sowing as P moves very little from its site of application. Seeds may be soaked or treated with P solution before sowing. Seedling roots may be dipped in P solution before planting. In drylands and acid soils P fertilisers should be placed near the roots.

Plants absorb K up to the harvesting stage but K fertilisers become available slowly. Therefore the entire quantity of K fertilisers should be applied at sowing. In warmer areas, on lighter soils, or on poorly drained soils, a high application of N and in coastal areas, K application in two to three splits gives good results. K fertilisers move slowly from their site of application, therefore, they should be applied in the rhizosphere.

9.17 DOSES OF MANURES AND FERTILISERS

Plant nutrients are lost or removed from the soil in various ways. For successful cropping in season after season, year after year and generation after generation, plant nutrients must be replenished in the soil in the right quantity, form, place and time. The quantity or dose of manures and fertilisers may be a) corrective dose, b) maintenance dose and c) productive dose. A soil may be deficient in one or more elements which may cause a drastic reduction in the quality and/or quantity of yield. The application of manures or fertilisers to correct the same deficiency is regarded as the corrective dose. The fertility of cultivated soils declines gradually. The application of manures and fertilisers to maintain their normal status is regarded as a maintenance dose. When any crop is produced it requires the application of manures and fertilisers for its expected yield. These additional nutrients are beyond the corrective and maintenance doses and are regarded as productive doses. In general, corrective doses are more than maintenance doses as correction is needed when the soil is depleted. When the application of a productive dose is delayed or becomes inadequate for any reason crop plants use nutrients depleting the maintenance level. Therefore, the maintenance dose is determined according to the target yield, the use efficiency of nutrients and other factors. The corrective dose of different fertilisers varies with the severity of depletion or deficiency. Any cultivated field soil under normal conditions should not show deficiency of major nutrients. The level of secondary nutrients should be at the maintenance level and ameliorative measures should be taken to maintain this level. The application of a liberal quantity of organic manures helps to maintain the maintenance level of soil fertility by supplying nutrients to crops, providing food to soil micro-organisms, by their buffering action and by reducing leaching and wash out of nutrients.

Productive doses are generally regarded as recommended doses. These doses are of four types: a) threshold, b) low, c) optimum, and d) superior recommended doses.

a) *Threshold recommended dose*: This refers to the minimum quantity of manures or fertilisers needed to get a positive response. With a lower dose there is no response, thus the use of less nutrients than this dose requires is useless with respect to an adquate economic return. For soil application of major nutrients the threshold doses has been found to be 20 kg per ha.

b) *Low recommended dose*: This dose provides a positive response and fetches a maximum profit per rupee invested for the nutrients but does not provide maximum profit per ha. This recommendation is mainly for the general cultivators who have limited resources to purchase nutrients or when the nutrients are limited in supply and a larger area is intended to be covered with the same quantity of nutrients with a rate above the threshold dose.

c) *Optimum or average recommended dose*: This is the quantity of nutrient applied to derive maximum profit per unit area from the application of nutrients but irrespective of the profit per rupee invested for the nutrients. It is also considered to be an economic optimum dose which is preferred by most of the farmers with all available resources.

d) *Superior recommended dose*: This refers to the quantity of nutrients applied to obtain the maximum possible crop yield within the limits of the nutrient dose. No financial benefit is considered in this dose except for yield maximisation. This is important under certain situations, for instance potential yield experiment, to meet targetted production, or when the profit is obtained from other enterprises using the yield as the raw material such as fodder for dairy farming. In general, yield increment is progressively less per unit increase of nutrient dose at higher levels of productivity. Therefore, the law of diminishing returns acts under such a situation.

Factors Determining Optimum Dose of Nutrients

The arbitrary use of plant nutrients not only incurs financial losses but may be harmful to the soil, the crop and the environment. It may permanently upset soil conditions or the soil may need conditioning which is time consuming and expensive. The soil must be tested to evaluate its initial fertility status and again at regular intervals to ascertain the extent of loss or gain of fertility after each cropping or within the crop duration. At the same time chemical, physical and biological properties of the soil are to be evaluated as they have contributing roles on nutrient availability. Soil reaction, moisture and temperature affect the responses of nutrients. The degree of soil disturbance due to tillage and other factors such as the extent of erosion or of leaching due to rainfall or irrigation and nutrient removal by weeds affect the fertility status and the soil thus requires a higher dose of nutrient application.

Only chemical analysis or biological evaluation of a crop other than the specific crop, does not always indicate the available nutrient stutus that is used to predict the yield of all the crops or even a particular crop. This is because each crop varies in the uptake of nutrients because of morphological, anatomical, physiological and agronomic differences. Besides these, conditions in the soil are always dynamic and thus a pre-sowing fertility evaluation of the soil may not show the real picture which will be experienced by the crop at its different stages of growth when rate and type of nutrient uptake varies considerably. Therefore, a soil-test crop response method has been suggested.

It has been observed that there should be a proportion of different nutrients (e.g. 1 to 1.2 : 1 : 1 for potato, 1.2 : 0.8 : 0.6 for rice for N, P, K respectively) for each crop variety grown under a specific cultural system, and in a specific season. The efficiency of some major nutrients may be limited because of the poor availability of any one of the micronutrients. The application of balanced fertiliser (a soil additive containing suitable proportions of each necessary mineral element to grow a crop) provides a greater stimulatory effect to individual nutrients than that of the application of each nutrient separately at different times. Some edaphic environmental or biotic conditions may not permit the availability of some nutrients even though they are present at a high level as per the soil fertility evaluation. In such cases a need-based application of nutrients as a corrective dose has been suggested.

There may be an interaction of plant nutrients with soil amendments, agro-chemicals and pesticides both in the soil and within the plant. Care must be taken to avoid such interaction.

Each field condition has its characteristic use-efficiency of nutrients (in wet land rice under normal conditions the use-efficiency of N, P and K nutrients from commercial fertilisers are only 37 per cent, 41 per cent and 36 per cent respectively; special measures such as split application and the use of slow release fertiliser can improve the use-efficiency of N upto 65 per cent, increasing the use-efficiency of N is most important as N is lost easily from the soil compared to P and K, the residual quantity of which can be used by subsequent crops). Special techniques improve the use-efficiency. Thus the recommended dose may be reduced for the targetted yield or greater yield or profit can be expected from the same quantity of nutrients when such techniques are adopted. Some crops such as potato, and maize need a higher level of nutrients in the soil as a luxuriant feeder but they leave a considerable quantity of nutrients as residue. Crops such as rice and maize have a tendency for luxury consumption (the absorption of nutrients by plants in excess of their current need for growth) of N and K nutrients in particular and therefore, scheduling a higher dose of such nutrients may affect the yield because of lodging, or insect-pest attacks of more succulent, soft, tender and leafy crops.

The recommended nutrients applied to soils may not be fully utilised by a single crop in one season and a considerable quantity of less mobile nutrients remain as residue in the soil. The applied nutrients are not the sole source of food for crop production, siltation from floods, rainfall and irrigation water, and the activities of soil microbes add a considerable quantity of nutrients which are utilised by the crop. Contributions from such sources are not always additive but they are most important under low levels of soil fertility. Specific soil treatments such as alternate wetting and drying, or the application of soil amendments increase the availability of fixed, occluded or unavailable nutrients.

Organic farming, chemicalisation, integrated nutrient management by combining organic manures, commercial fertilisers, biofertilisers, intercropping with legumes as a component crop or growing legume crops in rotation, green manuring, stubble farming and organic recycling provide a basis for the nutrient dose to be applied to an individual crop or to crops in sequence.

CHAPTER 10

Water Management

10.1 INTRODUCTION

Water management is the planned use of water for better utilisation in agriculture. It includes irrigation (the application of water to soil or crop plants to assist crop production) and drainage (the removal of excess surface or ground water from land by means of surface or sub-surface drains).

Water is one of the most important inputs essential for the production of crops. Plants need it in huge quantities continuously during their life.

An air-dry seed contains about five per cent water. Seeds imbibe water for germination and emergence from the soil. During this period the water content in the plant becomes positive instead of negative in tension. This is apparent in herbaceous plants as guttation. As the leaves develop the water pressure becomes negative (cultivated plants in general endure soil-water stress of about —10 to —16 bar while some xerophytes upto —80 bars). Thereafter the plant remains in negative tension except under certain special situations of calm, cold and humid weather. Plants show wilting when soil-water stress reaches the —15 bar. Such stress conditions affect the metabolic, synthetic and morphogenetic activities of the plant.

The soil needs the application of water to remove stress conditions, release nutrients in the soil solution for absorption by plants, leach or wash out injurious salts from the soil, prepare land for raising crops and maintain the temperature and humidity of the soil and micro-climate and the activity of soil microbes at the optimum level. On the other hand excess water needs removal for the normal aeration and functioning of roots and shoots of the plants except of aquatic or semi-aquatic crops which bear aerenchymatic cells which transport oxygen from the phyllosphere to the rhizosphere for root respiration. In such plants excess water causes poor tillering, poor culm strength, poor nutrient recovery and poor yield. Excess water creates unworkable soil conditions, impedes root respiration and activity, affects the development of underground storage tissues, produces toxic products on anaerobic decomposition of organic matters, increases loss of nutrients, decreases the efficiency of soil-applied nutrients and pesticides, affects crop management including harvesting, transport and cleaning.

Water acts as a carrier of nutrient ions from the soil to the plant system and photosynthate from leaves to different parts of the plant. Water moves in the plant body in liquid and vapour phases. The movement of water as a liquid phase takes place from the soil to the root hair to mesophyll cells in the leaves through the root epidermis, cortex, endodermis, pericycle, root-xylem, stem-xylem and leaf-xylem (veins). Thereafter it moves to the stationary air and ultimately to the turbulent air, as a vapour phase through pallisade (spongy) cells and substomatal cavity. From soil-water to root-cortex is said to be apparent free space where water moves by diffusion and potential gradient (osmotic and/or electrical gradient). Water movement in the vapour form is due to the driving force of vapour pressure gradient from mesophyll to atmosphere). The movement of the liquid phase of water from the root endodermis to the mesophyll cells takes place due to the diffusion pressure deficit created by the vapour pressure gradient in the intercellular spaces in the leaves. For these, water is absorbed by the plants from the soil solution. This water absorption depends on the diffusion pressure deficit of the plant and soil and the resistance imposed by the plant and soil systems. The plant resistance mainly depends on genetic factors, temperature, CO_2 concentration, water stress and metabolic inhibitors. The soil resistance is mainly regulated by capillary conductivity and the length and spread of roots per unit volume of soil.

The rooting-zone of the soil, the plant body and the lower layer of the atmosphere behaves as a continuum in relation to water transfer. Solar radiation is the primary source of energy for water transport in the soil-plant-atmosphere continuum, the sink being the latent heat change in the evaporation of water at the mesophyll cell surface and at the soil surface.

On draining excess water from the root-zone, the soil gets

aerated, roots of mesophytes can function normally, underground storage tissues get sufficient space for their normal development, aerobic decomposition of soil organic matter takes place and soil reaction reverts towards its original state.

10.2 IRRIGATION

For successful crop production plants must be supplied with water as required by them. The water requirement of crops is that quantity of water required by the crops within a given period of time for their maturity and it includes losses due to evapotranspiration plus the unavoidable losses during the application of water and water required for special operations such as land preparation, puddling and leaching. This water requirement of crops is partially met by *in situ* rain or ground water which are very often not in conformity with the requirements of the crops at different stages of their life span. Rains do not occur throughout the year in all arable fields and their intensity, duration and distribution also differ. A very high ground water-table is not always desirable. In such areas, subsurface drainage is needed for various reasons. The application of water is found to be essential for maintaining the optimum soil-moisture balance suitable for the crop variety, its cultural practices and for other practices such as the application of fertilisers and pesticides and for early ripening and ease of harvesting by digging. Irrigation is an important practice for higher production both in mono- and polycropped areas. Irrigation requirements which are a part of the total water requirement of a crop exclusive of effective rainfall and soil-moisture stored in the root-zone or that contributed from the shallow ground water-table are less than the total water requirement of a crop. The entire volume of water used for irrigation does not meet the purposes of irrigation as a considerable part of it is lost in conveyance and by deep percolation and seepage in the field and only a percentage serves the purpose in practice. This is termed irrigation efficiency (the ratio expressed in percentage of water stored in the root-zone depth of the soil to the water delivered in the field from the farm supply-source). Effective irrigation is the controlled and uniform application of water to crop land in the required amount at the required time, with minimum cost to produce optimum yields without the waste of water and any adverse effect on the soil in the form of soil salinity and water-logging problems.

Irrigation has a tremendous effect on productivity and production of a given area within a given time with higher efficiency of input and crop insurance against short duration droughts. All land areas are not irrigable. Irrigable land is the land under existing or potential irrigation development which by reasons of topography, quality of land and other characteristics is physically suitable for sustained irrigation and for which an adequate and suitable water supply can be provided at a reasonable cost. The basic features of irrigable land are irrigable terrain, potentially fertile soil, a climate in which the crop can thrive and a reliable cheap source of water of consistent quality. In a potentially irrigable area multiple cropping with most remunerative crops can be adopted provided other physics, chemical, biological, climatological and technical factors of production are not limiting.

Water is a dearer input and difficult to receive, retain and release whenever it is required and in whatever quantity it is needed. Irrigation is a costly proposition. For surface Water resources the construction of dams, reservoirs, lakes, tanks and ponds and for ground water resources the installation of shallow and deep wells by digging, drilling or sinking and the construction of conveying channels, pipes and sluice gates from the storage or source points to the fields, installation, operation and maintenance of water lifting devices and conveying systems involve a huge cost. The misuse of water leads to the problem of water-logging and salt imbalance thus rendering agricultural lands unproductive or less productive or problematic. Besides these, other problems of irrigation are:

a) proper utilisation of water with proper discharge with a reasonable flow according to plot size, soil texture, type of crop variety and management practices;

b) creation of a proper water development system with point to point irrigation (carrying water through underground channels and discharging it at the point where and when it is needed in the required quantity);

c) proper levelling of the field for uniform and easy movement of water to each part and the construction of a water management zone in between seeding zones which are mostly ridges, to get rid of anaerobic conditions and other difficulties;

d) preparation of stable boundary bunds, temporary field bunds, plot bunds, levees and channels for irrigation and drainage which not only involves costs but reduces the effective sowing area;

e) erosion of soil and plant nutrients. If irrigation channels are along the slope there may be a greater shifting of soil and the formation of gullies and if across the slope much water is wasted;

f) rat holes, bores, leaks in the bunds need plugging; repair of bunds and channels needs time and involves expenses;

g) in deep clay soils a huge quantity of water is lost in filling up cracks before it reaches saturation level and in flowing further to cover the entire area; on the other hand, in very loose and friable soils there is the problem of deep, percolation. Under such situations the frequent application of light irrigation may be needed and this involves additional cost;

h) irrigation at the appropriate time and interval with a judicious amount of water depends upon the weather, the growth stage of the crop, the type of soil and other factors and therefore needs adequate experience;

i) the compaction of the soil if required, as in paddies needs to be done properly. This involves additional tillage;

j) the choice of crop varieties are limited, as in free flooded areas, rice is the only crop which can be grown throughout the year; no rotational practice can be adopted although there is a considerable wastage of water in wet land rice cultivation and water use-efficiency of rice is also considerably less than maize and wheat;

k) the adoption of green manuring *in situ* is not a worthwhile proposition because of the waste of time and water involved in it;

i) the harvesting of the natural gift of water and its reuse is restricted;

m) root growth is restricted in irrigated soils and therefore, nutrients that leach downward are not extracted by the crops;

n) in potentially irrigated areas, the ground water-table becomes shallow which restricts leaching requirements and seed bed preparation while on the other hand, drainage becomes a problem;

o) the irrigation system of an area is not always under the control of the individual farmer especially in India where there is subdivision and fragmentation of holdings of small units. Under such situations there is a restricted choice in the time, method of application and quantity of water to be used. Again, there may be rainfall, squalls, gales or cyclones immediately after irrigation which seriously affects the standing crop and its yield;

p) the irrigation system may not be truly dependable because of poor or delayed rainfall or its early retreat in the catchment areas of torrential river valley projects; or there may be mechanical or organisational breakdowns or the failure of the power supply, or breakage and siltation of channels and leakage or spillage from pipes. All these are unforeseen hazards which may seriously affect the standing crop;

q) an ineffective or poor drainage system along the irrigation system seriously affects crop cultivation. It may cause the prolonged submergence of a vast area with a varying depth of water, poor crop growth but profuse weed growth, and the upward movement of injurious salts;

r) the requirements of costly inputs such as fertilisers and tillage are considerable in irrigated agriculture because without an adequate supply of fertilisers plants transpire water without much gain of biomass. The application of nitrogenous fertilisers with nitrification inhibitors becomes essential under specific conditions. Irrigated soils become more compact on drying and thus tillage requirements are high. Implements and power for tillage should also be heavier;

s) irrigation water often becomes acidic or alkaline and contains injurious salts, impurities and weed seeds that affect crop production;

t) insect-pests, pathogens, parasites and weeds are greater in irrigated areas. They need to be controlled properly;

u) cultural practices, the cropping system, the cropping pattern and even the farming system are mainly decided by the source of water. systems and methods of irrigation.

10.2.1 Movement of Water into the Soil

Immediately after rain or irrigation on the surface of the soil a considerable quantity of water moves downwards (percolation) and sideward (seepage) through the soil which becomes soaked. In the meantime, part of the water is lost in the form of vapour (evaporation). If the supply of water is more than the intake capacity of the soil within the specified period, it flows from the surface (as run off or surface flow). If the infiltration rate (the volume of water passing into the soil per unit area per unit time) of the soil is poor, there is more surface flow even with less water supply. When the movement of water through the macropores is stopped, a considerable quantity of water is retained by the micropores in the soil for the time being.

When the soil-moisture status becomes within the range of available water (0 to —15 bar tension) a considerable quantity of this water is absorbed by plants (crops and weeds) and used for metabolic activities, transpiration and guttation. At the same time a part of this water is lost by evaporation. As a result, the soil dries up first on the surface and gradually at lower depths. If there is mulch or vegetative cover over the soil surface evaporative loss of water is reduced considerably. When the soil moisture is depleted due to continuous plant absorption and/or evaporation a small quantity of water moves in the opposite direction, i.e. upward and sideward to recharge the moisture level of the root-zone. These are the sources of water for the plants in between two effective periods of rainfall or irrigation.

As the soil-moisture content decreases, only the micropores are available for water flow and hence, the water transmissibility of the soil decreases. There is a diurnal variation in evapotranspiration and water absorption. At night, in most plants, the stomata close and no evapo-transpiration takes place but water absorption by the plant continues in an attempt to eliminate the potential difference between the leaf and the roots. During the day the moisture content of the plants and the soil decreases and therefore the equilibrium potential gradually decreases as plants show signs of temporary wilting. If this condition is allowed to continue further, the soil moisture content will be insufficient to maintain the equilibrium potential and to enable the plant to survive the water deficits occurring thereafter. During this stage plants wilt permanently (plants do not regain turgidity when water is added to the soil; it occurs at a suction of 15 bars for most cultivated crops; growth completely ceases during the period). Water must be applied well before this point is reached if crop yields are not to be seriously reduced. A practical limit of between one and two bars, soil suction in the soil at the level of densest root growth is safe for most plants. Most soils release at least 50 per cent of their available water by the time this suction is reached. However, the economics of water application and moisture sensitive phases of plant growth must be considered. Soil type is also an important factor. In swelling clay soils, it is difficult for water to infiltrate unless the soil-mass is in a cracked condition which does not occur until most of the available water is depleted. A sandy soil, on the other hand, may release 80 per cent of its available water at suctions less than one bar. However, the reserve of moisture at this stage is very small and thus 50 per cent moisture depletion can be regarded as a thumb rule for re-irrigation.

The other consideration is that rooting depth differs with the progress of crop duration except for perennial and ratoon crops. Roots do not penetrate a dry soil and therefore, sufficient water should be applied to wet the soil beneath the existing root-zone during the period of root growth. If the clay soils are not irrigated immediately after the development of small cracks, these cracks deepen and widen resulting in tearing of roots and accelerate evaporation from the manifold area exposed due to vertical cracking.

If infiltration is induced by preventing run off and impounding for a longer period, the ground water-table recharges and comes up to the surface. If the water or soil contains a huge quantity of defloculating agents such as Na, K and Mg or the water is charged with dispersed or suspended substances such as clay particles, infiltration is seriously impeded. Infiltration is affected by the moisture content of the soil; the drier the deeper layers of soil, the larger the potential gradient between the wet front and the soil mass beneath, and hence, the more rapid the intake. A water-table near the soil surface reduces the infiltration rate because of the distribution of moisture in the adjacent soil layers. The presence of hard pan or an impervious layer due to puddling or heavy traffic affects the infiltration rate, the reduction depends on the depth of the layer below the soil surface. The passage of water over the soil causes the movement of the smaller particles which may block the pores in the surface layer. The development of roots in the top layers of the soil benefits the infiltration process. The addition of organic matter to sands binds the particles together and causes a reduction in intake rate; in clays, a structural break-up is encouraged, both of these effects are beneficial as they increase retention and release water for plant use.

10.2.2 Soil Texture and Structure and Moisture Retention Capacity

A sandy soil, by virtue of its relatively large pores cannot exert sufficient force to prevent gravity draining a large amount of water held at saturation: the water holding capacity is therefore, low but the remaining moisture for plant use is easily removed by the roots. Clay soils, having a large number of small pores, possess the ability to hold a large amount of water, but moisture extraction by the plant is resisted by greater forces. Loam soils possess good moisture holding properties but release their moisture at low suctions and have good internal drainage and aeration.

10.2.3 Sources of Irrigation Water

Though the primary sources of water for irrigation are rainfall and snow they are not available throughout the year in all the arable areas and in appropriate quantities required by each crop variety at the different stages of its growth. These two sources are almost replenished annually, therefore may be used as resources of irrigation water for an indefinite time. They recharge the other water bodies in the earth, both at the surface or underground and are utilised by diverting or lifting water for direct use or temporary storage and then by channelising it to other sites at different times for successful cropping in vast irrigable areas throughout the year or when water application is needed most.

a) Surface water resources

Ever flowing and torrential rivers, tributaries, rivulets, streams, brooks, springs, estuaries and canals with flowing water and lakes, tanks, ponds, pools and artificial reservoirs such as dams, barrages and diversionary bunds with different storage capacities for different durations are the major sources of surface water for irrigation. Fluxial and phreatic waters are also used to recharge some of these surface water bodies. Some of these have irrigation potential round the year over a large command area while some have potential for a limited period over a small area. Except for the snow fed rivers all water courses fluctuate seasonally and silt enormously. The nature and extent of flow in these courses depend on the pattern of rainfall in their catchments, therefore, water needs to be stored for future use. All types of reservoirs need to be filled by harvesting water from their catchments or by connecting them with some water courses in addition to the *in situ* rainfall particularly in medium to low rainfall zones with seasonal rains. All these water bodies need renovation and desilting at least once in a decade to maintain their storage capacity. The cost of such operations may be derived from other forms of economic returns, for instance pisciculture or afforestation in and around these areas.

b) Ground water resources

An enormous quantity of water is reserved underground enriched by percolation of the surface water through the porous strata of the earth's crust. Some areas have unconfined aquifers (the permeable beds only partly filled with water and overlying the relatively impervious layers. Their boundaries are formed by the free water table or phreatic level. They have seasonally variable water tables from where water needs to be drawn to the surface for irrigation), some have confined aquifers (the beds confined between two impermeable layers above and below the beds; thus they are not exposed to the atmospheric pressure and maintain a piezometric level on sinking a well. They are also known as artesian aquifers.) and some have perched aquifers (in some localised zones there are unconfined aquifers above the main water-table separated by impervious stratum preventing percolation and causing ground water to accumulate in a limited area above the stratum).

The rate of replenishment of the ground water reservoirs varies with the pattern of precipitation, surface water, stream flow and the permeability of the soil and other earth materials in the passage of percolation to the reservoirs. The rate of pumpage should not be more than the replenishment rate in the case of unconfined aquifers while it may exceed in artesian aquifers.

Ground water is lifted to the surface after digging shallow and large diameter dug (open/percolation) wells or drilling shallow or deep tube wells according to the availability of aquifer. Shallow wells recharge through seepage, percolation, and the high water-table while deep tube wells depend on the aquifers which recharge by surface water perhaps some distance away.

10.2.4 Planning Development and Management of Irrigation Water

Irrigation water may be trapped from rivers, tanks or wells or by a combination of them.

a) River irrigation

Different river irrigation projects are working in our country. River water is diverted by constructing barrages, dams, diversion bunds, check bunds and spill-ways provided with sluice gates to regulate the rate of downstream flow and the height of upstream flow, channelised through mains, distributaries or feeder channels, forming a network of channels distributed in the command area. Water is allowed to flow along the slope by raising the water level in the river bed or reservoirs.

As a minor irrigation project (river lift irrigation) the water of the river or stream is lifted to irrigate a limited command area adjacent to such water courses.

b) Tank irrigation

A large number of small, medium and large size tanks have been dug for storing water for multi-purpose local use. Most of them supply water for irrigation. Some of them are used exclusively for irrigation of a definite command area with fertile soil. A number of crops are grown in these tank-based farming systems. Such tanks are recharged with catchment flow in addition to *in situ* rainfall. Water is either released through gravity flow or lifted for irrigation.

A considerable quantity of water in these surface water reservoirs is lost through percolation, seepage and evaporation. Aquatic weeds and siltation affect the storage capacity.

c) Well irrigation

Different types of wells are used for drawing water for irrigation. Water is lifted by different sources of power such as manual, animal, wind, solar, fuel oil and electric.

d) Combined use of surface and underground water for irrigation

The excess quantity of pluvial, fluxial or phreatic water of a locality can be used for irrigation of other areas or the same area during a scarcity period by storing them in reservoirs or recharging the ground water or root-zone. Ground water can be used to supplement surface water for irrigation which supplements rainfall and water conservation techniques (for instance, the use of mulches reduces the irrigation requirement). Combined use helps to utilise whatever water there is, whenever it is available, for crop production by enriching soil-moisture in the root-zone and surface and underground water reservoirs for subsequent use.

Water can create, preserve and destroy the life of the earth and therefore water must be used in a proper way, with due care and caution as a lack of it may create dryness and too much causes devastating flood; both may cause loss of lives.

10.2.5 Water Conveyances

Water can be conveyed from the storage point or the farm inlet to the point of use through earthen or lined water courses or through metallic, canvas, cement, or plastic pipes, hoses or tubes. Several lining materials of varying cost and durability are now available. These are mud, cement, bitumen, asphalt, bricks, plastic membranes and sodium compounds. Conveyance losses are high in light soils, and lining is essential for using water economically. Lining decreases conveyance-seepage losses, provides safety against breaks and burrows, prevents weed growth, decreases erosion of beds or sides or both from high velocities, decreases siltation, cuts down maintenance costs, reduces drainage problems and increases conveyance and distribution efficiencies of open channels. The water courses should be at a higher elevation than the field level.

The conveyance lines should have water control structures such as water gates, diversion boxes, drop structures, plugs, and outlets.

10.2.6 Layout of the Farm Fields

After assessing the nature and extent of water availability (quantity per day, turn period or rotation, flow rate, system and method of irrigation, water quality), soil (texture, structure, depth of water table, infiltration and permeability, soil depth, slope of land) and crops (type, depth and spread of root system, life span, consumptive water needs, critical periods with respect to moisture; acreage, market value, yield response, drainage requirement), season (warm-wet, cold-dry or hot-dry) and the topographic situation (direction and extent of gradient and other factors), a farm is divided into a convenient number of blocks, fields and plots. Each irrigation and drainage channel should connect two contiguous blocks alternately.

One side of the irrigation channel may be extended for an

approach road connecting two blocks. If needed siphons, flumes and culverts may be provided.

Individual crop fields may be laid out by dividing and sub-dividing with ridges, cross-ridges, furrows, cross furrows, check bunds, dykes and levees along with drains and outlets. The ultimate irrigation beds or furrows should be graded (with 0.3 to 1.5 per cent slope) along and levelled across the run for the easy, smooth and quick movement of water and uniform wetting of the root-zone. For wet land paddies where standing water is maintained at different depths, the entire field should be levelled uniformly. In areas where the ground has a mild gradient (less than 0.1 per cent, is undulating), and the slope is two or more directional, the land should be shaped for contour checks with flat basins. In steeper (more than three per cent slope), narrow basins the land must be formed into a series of terraces inter-spersed with flat basins of moderate width.

There should be the minimum wastage of land under field bunds and channels. The water flow in the channel should be non-silting and non-erosive. The width and depth of water courses should be according to flow size. Field channels should be as straight as possible and durable.

10.2.7   Suitability of Soil for Irrigation

For most crop plants, except rice, the ideal soil for irrigation is that which is deep, without any water-table, has high water-holding capacity, infiltration rate and permeability, and low salt content. The loams and clay-loams are generally good soils for irrigation, since the run off, the number of irrigations necessary and the investment for drainage are low as compared with those in the case of other soils. Any soil can be put under irrigated agriculture permanently but only with due care and caution which, on certain soils, may be beyond economic limits. The heavy soils often need surface as well as subsoil drainage. The light sandy soils involve a high application of costly inputs besides a considerable wastage of water.

In the case of rice, the soils with low percolation rates are ideal for economising on water (Oastane, 1980). For rice, deep clay soils having a high water-table are suitable for maintaining submergence for a prolonged period.

10.2.8 Quality of Irrigation Water

Irrigation water always contains some soluble salts in it. Apart from the total concentration of the dissolved salts, the concentration of individual salts especially the injurious ones is important in determining the suitability of water for irrigation. In general, irrigation water should not have more than low to moderate electrical conductivity for the total dissolved salts and sodium absorption ratio. It should not contain more than 0.5 ppm boron and safe to marginal level of residual carbonates and sulphates. Some salt tolerant species such as barley, sugar-beet, rape, cotton, Rhodes grass and *dhaincha* and semi-tolerant species such as rice, sorghum, maize, sunflower, lucerne, berseem, safflower, onion, spinach, lettuce, carrot, cluster bean, wheat, *bajra* and grasses can be grown successfully with water containing moderate to low concentration of salts. Boron tolerant species such as sugar-beet, lucerne, onion, turnip, cabbage, carrot and lettuce and semitolerant species such as wheat, maize, barley, cotton, sunflower, potato, tomato, peas, beans and sweet potato can be grown with irrigation water containing more than 0.5 ppm boron.

The continuous use of poor quality water for irrigation may permanently damage the soil. Thus it is better to avoid poor quality water for irrigation, however, dilution with good quality water, the application of sulphuric acid or gypsum to irrigation water or soil or the incorporation of a liberal quantity of organic matter to the soil to improve leaching of salts ameliorates the possible hazards from irrigation water with high pH value.

In medium to high rainfall areas even less satisfactory water can be used with advantage as the number of irrigations will be less and high rainfall will have a moderating effect by leaching salts from the soil.

During salt-sensitive critical stages such as emergence, growth of the seedling and flowering of most crops frequent irrigation with water in excess of the storage capacity of the soil helps to maintain stress-free conditions and accelerate leaching of salts.

Root crops should be grown near the furrow bottom where salt concentration is low. A higher plant population than usual helps to compensate the poor production from individual plants.

Split application of calcium based nitrogenous fertilisers improves the leaching of sodic soils. The provision of good drainage to lower the water-table improves the leaching of salts.

Organic irrigation with sewage effluent is a good practice if suitable precautions are taken and due care is given to possible long term effects on the total water supply from the accumulation of salts, nutrients and trace elements. The use of the effluent usually assumes at least a primary treatment (separation of solids from liquid) and usually a secondary treatment (separation, aeration and digestion, and discharge of clear stable liquid) (Michael, 1978).

Liquid manure from livestock housings is passed to a slurry tank where it is mixed and agitated to prevent settlement. It is drawn from the tank by a pump designed to chop straw and other semi-solid materials and diluted to discharge with fresh irrigation water.

Such organic irrigation water has some problems due to the physical, chemical and biological characteristics. The physical characteristics include the total suspended solids, colour temperature and odour. Special treatment and dilution before irrigation and drying the irrigated area between irrigation and cultivation help to alleviate such problems. The toxicity of trace elements may be the chemical characteristic. An assessment of trace elements may be made prior to irrigation use. Biological characteristics are concerned with the presence of pathogenic micro-organisms that may damage the crop or its consumer particularly when such a crop is used raw. Usually vegetables, root crops and creepers the consumable part of which remains in or near to the soil should not be irrigated with such water.

10.2.9 Basic Principles of Irrigation

The basic principles of irrigation are as follows:

a)   Maximisation of use-efficiency of irrigation water for crop production so that a given quantity of water can be used to irrigate the maximum possible area and/or period to increase productivity of the irrigable land and crop either horizontally or vertically or both.

b)   Except for paddy fields, other fields should be kept

moist close to field capacity to maintain a sufficient level of available water for crops. Crop plants should never be allowed to extract more than 60 per cent to 70 per cent of the available water.

c) Except for rice, standing water even for an hour should be avoided as this condition becomes intolerable for most crops.

d) The application efficiency of irrigation should be at the maximum. Levelling, bunding, special shaping of beds, plots or fields with minimum wastage of cropped area may be made for uniform distribution of water in the root-zone. Loss of topsoil and leaching or washing out of plant nutrients should be at the minimum.

e) Irrigation water should soak the root-zone and water management zone and the loss of water after irrigation should be checked by adopting conservation techniques such as mulching and weed control.

f) Irrigation interval and intensity should be based on soil, crop, cultivation practice and environmental conditions.

g) In irrigation practice, soil-moisture status should be allowed to fluctuate within the limits of available water. Such fluctuation helps soil aeration, nutrient availability, multiplication and activities of beneficial soil organisms including symbiotic organisms, growth and distribution of roots, proper development of underground storage tissues and the reduction of the activities of pathogenic and parasitic soil organisms.

h) The application of fertilisers, pesticides and growth regulators may be simultaneous with irrigation to reduce the cost of their application, distribute them uniformly and increase their efficiency.

i) Irrigation to meet special purposes such as land preparation, ameliorating the injurious effects of salts and transplanting of seedlings should be judicious.

j) Irrigation should be considered as an agricultural practice mainly designed to supplement the deficiency of climate in supplying soil moisture with quality water.

10.2.10 Scheduling Irrigation

For scheduling irrigation due consideration should be given to soil and its condition, crop variety, type of culture, growth stage of the crop and atmospheric conditions. Different irriometers (tensiometer, gypsum block etc.) are used to determine moisture content in the root-zone of soil. These are not always applicable for extensive field use. The use of indicator plants which show symptoms of wilting earlier than the plants growing in the field, tissue testing, crop logging and soil moisture determination are also difficult from the farmers' point of view. Meteorological data with an open pan evaporimeter indicates the consumption of water and thus helps to determine the time of irrigation. All these need standardisation with respect to soil, season, crop variety, cultural practices such as the extent of fertiliser use, the crop density and crop geometry, along with the intensity of previous irrigation or precipitation and other local effects. Practical experience must be gained before scheduling irrigation. Wet land rice needs irrigation when the soil surface shows hair like cracks or even before this stage is reached. Once the cracks deepen the objective of puddling is nullified as the cracks make the impervious layer porous. Well fertilised soils, light soils, dense, leafy, shallow rooted crops and the crops with underground storage tissues require light but frequent irrigation. Deep rooted crops in heavy soils require heavy irrigation with wide intervals. Heavy soils with shallow rooted crops require light but frequent irrigation. Light presowing irrigation helps uniform seed germination and crop stand. Light but frequent irrigation helps to establish seedlings, endure moderate vagaries of weather including light showers or high wind speeds that prevail immediately after irrigation. The depth of irrigation may be increased with the progress of crop growth up to a certain limit. Light irrigation at wider intervals helps the uniform and early ripening of the crops. Shallow but repeated irrigation reduces the injurious effects of salts.

Irrigation should be scheduled when the soil temperature is moderate. There is a greater loss of water through evaporation if irrigation is applied to very hot soil. The activities of roots may collapse with a sudden drop in soil temperature and/or the presence of warm water that receives heat from the soil. Irrigation on cold soil may injure some crops. Moreover, during the cold period the evapo-transpiration rate is less and water use-efficiency is reduced. Therefore, in the hotter season, it is better to irrigate the field during the late afternoon or at night so that both soil and plant get sufficient time to absorb water before dry and desiccating weather prevails the next day. If irrigation is utterly necessary during the very cold season, water should be applied during the morning hours when the soil starts warming up.

For the first irrigation to a grown up crop, water should not be applied when it is windy as this may cause root and shoot lodging of crops particularly when the flood method of irrigation is adopted. For the furrow method, the first irrigation should be with as little water as possible to stabilise the soil in the ridge. For the overhead system, the first irrigation should be with very little water to stabilise the surface soil.

Frequent irrigation increases application loss and cost but it helps to increase use-efficiency of fertiliser and other inputs. However, how much water should be applied in each irrigation depends on the need of the crop for water, the availability of water for irrigation, the capacity of the root-zone soil to store water, the season of raising the crop, the cultural practices and the anticipated rainfall.

Growing crops use water continuously but the rate of use varies with the kind of crop grown, the age of the crop, the temperature, atmospheric conditions, special treatment to the crop (use of anti-transpirants and growth modifiers) and the moisture status of the soil. It is not possible or desirable to supply water uniformly throughout the life span of the crop. There are certain critical periods of water requirement for each crop. If the crop experiences stress conditions during these periods there is a drastic reduction in yield. These periods are mostly specific. Therefore, irrigation should be scheduled centering around such critical stages of water requirement of the crop provided there is no other source to recharge the root-zone with moisture.

Different growth stages of different crops have been recognised as critical stages of water requirements. They are as follows:

Crop	Critical stages
Rice	: Tillering initiation, flower primordia and flowering.
Wheat	: Crown root initiation (CRI), milk.
Maize	: Silking, cob development.
Sorghum	: Seedling, flowering.
Barley	: Early tillering, boot, grain filling.
Oat	: Ear emergence.
Pearl millet	: Flowering.
Finger millet	: Flowering, grain formation.
Soybean	: Early seedling, flowering, pod development.
Peas	: Start of flowering, pod formation.
Red gram	: Flower initiation, pod filling.
Gram	: Pre-flowering, flowering.
Sunflower	: Flowering, grain filling.
Safflower	: Flowering, branching development.
Rape and mustard	: Pre-flowering, capsule development.
Linseed	: Pre-flowering, capsule development.
Groundnut	: Pegging to pod formation.
Potato	: Stolonization, tuberization, and bulking.
Sugar-cane	: Emergence, tiller formation and elongation.
Cotton	: Commencement of sympodial (fruiting) branches, flowering and boll formation.
Tobacco	: Topping.

At each irrigation a volume of water sufficient to supply the needs of the crop for a period varying from a few days to several weeks is stored in the unsaturated soil in the form of available soil water. Therefore, how frequently water should be applied to soils of different properties and management practices in order to best supply crops needs is a question of real and practical significance.

10.2.11. Systems and Methods of Irrigation

Several systems of irrigation are in vogue to suit different crops, topography, soil types, water resources, climatic conditions and costs. These systems are:

a) surface irrigation system;
b) subsoil irrigation system;
c) sprinkler irrigation system; and
d) drip irrigation system.

a) Surface irrigation system

In this system, water is directly applied to the surface of the soil from a conveyance system, channel or pipe located at the upper reaches of the field and spread by gravity flow incidental to the slope of the land. There are several methods in this system, the commonest being free flooding from a ditch, check flooding, basin, border strip and furrow.

For irrigation with the surface system, fields are laid out every time before the crops are sown, since these layouts are destroyed during preparatory tillage. In some instances the same layout may be used for irrigating the subsequent crop. However, the field must be levelled well to achieve a higher water application efficiency.

1) *Free or wild flooding*

In this method water is allowed to flood the entire field in an uncontrolled way. This method is practised largely where irrigation water is abundant, inexpensive and harmless to the soil and crop. This method is adapted in wet land rice where water may be applied as a continuous flood, rotational flood or as intermittent flood (De, 1979).

Except for levelling and preparation of impervious boundary bunds no other field layout becomes necessary. No land area is utilised for water distribution. Supervision of water application may not be required. The main drawback of this method is that a huge quantity of water is lost. All types of field management practices specially the nutrient management practices are very poor.

2) Check flooding or check basin

A basin is a flat piece of land surrounded by checks or levees On essentially level ground the shape and size of basins depend mainly on the soil characteristics, the available stream flow and the cultivation practices. Checks, rectangular or square, with sizes varying from about 10 to 100 sq. m, or even more are used. The basins are levelled in both directions. A supply ditch is aligned aloag the highest contour. Individual flat basins are connected one after another with this ditch by breaching or by setting portable siphons. After irrigation the breached bank is blocked and patched up or the siphon is removed.

This method is well suited to all irrigable soils and to a variety of crops. Crops, susceptible to complete saturation of the root-zone such as potato, tobacco, maize and chillies cannot be irrigated by this method.

Slopes upto two to three per cent can be irrigated by using this method with a good control on irrigation water and high water application efficiency. On steeper slopes, this method can be used after proper terracing.

The limitations of this method are that it has too many ridges which occupy a larger area of land and the repair of ridges and careful supervision during irrigation are needed.

3) Basin or ring basin

This method is commonly used for widely spaced crops to irrigate an individual plant or plants grown in pits or pockets. A ring or flat basin is made around the plant and a number of plants or pockets are connected with a ditch passing between the rows of plants. The soil around the plant is soaked with irrigation water and not the entire land. This method considerably economises water. Crops, such as sweet gourd, pumpkin and ash gourd are irrigated by this method. This is a very common method of irrigation of orchard crops.

4) Border strip

In this method the field is divided into a number of long

narrow strips, with small parallel ridges on the sides. The strips are 2 to 10 m wide. The length of a strip ranges from 10 to 300 m or more, depending upon the slope. Individual strips are levelled perfectly and connected with the supply ditch laid out at the upper elevation. After irrigation they are disconnected.

This method is suited well to all irrigable soils and to closely spaced row crops and even to pasture crops. Wheat, barley, rape and mustard, peas, beans and grams are irrigated by this method. Slopes upto seven per cent can be irrigated when the pasture crops are grown. On steep slopes, this method can be used by proper terracing or trenching along the contours to avoid the shifting of soil.

For laying out border strips, the land needs to be graded uniformly to achieve a high water application efficiency. Larger flows are required for this method. Repair of ridges and supervision during irrigation are needed.

5) *Furrow method*

In row crops furrows are made between the two ridges. Water is applied to the furrow and the top of the ridge is not directly wetted. The furrow can be made along the slope when the level of the land is sloping gently up to three per cent. When the slope exceeds three per cent and is up to about 15 per cent, the furrows are laid out on graded contours or as cross-slope furrows. Water distribution can be controlled well to achieve uniform application and the consequent high efficiency. The length of the furrows varies with the soil type, the slope and the quantity of water to be applied and may vary from 10 to 1000 m in different situations. The depth of the furrow should be such that water movement within the soil is predominantly horizontal into the plant root-zone.

Crops, sensitive to the saturated soil condition in the root-zone are irrigated by this method as a considerable number of roots remain above the saturated soil after the application of water in the furrows. Crops which have underground storage organs such as potato, sugar-beet, sweet potato and radish get sufficient space for the development of such organs above the wetting zone in the soil. Wide spaced crops such as sugar-cane.

maize and cotton are irrigated by this method. Under conditions of limited water resources, alternate furrow irrigation is adopted.

Water is applied either by breaching or with flexible siphons that connect the head ditch with the furrow. Individual furrows may be connected separately or a number of furrows may be connected by a common passage made at the upper end of the furrow or at a convenient location.

If the field is quite level, water may be applied by check and furrow method. In this method, after completion of the run of water in the first furrow the flow is continued to the second furrow through a breach in the ridge at the terminal end of the run. After completion of the run in the second furrow the terminal end of the flow is continued to the third furrow through a breach. A number of furrows are irrigated with a single run of water that moves in a zig-zag way through furrows.

Small furrows known as corrugations are sometimes prepared in the case of the border strip method to increase the efficiency, the uniformity of flooding from irrigation and also to facilitate uniform and quick surface drainage required due to overflooding from irrigation or rain water. This method can be adopted even for close growing crops grown on rows such as rape and mustard, groundnut, Bengal gram, peas, soybean and linseed, on medium types of soil with rather uneven topography.

The cost of preparing furrows is more but in this method water is applied uniformly. Because of less open water surface there is less evaporative loss from furrows, the risk of puddling clay soil is reduced, men and machines can work in the field sooner after the end of water application (Withers and Vipond, 1974). Due to separate water application zones a considerable volume of soil is not compacted and soil aeration is not completely interrupted, weed infestation and the requirement of interculture and proportionately less than in flooding methods of irrigation.

b) Subsoil or subsurface irrigation system

In the subsoil irrigation system, water is applied into a series of field ditches deep down up to the impervious layer. It then moves laterally and then vertically through capillaries and satu-rates the root-zone. A continuous supply of moisture in the crop root-zone is, thus, assured from the artificial water-table conditions created by the ponding of irrigation water on the impervious layer. In artificial sub-irrigation, perforated or por-ous pipes are laid underground in the vicinity of the root-zone and water under pressure is distributed through these pipes. An impervious subsoil at a depth of two metres or more, a highly permeable loam or sandy-loam surface soil, uniform topographic conditions and moderate slopes favour sub-irrigation. Under such conditions, proper water control to prevent alkali accumu-lation or excess waterlogging usually results in the economical use of water, high crop yields, and low labour cost in irrigation (Israelsen and Hansen, 1962).

The initial investment and maintenance costs are prohibitive. The system is, however, very efficient, since the water losses through evaporation from the wet surface can be minimised. This system is practised at present in Gujarat and Kashmir for growing cash crops on sandy-loam soils (Dastane, 1980). Small grains and root crops can be irrigated by this system provided there is no development of salinity or alkalinity due to the upward capillary water flow from the shallow water-table.

C. Sprinkler or overhead irrigation system

In this system water is applied to the surface of any crop or soil in the form of a thin spray from above. A typical sprinkler system consists of a pump to lift and convey water under pres-sure, pipes or tubing for the conveyance of water, the sprinkler heads or nozzles and the risers which connect the sprinkler heads with a pipe line. Based on the equipment with which spraying is done, the sprinkler system has been classified as the rotating head type and the perforated pipe type. The sprinkler system is also classified on the basis of portability of equipment as (1) portable (2) semiportable, and (3) stationary or perma-nent. There are self-propelled sprinkler systems which move laterally or radially around a central pivot feeding-line. These portable systems can be designed to cover any area ranging from three to four hectares to fifty to sixty hectares.

This method is advantageous, as water can be applied at a controlled rate and a uniform distribution and high efficiency can be ensured. This method can be adopted in the case of almost all crops and is very popular in the case of cash and some orchard crops and in all types of nurseries. This system is specially suited to shallow sandy soils of uneven topography, where levelling is not practicable, and in areas where water and labour are scarce. On some soils with salinity problems, this system is advocated for the leaching of salt more effectively, for emergence and to secure the quicker and better growth of plants. The pesticides, including herbicides and fertilisers have also been applied successfully by adopting this system. The sprinkler system is also used for cooling the crops during high temperatures and for frost control during freezing temperatures (Dastane, 1980).

In temperate and humid climates evaporative loss from a spray is much the same as the loss from open water in surface irrigation; but in hot, dry climates evaporative loss can be excessive during the summer months and night irrigation alone is advisable. In less extreme arid conditions, high precipitation rates and sprinklers with large drop sizes solve the problem if they can be tolerated by crop and soil (Withers and Vipond, 1974).

In some instances, water is carried to the open ditches laid at five to eight metres apart. The water is applied to the crop of low height or bare field by splashing with plates. Such a splash method of irrigation helps to moisten the dry surface soil for seed-bed preparation, leads to the early and uniform emergence of seedlings and softens the surface crust if formed due to rainfall immediately after sowing seeds. In potato crops, the first irrigation before ridging and earthing up is mostly applied by this method.

D. Drip or trickle irrigation system

This system involves the slow application of water, drop by by drop, to the root-zone of a crop. The equipment consists of a pumping-unit to create a pressure of about 2.5 kg/sq. cm, pipe lines which may be of PVC tubing with drip type nozzles or emitters, and a filter unit to remove the suspended impurities in the water. The amount of water dripping from the nozzles can be regulated, as desired, by varying the pressure at the nozzles, and the size of the orifice of the nozzles. Water may be lifted and distributed through overhead pipe lines which are fitted with nipples to drop or trickle water to the desirable site or water may be held at a certain height from where it passes to the orifice which is impregnated into the root-zone. Water supply may be continuous or intermittent.

In this method, water is used very economically, since losses due to deep percolation and surface evaporation are reduced to the minimum. This method is, therefore, very suited to arid regions. The successful growing of orchards even on saline soils has been made possible by the drip system of irrigation. The system can be used for applying fertilisers in solution.

The initial high cost of the equipment and its maintenance are the major limitations of this system. However, it may be cheaper than the sprinkler system, especially for orchards and other widely spaced crops.

Porous pitchers may be inserted near the pit or pocket or base of the plant. The pitcher is then filled with water and covered with a lid. Water that comes out of the pitcher irrigates the root-zone of the crop. The pitcher is then refilled. Such pitcher irrigation is a most important and easy method of applying water in arid and semi-arid zones with light soil and wide spaced crops such as cucurbits or orchards.

10.2.12 Water Lifting Devices

There are several devices for lifting water from low level water reservoirs, such as wells, lakes and rivers.

Earthen pots, buckets, swing baskets, scoops, *picota*, and the Archemedian screw are some of the ancient devices worked with manual labour. They are suitable for low lift with discharge varying from 6,000 to 12,000 l/hr. A very limited area may be irrigated with such devices. The Persian wheel and *mhote* are operated with animal power and can lift water from depths of 20 m or so, with a discharge varying from 4000 to 12000 l/hr. A limited area can be covered with irrigated crops with these devices. Wind mills are not dependable devices as the wind does not blow at the required speed throughout the year.

Horizontal and turbine pumps worked with petrol, diesel or electric motors, are the most popular water lifting devices of today. They are of various capacities, up to 2,00,000 1/hr or more. A large area can be commanded with pumps having such capacities.

Irrigation practices comprise three interlinked components: the quantity of water at each irrigation (which is normally five to ten cm), the interval between two irrigations (which is normally ten to twenty days) and the total number of irrigations in the life span of a crop (which varies from one to fifty or more).

The total number of irrigations received by a crop is governed by the availability of water, the needs of the crop and the profitability of the crop (Dastane, 1980).

10.3 DRAINAGE

The provision of drains is a must on every irrigated and rain fed farm for sustained farming. Excess water even for a brief period may cause severe damage to crop and soil. Sometimes drainage becomes a greater problem than irrigation, specially in low lying areas.

In heavy rainfall areas, there should be adequate arrangements for surface drainage for the speedy disposal of water. In areas with more phreatic water both surface and subsurface drains (trenches) should be provided. In high water-table areas, there should be subsoil drainage. The spacing between the drains and the size of individual drains depends upon the permeability characteristics of the soil, the gradient of the land, excess water tolerance capacity of the crop grown, and the quantity of water to be disposed of and its speed.

The drains may be open or closed. The open drains are easy to construct and maintain, but they occupy considerable land. The closed drains are costly, but save land and hence, are desirable where land is scarce. The closed drains may be of mole or tile type. A mole drain is an unlined channel in the subsoil made with a tractor-drawn mole plough. This type of drainage is suitable only on heavy clay soils. These drains last for two to three years.

Tile drainage is rather an expensive system but it lasts

several decades. A tile is made of clay or concrete and it is about 30 to 50 cm in length and 7 to 12 cm in diameter. These are placed end to end, with a gap of two to three mm between them. Excess water enters the system through this space and is conveyed along the gradient (Dastane, 1980).

In some locations, surface or subsoil drainage should be accompanied by pumping or siphoning. With proper planning and designing the drainage water of one field may be used as irrigation water for another. On a mild gradient (about 0.1 to 1 per cent) a shallow common channel may be constructed for irrigation and drainage purposes. Such channels are very accessible to mainstreams and without much effort can increase the level of water for discharge. If the level of water in the channel needs to be raised for irrigation the banks of the channel may be raised accordingly. If required, accumulated drainage water may be lifted by a suitable water lifting device for irrigation purposes.

CHAPTER 11

Dry Farming

11.1 CONCEPT

Dry farming or dryland farming is the practice of crop production entirely with rain-water received during the crop season or on conserved soil moisture in low rainfall areas of arid and semi-arid climates and the crop may face mild to very severe moisture stress during their life cycle. A dryland crop refers to a crop grown on well-drained soil where the ground water throughout the year lies well below the soil layers occupied by the crop roots. The crop water requirements are thus satisfied solely by natural rainfall (i.e., surface soil moisture from precipitation is the primary source of moisture for crop production). The term is synonymous with a pluvial crop and excludes phreatic or fluxial crops under rain fed farming systems.

Drought is an occasion when the rainfall for a week is half of the normal or less, when the normal weekly rainfall is five mm or more. Agricultural drought is a period of four consecutive weeks in the period from May to the middle of October or six consecutive weeks during the rest of the year. *Seasonal drought* occurs when the actual seasonal rainfall is deficient by more than twice the mean deviation.

Subramanyam and co-workers made use of the *aridity index* of Thornthwaite and drought years were classified as moderate, large, severe or disastrous according to whether the departure of the yearly aridity index from the climatic normal value was less than 1/2 of the standard deviation (σ), between 1/2 σ and σ, between σ and 2 σ or above 2 σ respectively.

The Indian Meteorological Department (IMD) at Poona, has tried to evolve a drought index on the basis of rainfall departures, monthly rainfall deficits and water periods. Employing Thornthwaite's water balance technique and using potential evapo-transpiration (ET) values computed for 300 stations from the Penman formulation, areas of arid and semi-arid climatic zones were demarcated.

In the arid zone there is a constant water deficit Conditions very close to aridity prevail in the rain shadow tract. Semi-arid regions comprise areas where precipitation meets one-third to two-thirds of the ET needs. The drought climatology of the country shows that on an average, drought may be experienced over large areas on 20 to 25 per cent occasions in each of the months of the *kharif* season. For identification of drought areas IMD has defined drought as a situation occurring in any area in a year when annual rainfall is less than 75 per cent of the normal. When deficiency of rainfall is above 50 per cent of the normal it is termed as severe drought. However, agricultural drought is the moisture deficit which results when the amount of water available in the soil is not sufficient to meet the demands of potential ET.

Areas where drought has occurred in 20 per cent of the years during the period are considered *drought areas* and where it. has occurred in more than 40 per cent of years, *chronic drought areas*. Using the annual and southwest monsoon rainfall departures from normal during the period 1901 to 1960 for about 500 stations, drought areas have been identified by the IMD. The irrigation Commission accepted the approach and in their Report (1972) indicated the drought and chronic drought areas as follows:

11.1.1 Drought Areas

(20 per cent of the probability of rainfall deficiency of more than 25 per cent of normal): Gujarat, Rajasthan and adjoining parts of Punjab, Haryana, west Uttar Pradesh and west Madhya Pradesh, madhya Maharashtra, interior Karnataka, Rayalaseema, south Telengana and parts of Tamil Nadu; small portion of northwest Bihar and adjoining east Uttar Pradesh; a small portion of northeast Bihar and the adjoining portion of West Bengal.

11.1.2 Chronically Drought Affected Areas

(40 per cent probability of rainfall deficiency of more than 25 per cent of normal): west Rajasthan and Kutch.

The most of the areas identified as susceptible to dronght

fall within the arid and semi-arid zones. Chronically drought affected areas are identical with the intensely arid zone. All the districts which comprise the drought zone are not equally vulnerable to crop failures as *protective irrigation* has been developed in some of the districts or taluks. As such those of the districts/taluks which enjoy a minimum percentage of irrigation should appropriately be excluded from the list of drought affected areas (NCA, 1976).

In semi-arid regions, rainfall occurs only during a well defined wet season (when total precipitation exceeds total evapotranspiration) after a dry season and tends to vary strongly from year to year. Since ET is low during the winter season, the rainfall may accumulate and consequently be more efficient than summer rains (*kharif* season). The minimum effective rainfall (that fraction of total precipitation which forms a part of the crop consumptive use) for producing a crop in dry regions is found to be 200 to 300 mm in winter rainfall areas (southwest monsoon) and 400 mm in summer rainfall areas (northeast monsoon).

Soil is another important factor determining productivity of farming. The chemical capacity, water balance and workability of the soil are the deciding factors for its use for arable purposes. With respect to relationships between the crop plant, climate and soil there may be four situations:

a)   optimal climate and soil conditions;
b)   optimal climatic conditions and limiting soil conditions;
c)   limiting climatic conditions and optimal soil conditions; and
d)   limiting conditions for both climate and soil. Dry farming is mostly restricted to both (b) and (c) conditions and thus it consists of farming under very variable conditions of soil and climate of semi-arid regions, i.e., from the least productive to highly productive soil, over a wide spectrum of soil moisture (from below-average to above-average rainfall) without irrigation from sources other than rainfall in the locality.

Irrigation water is not always available to supplement the whims of rainfall. Irrigation is a costly proposition and it is not always practicable even if sufficient water is available. The causes may be the poor quality of the water, soil, climate and crop, the land situation (irregular physiography), low irrigation efficiency, a high rate of shifting of soil and plant nutrients and the upward movement of injurious salts. Under such conditions special measures are needed to utilise whatever rain-water is available by manipulating the soil and crop environment.

Climatic water availability is independent of the soil in low rainfall conditions (less than 250 mm). In medium rainfall seasons, the soil type plays a definite role and in relatively high rainfall seasons again stress-free periods for crop growth are independent of the soil type.

Dryland farming is the only way to utilise a vast geographical area with abundant sunshine and moderate type of soil which has not been explored properly and thus causes regional economic and soil inequalities among the farming community. The whole philosophy of dryland farming revolves around the principle that water is a limiting factor in these areas and one needs to maximise the efficiency of natural rain-water for crop production. The need for a scientific approach towards farming in rain fed dryland areas was felt with the realisation that the occurrence of drought is more or less inevitable. In such areas emphasis is to be given on matching the crop to the soil and water availability and not vice-versa as it is with irrigated farming.

Dryland farming has two dimensions, e.g.,

a) growing and managing crops that can be profitable under the rainfall deficient years, during which drought tolerance and efficient water use are the main requirements, and

b) growing and managing crops that are capable of making the best and efficient use of the favourable environmental conditions provided during the good rainfall years.

Even under favourable seasons soil moisture in the root-zone usually fluctuates in a range between field capacity and permanently wilting point. Therefore, more or less severe water stress may be experienced at any stage of development (Arnon, 1975).

Three types of agriculture are possible in such areas. They are (1) crop production, (2) animal husbandry including pasture management and (3) agro-forestry.

11.2 IMPORTANCE OF DRYLAND FARMING

In India rain-dependent areas are vast and they have a great contribution to make in agricultural production. With just one-fortieth of the world's land, India supports over half of its buffaloes and over a seventh of its cattle and goats which also share land utilisation with humans and add to the national wealth. The second largest number (after Africa) of drought victims of the world live in India.

Out of 143 million ha cultivated land about 37 million ha are under irrigation and the rest of the area (about 75 per cent) is rain fed but all is not too dry. About 33 million ha are in the high rainfall (above 1150 mm) region, and about 37 million ha are each in the medium (750 to 1150 mm) and low (less than 750 mm) rainfall areas.

About 90 per cent of hot desert is located in the northwest part out of which 60 per cent is located in Rajasthan and sustains a human population of 20 million and a total livestock population of 23 million.

The dryland areas contribute about 42 per cent of food-grains, almost all of coarse grains and about 75 per cent of pulses and oil-seeds of the total production of the country. About two-thirds of rice and rape seed-mustard and one-third of wheat are grown in rain fed areas. A large portion of industrially important crops such as cotton, groundnut and castor are cultivated under dryland conditions. Such an area has a great contribution to make to the production of food, feed, fibre, fuel and furniture timber.

Stretching over 75 per cent of the arable area in the country the dryland areas generate enough employment and energy in a highly populated developing country like India.

Efforts are being made to bring more areas under irrigation and thus increase the cropped areas. But, even when we achieve the target of 113 million ha under irrigation by the year 2000 A.D. we would still have about 45 per cent area under rain fed agriculture. Intensive irrigated agriculture is imperative for survival but improved dryland agriculture is necessary for production stability and equity (Gautam, 1987).

Unless dryland agriculture is streamlined scientifically and production in rain fed regions is elevated and stabilised there will remain the problem of regional inequalities between the irrigated and rain fed areas in the country.

Dryland farming also contributes a considerable amount of milk, meat, wool, hide, bone meal, animal power and other products from animals and birds either raised in this region or in other regions depending on the produce of this region.

There is also a great potential for large-scale production of food, feed, fibre, fuel and furniture timber from agro-forestry.

The adoption of improved dryland farming technology will increase cropping intensity from 116 to 130 per cent which will increase the labour requirement from 64 to 84 man days per ha per year by 2000 A.D. and will provide employment of an additional 11 million man years.

The National Commission on Agriculture, 1976 estimated that by 2000 A.D. the country would need 225 million tonnes of foodgrain, 10.2 million tonnes of oil and 17.2 million bales (180 kg each) of cotton which can be achieved only through the development of both irrigated and rain fed farming (Singh and Seth, 1983).

Table 11.1 · Percent area and production of different cropt in irrigated and unirrigated areas

Crop	Per cent area		Per cent production from unirrigated area
	Irrigated	Unirrigated	
Rice	40	60	40
Wheat	58	42	29
Barley	51	49	29
Jowar	4	96	83
Bajra	4	96	81
Maize	15	85	59
Ragi	13	87	66
Small millets	–	100	100
Pulses	8	92	92
Oil-seeds	8	92	92
Cotton	16	84	68
Fodder (dry) except wheat	3	97	93

Source: ICAR, 1979, 50 years (1929-79) of Agricultural Research and Education.

The seriousness of human sufferings caused by frequent failures of crop drew Governmental attention as early as the eighties of the last century when the Famine Commission recommended protective irrigation.

The Royal Commission on Agriculture (1921) realised the importance of millets and rain fed agriculture in the economy of the country and emphasised the need for developing Dryland Agriculture on scientific lines. In 1923, Dr. Mani, Director of Agriculture, Bombay in consultation with Mr. Mehta, Minister of Agriculture, initiated systematic research on dryland farming.

Though V.A Tamhane and K.V. Keintkar made pioneering efforts in developing dry farming at the beginning of the third decade of the century, the organised research efforts to improve crop production on drylands were initiated in 1933-34 by the ICAR when it sponsored five Dry Farming Research Schemes located at Rohtak, Solapur, Bijapur, Raichur and Bellary.

The establishment of the Soil Conservation Research Centres in the mid-fifties provided further needed information on the factors of production (land-use classes, rainfall pattern, run off collection and fertiliser use).

The Desert Afforestation Research Station was established in 1952. In 1957 it was registered as the Desert Afforestation and Soil Conservation Station and as a full-fledged Institute on Arid Zone Research in the year 1959 (CAZRI, 1982).

The All India Coordinated Research Project (ICAR) for Dryland Agriculture was formally launched in June, 1970 in active collaboration with the Government of Canada (Research Branch, Canada Agriculture). It is a multi-disciplinary research unit with 23 Cooperating Research Centres located in typical agro-climatic regions of India (Hyderabad, Anantapur, Bellary, Bangalore, Kovilpatti, Bhubaneswar, Ranchi, Varanasi, Rewa, Jhansi, Dehradun, Ludhiana, Rakh Dhiansar, Hissar, Agra, Jodhpur, Udaipur, Rajkot, Anand, Indore, Akola, Solapur and Bijapur).

With the gain in experience over the years the activities at the Headquarters have expanded to include Agro-economic research at 16 centres, pasture research and development at five centres, Operational research programmes at five centres, Krishi Vigyan Kendra for Dryland Agriculture at one centre and a Trainers' Training Centre at the Co-ordinating Cell at Saidabad, Hyderabad.

Pilot Projects on Integrated Dryland Development were

launched at 24 locations in close proximity to Dryland Research Centres by the Ministry of Agriculture which operated till 1978-79. Thereafter these were transferred to the State Sector as decided by the National Development Council. The main objective of this project was testing research findings under cultivators' conditions. The Projects were more research oriented, demonstrative and probative in nature. The demonstrations indicated that production can be increased several times and an overall increase of 50 to 100 per cent in dryland areas can conveniently be achieved (Annual Report, Ministry of Agriculture, 1978-79).

The Indo-British Dryfarming Operational Research Project at Indore was launched by J.N. Krishi Vishwa Vidyalaya, Madhya Pradesh in collaboration with U.K. in 1973-74 but the actual work started in 1975. The projected area covers 2712 ha of cultivated land. The main aproach was to convert the maximum possible *kharif* fallow area in *kharif* cropped area and to improve and stabilise production by the scientific management of natural resources of land and water and the adoption of improved agronomic practices.

Considering the importance of farming in drylands distributed worldwide, the International Crop Research Institute for the Semi-Arid Tropics (ICRISAT) was established on October 11, 1972 near Hyderabad. ICRISAT has co-operation with five Agricultural Universities in India. They are Bhavanisagar, Anantapur, Dharwar, Gwalior and Hissar with stations on their campuses to test the performance of breeding material under various climatic conditions and latitudes.

Subsequently, the International Centre for Agricultural Research in the Dry Areas (ICARDA) established in 1977, is one of the newest links in a world-wide network to improve and increase food production. The Centre's principal geographic area of concern involves 22 countries of the Near East and North Africa with a population of 300 million people. The five main operational programmes of the Centre are Farming Systems, Cereal Crops Improvement, Food Legumes Improvement, Forage and Pasture Crop Improvement and Training and Communication (ICARDA, 1981).

11.3 RAINFALL PATTERNS OF INDIA AND ITS DISTRIBUTION

Rainfall pattern and distribution in a region is a good index of its water resources. A considerable portion of the Indian subcontinent belongs to the subtropical zone. However, the region as a whole shares the characteristics of a tropical climate. Literally, the word monsoon means a wind system which undergoes a seasonal 180° reversal of direction. In India two factors make it unique. One is the continuous and high mountain mass in the north which forms an effective barrier to the air movement across it. The second is the peninsula shape of the subcontinent with its land in close proximity to the ocean, thereby providing a rich source of moisture.

The rainfall in the country is primarily orographic, associated with tropical depressions originating in the Bay of Bengal and the Arabian Sea. The moisture-laden summer monsoon, accounting for the bulk of rainfall in the country, originates from the vast expanse of the Hind (Indian) Ocean and enters the Indian subcontinent from the southwest as a southwesterly current. The physiographic features of the Indian Peninsula and Western Ghats divert the monsoon into two branches: the Arabian Sea branch and the Bay of Bengal branch. The Arabian Sea branch strikes the Western Ghats from Kerala to Gujarat between the last week of May and the first week of June. After surmounting the Ghats the southern part of the current blows across the Peninsula as a westerly or, in places, as a northwesterly wind. The northern portion of the current which crosses the Saurashtra coast blows across Rajasthan as a southwesterly wind and provides rain mostly in the coastal districts near the Aravalli Hills and the Punjab Kumaon Hills, but very little in the plains of Rajasthan.

The Bay of Bengal branch turns north and enters Bangladesh, Assam and West Bengal in May as a southerly current. With the progress of summer, this current moves northwards and by about June it strikes the Himalayas, it turns westwards and enters Northern India as an easterly current. This occurs in the whole country, except in the western part of Rajasthan, coming under the influence of the monsoon by about the first week of July. However, the advance of the monsoon is not regular every year and is governed by the movement of depressions and low pressure waves travelling from the Bay of Bengal across the country. Whereas Assam, the western coastal plains and the Western Ghats receive heavy rainfall almost every year, rainfall in other parts of the country is dependent on the number of depressions in the Bay of Bengal. Therefore, the annual monsoon variation in the central, northern and the northwestern parts of the country is much more pronounced than in the regions of abundant rainfall.

The monsoon starts withdrawing from Punjab downwards up to Saurashtra and Gujarat in the first week of September and disappears by the middle of November except in the extreme south and southeast, where the receding monsoon gives rain as the northeast monsoon from the Bay of Bengal.

During winter, the northern part of the country gets some rainfall from western disturbances, but these are irregular and not reliable compared with the southwest monsoon. Also severe cyclonic storms experienced during the transition months of April to June and October to December cause some precipitation. Some parts of the country receive hot weather rainfall between March and May mainly owing to large-scale thunderstorms called Norwesters. These rains are of substantial importance in West Bengal and Assam.

The annual average rainfall in India is about 1200 mm i.e., slightly more than the global mean of 990 mm. The more important fact is its extreme spatial and temporal variation. The regional variation is so large that the Khasi Hills of the northeast get an annual average rainfall of more than 10,000 mm whereas the average annual rainfall received in parts of the Rajasthan desert is even less than 150 mm.

Out of the total rainfall received annually, about 75 per cent is received in the four-month period from June to September in northeast India and 60 per cent in the northern belt during the same period where as, in the southwest region 60 per cent rainfall is received between October and December. On an average, more than 60 per cent of the annual rainfall occurs during the monsoon (June to September) throughout India except for Jammu and Kashmir, coastal Andhra Pradesh and Rayalaseema; more than 20 per cent occurs during the post-monsoon period (October to December) in Tamil Nadu,

coastal Andhra, Rayalaseema, Karnataka, Arabian Sea Islands,
Kerala and Bay Islands and more than ten per cent in
Orissa, Marathwada, madhya Maharashtra and Telengana
regions.

During winter (January-February) 19 per cent rainfall occurs

Table 11.2: Annual total rainfall and evaporation (mm) in different
regions of India

Region	Rainfall	Evaporation
Arunachal Pradesh	4142.1	
Coastal Karnataka	3264.8	
Sub-Himalayan West Bengal	3126.8	
Kerala	2996.1	1150.0
Bay Islands	2994.5	
Konkan	2872.0	
Assam and Meghalaya	2752.0	666.0
Nagaland, Manipur, Mizoram, Tripura	2434.6	
Uttar Pradesh (west) Hill	1649.2	
Himachal Pradesh	1602.6	
Arabian Sea Islands	1572.4	
Orissa	1482.2	2120.0
Gangetic West Bengal	1435.2	1355.0
Bihar plateau	1372.0	1507.0
Kerala (south)	1244.9	1801.0
Bihar plains	1202.9	
Vidarbha	1099.6	2732.0
Madhya Pradesh (west)	1044.9	
Coastal Andhra Pradesh	1008.3	1085.0
Tamil Nadu	1008.1	1900.0
Uttar Pradesh (east)	1007.7	
Jammu and Kashmir	994.6	
Gujarat	976.5	
Telengana	926.5	2619.0
Madhya Maharashtra	920.7	3648.0
Uttar Pradesh west (plains)	830.8	2909.0
Marathwada	773.6	
Rajasthan (east)	704.1	2835.0
Rayalaseema	677.8	2892.0
Interior Karnataka (north)	675.0	
Punjab	601.9	
Haryana, Delhi and Chandigarh	539.2	
Saurashtra and Kutch	482.6	3133.0
Rajasthan (west)	311.4	3448.0

in Jammu and Kashmir whereas, Punjab and Himachal Pradesh receive nearly ten per cent annual rainfall

During summer (March to May) more than 20 per cent rainfall occurs in Assam, Meghalaya, Arunachal Pradesh, Nagaland, Manipur, Mizoram, Tripura and Jammu and Kashmir whereas, more than 10 per cent but less than twenty per cent rainfall occurs in the Bay Islands, both sub-Himalayan and Gangetic West Bengal, Himachal Pradesh, Rayalaseema, Tamil Nadu, north and south interior of Karnataka, Kerala and the Arabian Sea Islands.

As a characteristic feature of a monsoonal climate, the rainfall in most parts of the country is subject to uncertainty of occurrence, marked by prolonged dry spells and aberration in the time of commencement and withdrawal and the total amount received, as there is variation with respect to onset, duration, distribution, intensity and retreat of the monsoon.

11.4 DISTRIBUTION OF DRYLAND FARMING REGIONS IN INDIA

The climatological data of the dryland farming regions of India have been analysed by Sarkar and Biswas (1980). The entire dry farming areas have been divided into four zones and the crop potential have been described as follows:

Zone D: This zone has the lowest crop potential. It occurs at three places. The first extends from the Jamnagar district of Gujarat to Firozepur in Punjab; the second includes parts of Ahmednagar, Pune, Satara, Solapur and Sangli districts of Maharashtra, and the third includes portions of Bijapur, Raichur and Bellary districts of Karnataka and Kurnool and Anantapur districts of Andhra Pradesh. In this zone, there may be a break of the 'optimum moisture availability index (OMAI). [Moisture availability index (MAI) is the ratio between rainfall (weekly/monthly) at 50 per cent probability level to potential evapotranspiration (PET) of the corresponding period. Hence the use of assured rainfall at a 50 per cent probability level really indicated a chance of agricultural success higher than 50 per cent. The MAI obtained by using assured weekly rainfall at 50 per cent is considered as OMAI] of one week duration and in many cases may extend up to four to five weeks. Accumulated assured rainfall (AAR) is 80 to 100 mm in the western parts of the tract in Gujarat and 200 to 225 mm in Punjab at 50 per cent probability. Crop production without irrigation is almost a speculation. However, at some places where AAR is 200 to 225 mm and there is hardly any break in OMAI, a short duration crop may be grown. Since soil moisture availability is extremely limiting, commercial crops should not be grown in this region. Where the break is more than two weeks, pasture development and cattle rearing are recommended.

Zone E: This area extends from Rajkot in Gujarat along the east of zone D up to Punjab through Rajasthan and Haryana. This area also extends from Ahmednagar (Maharashtra) to coastal areas of the Cudappah (Andhra Pradesh) through Satara, Pune, Solapur and Sangli districts of Maharashtra and Bijapur, Bellary and Tumkur districts of Karnataka. The AAR is about 200 mm in Sirohi district and adjacent areas of Rajasthan and 350-375 mm in Karnataka and Andhra Pradesh at 50 per cent probability level. A short to medium duration crop may be raised at most of the places. On the other hand, at 40 per cent level of probability, medium to long duration crops may be raised in this area, as crops can thrive on stored soil moisture for a few weeks after the cessation of rains. But at 30 per cent probability, long duration crops may be grown at most of the locations in this zone once in three years.

Zone F: This zone is confined to two localities in the country. In the north, it comprises vast areas of Gujarat, Rajasthan, Uttar Pradesh and a portion of Haryana and Punjab. In the south, this zone stretches from the Nasik district of Maharashtra to Kanyakumari in Tamil Nadu. A large portion of Karnataka and Andhra Pradesh also fall within this zone. As there is hardly any break of OMAI and AAR ranges from 230 to 450 mm, medium duration crops may be grown at most of the places once in two years. On the other hand, two short duration crops or mixed crops may be raised at 40 per cent probability level, whereas, at 60 per cent level only one short duration crop can be grown at locations having 225 to 250 mm AAR.

Zone G: This zone has the highest crop potential. This

area consists of small portions of Uttar Pradesh, Madhya Pradesh, Gujarat and Tamil Nadu and considerable parts of Maharashtra and Andhra Pradesh. A part of Tamil Nadu receives the northeast monsoon and the growing season, therefore, differs significantly from the rest of the localities. At 60 per cent probability level, most of the places have the potential to grow a medium to short duration crop as AAR ranges from 180 mm at Dharmpuri (Tamil Nadu) to 380 mm at Nanded (Maharashtra). At 70 per cent probability, short duration crops can be raised in seven out of ten years, whereas at a 40 per cent level of probability, the crop prospects are very high in this zone (De and Singh, 1983).

11.5 DEPENDABILITY OF RAINFALL

According to the variability of rainfall four classes have been identified. They are:
  a) abundant rainfall (double the average),
  b) average or above average rainfall,
  c) dry or below average, and
  d) killer years with extremely low to nil rainfall.

11.6 ABERRANT WEATHER CONDITIONS

Aberrant weather is the most important vagary of the monsoon which can be considered as the 'core' input for dryland areas. The farmers have to meet the following types of aberrant weather conditions that arise in different years and impair crop production.
  a) Lower than normal rainfall with equal distribution.
  b) Early commencement of normal rains followed by long drought.
  c) Very late onset of rains with normal intensity.
  d) Early onset of normal rains with abrupt retreat.
  e) Normal in onset followed by intermittent drought with normal or above normal rains before retreat.
  f) Late onset with late retreat with above normal rains.
  g) Late onset with abrupt finishing.
  h) Late onset with above normal rains followed by mid-term drought.

i) Early onset with lower intensity followed by normal or above normal rains.

j) Abnormal rains with respect to onset intensity, duration, distribution and withdrawal with spatial and temporal aberrations.

11.7 SOIL CONDITIONS IN DRY FARMING AREAS OF INDIA

Five major soil groups are predominant in these areas. They are black, alluvial, red (including laterite and lateritic), sierozems and submontane soils. In the north and northwest of the country, alluvial, sierozemic and submontane soils predominate. In central and south India black and red soils occupy the highest area. In the higher rainfall regions and coastal areas, the laterites and lateritic soils predominate.

11.7.1 Black Soils

The black soil regions comprise Akola, Bellary, Bijapur, Indore, Kovilpatti, Rajkot, Rewa, Solapur and Udaipur. The black soils are deeper, clay to clay loam and are characterised by low permeability and high water holding capacity. These soils contain a high percentage of clay (more than 30 per cent) and the clay and silt fractions together make up more than 80 per cent of the soil. The dominance of montmorillanite type of minerals and the consequent cracking due to enormous shrinkage on drying are other features associated with this group. Low infiltration rate, high plasticity and stickiness, low organic matter content, high CEC, the calcareous nature and slightly alkaline reaction, pose problems of management practices, Vertisoils, when kept fallow during *kharif*, are exposed to soil erosion hazards. On deep black soils with low infiltration rates on flat lands with less than 0.3 per cent slope, waterlogging is a more serious problem than erosion. The soils of the Deccan Trap region are usually undulating and shallow, medium deep and deep black with certain unfavourable physical characteristics. Small showers of rain do not penetrate to a great depth and are of not much use. Most of the soil moisture is held back by the soils due to the high moisture content at wilting point.

11.7.2 Red Soils

The regions under red soils comprise Anantapur, Bangalore, Bhubaneswar, Hyderabad, Jhansi and Ranchi. The red soils are light textured, shallow to medium in depth and usually underlain by compact subsoil, fairly porous and low water holding capacity. Soils are prone to erosion and surface crusting. At Anantapur the soils are shallow, the depth seldom exceeds 15 cm. Moisture storage capacity is only about 30 mm over a 30 cm depth. There is a compact subsoil, which allows water movement but impedes the root penetration of most crops. Soils of Bangalore are moderately deep to deep, coarse to fine textured and well drained. The profiles contain a quartz gravel argillic horizon. Soils on slopes are susceptible to severe sheet erosion. The annual loss of rain water in terms of run off and percolation could be 24 to 36 per cent. The available water holding of the soils ranges from 80 to 90 mm/m depth of soils. In the Jhansi region, both red (uplands) and black soils (low lands) occur. Red soils are loamy sands to loams with low water retentivity and form the surface crust. Black soils are clay loams to clays. Both in Ranchi and Bhubaneswar regions a regular toposequence of soils is met with. The upland soils tend to be lighter and shallow and are prone to crust formation. On the slopes, they are medium deep and relatively heavy. In the valleys, the soils are light grey, heavy and very deep. The soils of Hyderabad are both Alfisols (red soils) which are light and drought prone and Vertisols (black soils) which have great water holding capacity.

In red soils, Kaolinite is the dominant clay mineral and as such the capacity of retention of cations such as K, Ca, and Mg is very small. The fixation of applied P_2O_5 occurs readily and with increasing acidity the fixation capacity increases.

Because of crust formation (during the non-crop season) run off in Alfisols is more than in Vertisols. Crusting just after seeding results in the poor emergence of seedlings, particularly in the case of small seeded crops such as finger millet and pearl millet.

On lands which have a three per cent slope, about 15 per cent of the rainfall is lost as run off, and the soil loss is 2.5 t/ha/year.

11.7.3 Alluvial Soils

The dryland farming centres with alluvial soils are Agra and Varanasi, extensive areas of Punjab, Rajasthan, southwest Uttar Pradesh and northwest Madhya Pradesh. These soils are fairly level, deep, light to medium in texture with favourable physical characteristics and good permeability. Small showers are useful and there is the utilisation of most of the water held by the soil due to low moisture content at wilting point.

The soils of Agra are deep sandy loams; $CaCO_3$ concentrations at places, limit the effective soil depth. In Varanasi soils are sandy loams to clay loams, slightly acidic to neutral and have saline patches in low lying areas. A distinct toposequence of soils exists. Crusting is a problem in uplands.

11.7.4 Sierozemic Soils

The dryland farming centres under these soils are Dantiwada, Hissar and Jodhpur. Sandy, loamy sands and sandy loams are found in Jodhpur and Hissar regions. Very deep alluvial sandy loams are met with in the tracts of Dantiwada. Low soil moisture storage, instability of soil structure, and poor soil fertility are the major problems of soil management in the desert ecosystem. High wind velocity leads to severe wind erosion. Soil drifting leads to soil and nutrient losses.

Surface crust formation after sowing following light showers limit the desirable crop stand. In an arid environment, high evaporation and low moisture storage capacity limit the availability of rain water for plant use. Soils are light, loose and low in bulk density leading to deep percolation and high evaporation of rain water.

11.7.5 Submontane Soils

Such soils are distributed in the dry subhumid environment

of Hosiarpur in Punjab and Rakh Dhiansar in Jammu and Kashmir and in the humid tract of Dehradun. The lands are sloping, the soils range from loamy sands to sandy loams, silty loams and clay loams with soil moisture storage capacity improving in that order. In Dehradun tracts, the land is undulating with regular toposequences and the soils are silty loam to loam in texture leading to heavy loss of soil due to erosion. Losses of soil were found to be the highest on sloping terrain and also on bare fallow and cultivated fallow lands followed by soils with non-legumes and legumes.

Soil crusting occur in soils of dry and sub-humid regions. Low moisture retentivity and low fertility status are the main problems of loamy soils (Randhaw and Singh, 1983).

11.8 MOISTURE STORAGE CAPACITY OF THESE SOILS

The water holding capacity depends on the soil texture, the degree of porosity, the organic matter content, soil mineral constituents, underlying permeable rocks (e.g. calcareous rocks), and the relief factor. Effective rainfall forms the soil moisture. The amount needed for this has been estimated as 15 to 20 mm from a single rainfall which penetrates to a depth of at least 10 to 12 cm of soil. The infiltration rates vary within a wide range between 25 mm/hour in light soils to 0.25 mm/hour in heavy clay soils. On steep slopes there is less infiltration and much run off. Surface run off occurs whenever the rainfall intensity on a single occasion exceeds the rate of infiltration and evaporation. Rainfall on full leaf vegetation (where a larger proportion of rainfall is intercepted) results in the loss of water by ET before reaching the soil, but some crops enjoy some benefit out of it. The soil moisture is subsequently depleted by evaporation and transpiration.

Generally, seasonally dry areas receive most of the precipitation during one season of the year, while the rest of the year they remain more or less dry. Thus there are alternating wet and dry seasons. The season of the year at which precipitation occurs has a considerable bearing on rainfall efficiency and on the amount of run off.

In the absence of irrigation, crops have access to four sources of water: (1) fluxial water—water accumulated above the soil surface due to run off concentration, phreatic (interflow) concentration and pluvial water, (2) water stored in the part of the soil profile to which roots have access, (3) water provided to the rooting-zone from a lower lying water-table, and (4) water from rainfall or dew.

The available water varies according to the soil texture and organic matter content. Available water ranges from 40 mm (sands) to 250 mm (clays) per 100 cm soil depth. In addition, wet land soils can hold substantial water between field capacity and saturation just after surface drainage. This water varies from 300 mm (light soils) to 100 mm (clay soils) per 100 cm soil depth.

The ground water contribution depends on the depth of ground water over time and on the soil texture. True ground water permeates great depths and does not disappear in the dry season. Its fluctuation is controlled by the contribution from adjacent aquifers and the subsoil strata as well as that from rain received *in situ*. The rates of recharge and withdrawal determine the depth of ground water.

Perched ground water occurs in higher lying landscapes. It is generally separated from true ground water by unsaturated soil and usually disappears during the dry season. The perched water-table fluctuates more rapidly than the ground water-table because it depends on *in situ* rainfall and on localised lateral water movement in the shallow subsoil. The measured recession of perched ground water has been in the order of three to six cm per day during rainless periods and is greater in higher lying fields. At a ground water depth of one m below the root-zone, the contribution to the root-zone can be greater than one mm per day only for fine sandy loams. With ground water at 50 cm below the root-zone, contributions range from one to five mm for most wet land soils but can satisfy evaporative demands above ten mm per day in the case of fine sandy loams. The ground water contribution to the root-zone takes place over five to fifteen days when the ground water recedes through the soil layers upto a depth of one m below the root-zone (for some very fine sandy loams, contribution continues from ground water depth up to three m).

Table 11.3: Moisture storage capacities of broad soil groups
under rain fed areas

Broad soil groups	Sub-group based on soil depth	Moisture storage capacity (mm)	Rainfall (in mm)
1. Black	Shallow to medium (up to 45 cm)	135-145/45 cm	Rajkot (625), Solapur (722), Bijapur (680)
	Medium to deep (45-90 cm)	145-270/90 cm	Akola (830), Udaipur (635), Bellary (550), Rewa (1080)
	Deep (> 90 cm)	300/m	Indore (990), Kovilpatti (730)
2. Red	Shallow to medium (up to 45 cm)	(a) 40-70/40 cm (Sandy loam)	Anantapur (570) Hyderabad (770)
		(b) 70-100/45 cm (loam)	Jhansi (930)
	Deep (> 90 cm)	180-200/90 cm	Bangalore (890), Ranchi (1400), Bhubaneswar (1500)
3. Sierozem	...ium to deep (up to 90 cm)	80-90/90 cm	Jodhpur (380), Hissar (400) Dantiwada (815)
4. Submontane	Deep	(a) 90-100/m (loamy sand)	Hosiarpur (1000)
		(b) 100-140/m (sandy loam)	Dehradun (1750) Rakh Dhiansar (1180)
		(c) 140-180/m (loam)	
5. Alluvial	Deep	(a) 110-140/m (sandy loam)	Varanasi (1080)
		(b) 140-180/m (loam)	Agra (710)

Table 11.4: Length of effective cropping season in different centres

Category	Regional centres and effective cropping seasons	Cropping intensity
A. < 20 weeks	Bellary (8), Jodhpur (11), Anantapur (13), Hissar (17), Bijapur (17)	Sole cropping
B. 20-30 weeks	Jhansi (21), Kovilpatti (21), Hyderabad (22), Udaipur (22), Solapur (23), Agra (24), Anand (25), Akola (27)	Inter-cropping
C. > 30 weeks	Bhubaneswar (31), Hebbal (32), Varanasi (32), Hosiarpur (35), Indore (35), Rewa (36), Samba (44), Ranchi (45), Dehradun (51)	Sequence cropping

11.9 CONSTRAINTS ASSOCIATED WITH DRYLAND FARMING AREAS

It is a dream desire to eliminate drought from this region. There are many problems that are associated with dryland farming systems. The problems may be grouped into four major factors: climatic, edaphic, technological and socio-economic.

11.9.1 Climatic

a) Scarcity of rain-water which is low and unpredictable with respect to intensity and distribution;

b) High *ET* (nearly 3650 mm annually compared to total annual precipitation ranging from 350 to 1400 mm);

c) Extreme thermic values (upto 49°C with the mean annual temperature exceeding 18°C at some places);

d) High solar incidence (450 to 500 Cal cm² day^{-1});

e) High wind velocity with desiccating winds causing a high rate of *ET* and wind erosion;

f) Low relative humidity;

g) Extensive climatic hazards such as weather aberrations, drought, flood, frost, gale, cyclones and burning winds due to dry, deserted and denuded situations.

11.9.2 Edaphic

a) Poor and marginal lands with soils low in fertility and productivity;

b) Uneven topography with high erodability;

c) Difficulty in workability particularly in Vertisols;

d) Shallow or very deep in depth with extreme permeability;

e) Low moisture storage and release capacity particularly in Alfisols;

f) Presence of dissolved injurious salts in ground water;

g) Problem soils with respect to soil reaction (pH) and high concentration of soluble salts in the surface soils;

h) Waterlogging in level lands, flooding and breaking small field bunds resulting in poor conservation of soil and water;

i) Movement of sand and soil;

j) High surface crusting that leads to poor crop stand and high cracking (specially in Vertisols) on drying leads to a high rate of evaporation and mechanical injury to roots.

11.9.3 Technological

a) Limited choice of crop varieties matching the short moist period suitable for cropping as about 60 per cent of the area has two to four-and-a-half wet months and about 19 per cent of the area has less than two wet months per year;

b) Difficulty in designing a suitable cropping system, crop mixture and crop geometry considering soil type and other flexible components such as the moisture availability index, and the duration of the monsoon;

c) Difficulty in evolving static and adaptable technologies fitting into the prevailing as well as anticipated rainfall conditions on a regional basis. The use of high input based technology is risky;

d) Difficulty in seed production and multiplication specially in an on-farm basis;

e) Farming is mainly human and animal power-based which is labour intensive and land saving but costly and time consuming resulting in difficulty in covering the entire arable area under the brief period of favourable weather during the season;

f) Limited use of fertilisers, poor response of biofertilisers and poor availability of organic manures;

g) Poor land capacity and the scope to improve it is very restricted;

h) Limited scope of use of residual moisture and nutrients;

i) Difficulty in conserving moisture for proper and timely utilisation by crops;

j) Difficulty in reclaiming problem soils;

k) Lacking light and speedy implements with cheap sources of energy;

l) Profuse weed, pest, pathogen and parasite infestation;

m) Unpredicted heavy showers resulting in the accumulation of more than optimum moisture in the soil, impairing field operations including sowing, interculture and harvesting in time;

n) Limited scope of land improvement;

o) Water harvesting and recycling is the high investment proposition at the initial level;

p) Water damage and loss of run off (upto 50 per cent of incident rainfall) causing loss of soil (greater than 10t/ha) and formation of gullies resulting in degradation and unsuitability for arable cropping.

11.9.4 Socio-economic

a) Peculiar ecological and socio-economic settings;

b) Frequent failure of crops and unstable production rendering farmers poorer;

c) Low cropping intensity; low farm income, malnutrition and poor quality of drinking water;

d) Small size of farm holdings and high population pressure on land (1.7 persons/ha);

e) Poor quality of produce fetching lower price;

f) Lack of marketing facilities and market incentives;

g) Low level of literacy and poor resource base of the farmers though they are economic minded and rational;

h) Unemployment for most of the period of the year;

i) Farmers are to depend on the favours of the monsoon or on financing institutions which are reluctant to provide assistance as there is a more risk in the recovery of the released amount from farming;

j) Farmers of these regions are deprived of innovations, initiative, inspiration, aspiration and appropriate incentives and appreciations and they wait, watch and worry over their future.

11.10 MANAGEMENT PRACTICES OF DRYLAND FARMING AREAS

The success of crop production in dryland areas depends mostly on the monsoon rains which occur two to five months in a year. Intermittent drought of variable length aggravates the situation. The core problem of dryland farming is the scarcity of water as per the requirements of the crop and soil. Traditional systems of management of drylands are found to be less remunerative, conserve less soil and water and are of the subsistence type where every farmer tries to produce everything he needs; however, most of the larger farmers do produce for the market (Singh, 1983). The scientific management of drylands has brought out clearly that doubling the production on selected drylands through improved technology is a distinct possibility even if 50 per cent of the yield potential is realised, (Gautam, 1987).

Management practices of dryland areas differ from one location to the other as the agro-climatic situation of each location differs from the other in one or more ways. The factors relevant to management practices may be meteorological, edaphic, genetic resources, engineering, hydrologic, nutrient management, pest management and technological. A brief description of each of these will help the individual to understand how to schedule appropriate management practices for dryland areas.

11.10.1 Meteorological

The application of agro-meteorology is generally categorised into four types, e.g. (a) weather forecasting, (b) weather modifications, (c) micro-meteorology, and (d) interpretation of weather data.

The analysis of long-term rainfall records provides statistics for mean rainfall, standard deviation (*SD*) and coefficient of variation (*CV*) on a weekly basis. Results show that *CV* in weekly rainfall for a majority of locations even during the southwest monsoon period was appreciably high, ranging from 75 to 150 per cent. Weeks with *CV* values upto 125 per cent are dependable and weeks with *CV* below 100 per cent are highly dependable. Percentage probability for different amounts of rainfall (rain > 7 mm, rain > 14 mm, rain > 21 mm, rain > 28 mm and rain > 35 mm) on a weekly basis from standard week No. 1 to 52 for 23 locations were computed (Singh, 1982).

Other relevant meteorological information such as median date, mean date, *SD* and *CV* for the onset and withdrawal of the monsoon, the probability of the onset/withdrawal of the monsoon at different intervals for instance, seven days, ten days, and fifteen days is computed.

11.10.2 Edaphic

Land and soil types of different locations have been studied in detail. Their slope, configuration, physiography, erodability and physical, chemical and biological properties as well as workability have been studied. Problems and prospects of different monagement practices including soil and water conservation for seasonal cropping and sequential cropping have been studied.

11.10.3 Genetic Resources

A number of crop varieties have been identified/developed that are suitable for different locations at different seasons. Some of them are CSH-1, CSH-5, CSH-6, Swarna, SPV-245 (Sorghum); J-41, K-559, HB-3, Composite-3 (Pearl millet), Indaf-8 (Finger millet); S-3, Sharda, Pusa Ageti, Prabhat, UPAS-120, AS 71-73, AL-15, Pant, T-21 (Red gram—*rabi*); S-9, S-8, S-12, Pusa Baisakhi, K-851, T-44, SML-32, PS-18, ML-131, RS-8 (Green gram); S-1, T-9 (Black gram); G-2, C-13, C-20, C-152 (Cow-pea); Raj-1, FS-277, HG-75, HG-182, Durgajay, Ageta Guara-112, Malsonan (Cluster bean); Aruna, Bhagya, Soubhagya, Gauch-1 (Castor); Bragg, Punjab-1, Onkur JS 72-44 (Soybean); AK 12-24, Jyoti, JL-24, GAUG-10, DH 330, Kadiri 71-1, M-13, Punjab-1, TMV-3, RSV-87 (Groundnut); Pusa 1-1, Pusa 1-5 (Lentil); Sona barley, RDB-1, RD-31, Ratna, DL-3, PL 56, Azad, RS-6 (Barley); Durgamati, T-59, Pant toria-30, Pant toria 40, Pant toria-42, Suphala, RL-18, Bankura local (toria); B-54, Pendent type (rape) (Mustard); C-235, Dohad, Ujjain-21, Ujjain-24, RS-10, JG-62 (Bengal gram); N-62-8, JSD-1, Tara, Manjira, Annigeri, JSF 2 (Safflower); EC 68414, EC 68415, Modern, BSH-1 (Sunflower); G-5, G-2, Hunius, Diara, Sathi, Bassi selected (Maize); Laxmi, EL-156 E, PRS-72, Khandwa, Indore-1, Narmada, Banawar-1, H 777, H 999, H 974 (Cotton); Nila, Chambal (Linseed); N-22, TN-1, Bala, Kanchi, Saket-3, Ratna, MW-10 (Rice); S-308, Kalyan

Sona, RR-21, Narmada-4, D-134, N-112, Mukta (Wheat); Type-13, Pratap, B-67 (Sesamum), M.P. Chari (Fodder Sorghum), HFG 119, Guara 80 (Fodder Cluster bean).

Most of these varieties are of short duration, drought tolerant, tolerant to high population pressure, widely adaptable to variable conditions of soil, climate, rainfall, insect-pests and disease conditions, high yielding, suitable for intercropping and sequential cropping.

11.10.4 Engineering

There is a definite and positive relation between farm power availability and farm productivity. In India the available power with the farmers of drylands is low. In order to produce more from the drylands with limited power available, it becomes very important to make efficient use of power so as to increase the productivity of these lands.

In dryland agriculture, tillage, seeding, fertilising, interculture, plant protection and transportation are the major farm operations that are to be performed in time. Any delay in the operations due to limited power, non-availability of suitable tools/implements/machinery or aberrant weather conditions causes a reduction in the yield and the quality of produce. In drylands most of the farmers use human and bullock power and the average power available for these lands is nearly 0.22 hp/ha (compared to average farm power use to the tune of 0.35 hp/ha). This is very low considering the heavy demand of power over short periods to complete critical farm operations in the limited time available, because of the dependence on rainfall. This low power availability can be overcome either by employing more power (0.5 to 0.8 hp/ha) or by using improved, light and wide width implements.

In terms of physical output, the amount of work done by the power units depends upon the type of tool, implement or machinery engaged to carry out a given operation and also the prevailing soil moisture and weather conditions.

All tillage operations should be carried out at optimum soil moisture conditions when the resistance to tillage tools is low, resulting in lower draft and power requirements and better soil tilth. At higher as well as at lower moisture, the workability of the soil is very poor.

The development of minimum tillage is rewarding as there is a limitation of power. This helps in completing different operations (tillage, seeding, fertilising) in time.

Sowing crops in line helps in the efficient management of the crops. Weeding and hoeing become easy and quick compared to that of broadcast crops.

An improved *bakhar* blade provides for higher field capacity and field efficiency and requires a lower unit of draft power in comparison to local straight blades. The improved harrow (tilting hook type) provides a higher field capacity and requires lower draft power than that with a fixed hook. The use of improved tools and implements such as the wheel hoe, mould board plough, three-row seed drill, seed-cum-ferti drill and wheeled tool carrier considerably improve the working efficiency of man and bullock with less damage to dryland soil.

Cultivation practices such as wide-row cultivation, year round tillage, set-row cultivation and till planting have been found to improve the working efficiency of animal power as it either covers a larger area within a given time or distributes work over an extended period.

Multi-purpose implements permit farmers to carry out their basic operations of tillage, planting, fertilisation, weeding at the correct time and in a precise manner to increase productivity.

Proper crop planning for the year round and uniform use of available power will help in improving the timeliness of farm operations due to the distribution of work. It will then be possible to manage a greater land area within the right time with the same power resulting in improved power efficiency, low power cost and a high amount of produce with better quality.

11.10.5 Hydrologic

A water balance is the balance between the inlet and outlet of water in soil. In drylands, rainfall is the only source of soil moisture while evaporation, transpiration, percolation and run off are the major factors of moisture loss from the soil. Crop production requires the presence of the available moisture throughout the crop duration as well as for an additional period for land preparation and sowing. The amount of available soil moisture required for this varies greatly at different times. Thus the soil-water balance greatly determines the plant-water balance as the atmospheric-water balance has little contribution to dryland soils and crops. However, dew has a positive role in crop production where a considerable amount of dewfall occurs.

Crop growth is only indirectly related to the rainfall amount *per se*, and the soil moisture status in the root-zone, rather than the amount and distribution of rainfall is the main factor, contributing towards crop growth and yield. Water balance models have been developed by various workers in different parts of the world to facilitate the computation or estimate of soil moisture data from the readily available weather records. It was found that for operational purposes the water balance model of Thornthwaite could be used even in our tropical conditions for estimating soil moisture.

Using the water budgeting technique of Thornthwaite, the frequency of 'dry' and 'wet' spells in the Hyderabad region was worked out on a weekly basis for the cropping period (from week No. 21 to 45). The data was analysed in terms of percentage probability of different amounts of water surplus (WS), percentage probability of actual to potential evapo-transpiration (AE/PE) in different frequency ranges, and percentage probability of the occurrence of situations when the actual water supply (AE) falls below that of demand (PE) in varying degrees. The result shows week Nos. 26, 27, 28, 34 and 41 may be treated as 'dry' weeks: Nos. 30, 31, 32, 37 and 39 may be treated as 'wet' weeks: and the period from week No. 26 to 40 may be treated as a period of optimum moisture availability for crop growth.

Considering a period of 127 days (11th June to 15th October) as the cropping season for the Hyderabad region, daily water budgeting was done for the entire cropping period, with the object of working out the frequency and variability in 'effective rainfall' vis-a-vis rainfall, and also the frequency and variability in the number of 'dry days' during the cropping season. It was found that the mean 'rainfall' for the cropping period was 670 mm with SD of 164 mm and CV of 25 per cent;

the mean 'effective rainfall' for the period was 440 mm with SD of 58 mm and CV of 13 per cent; the mean number of 'dry days' was 34 with SD of 16 days and with CV of 47 per cent; and the most salient finding was that 'effective rainfall' in the semi-arid region of the country is far less variable than the 'rainfall'.

The water balance and aridity index ($Ia = WD/PE \times 100$ where WD is the water deficit and PE is potential evapo-transpiration) were worked out for individual years for different dryland locations.

The water balance and moisture index ($IM = P - PE/PE \times 100$ where P is precipitation and PE is potential evapo-transpiration) were evaluated for all the dryland locations for individual years to find out the fluctuations in different moisture regimes from year to year. The highest fluctuation was observed at locations falling in the sub-humid category of climate, thus indicating a high variability in crop yield or low stability in agricultural production in the sub-humid belt of the country.

The water balance analysis was carried out on a weekly basis with week-wise values of various elements of climatic water balance, for instance, P, PE, AE, WS and WD. This information was of specific importance in understanding the overall water economy at respective locations and run off collections for supplemental irrigation during the period of drought. Based on the AE/PE in different limits, the soil water availability periods were designated as excellent, good, poor and very poor and the total number of weeks in different categories were worked out for all the locations.

It was found that the modified Thornthwaite moisture index ($Ima = AE/PE$) was much superior to IM ($P - PE/PE \times 100$) in explaining the variability in crop yields from year to year. The Ima evolved by integrating data on climate, soil and crop was found to have additional advantages.

A water balance model of Baier and Robertson ($VSMB$ or VB) was tested for its reliability and suitability for tropical drylands of India. Good to very good agreement between observed soil water data and the soil-water estimated by VB on a daily basis was found for respective locations. From the computerized daily zone-wise soil-water, the following studies were undertaken:

a) Soil-water in the root-zone was represented in graphic form for the individual cropping season, showing the availability of soil-water at different depths at different periods of the season;

b) *AE/PE* for rainy and post-rainy cropping seasons were worked out and, thereafter, correction studies were made between *AE/PE* and yields of more important crops of these regions;

c) The aridity index was worked out for the cropping period to assess the magnitude of drought and information was provided on the efficient management of water;

d) The amount of surplus water (*WS*) available for run off and deep percolation was worked out during the cropping season. This shows a very high potential for run off water collection and its recycling.

In dryland agriculture, the catastrophic (famine producing) droughts and flash floods are weather events of specific importance which cause heavy reduction in agricultural production in different regions of the country. By analysing the long term weather data the 'risk' and 'return' periods are now being calculated (Singh, 1982).

The proportion of water in the living plant is far higher than in the soil. A water deficit occurs in the plant whenever transpiration exceeds water absorption; this may be due to excessive water loss, reduced absorption or both. The consumptive use (*CU*) of water (quantity of soil-water absorbed by crops and transpired or used directly in the building of plant tissues together with that evaporated from the cropped area) varies with crop varieties, the growing season and the stage of crop growth for instance, *CU* of modern wheat varieties ranges from 0.7 to 0.85 mm per day which increases with the advance of the season even up to four mm per day.

According to the degree of the internal water deficit and its duration the crop shows the signs of incipient, temporary or permanent wilting. A small too in turgor, causing incipient wilting, is an almost daily occurrence in warm, dry weather, even when the soil is moist. This does not produce visible symptoms of wilting. A more severe loss of turgor, causing leaf drooping, inward curling and subsequently spreading through the plant, will reduce growth. If plants regain their turgidity when the water supply is re-balanced, this is termed 'temporary wilting'. A longer period of dehydration causes 'permanent wilting' from which plants are no longer able to recover even in a saturated atmosphere.

Moisture stress does not affect all aspects of plant growth and development equally: some processes are highly susceptible to increasing moisture stress, while others are far less affected. The final yield of the crop will be the integrated result of these effects of stress on growth, photosynthesis, respiration, metabolic processes and reproduction (Arnon, 1975).

With the onset of dry conditions, a progressive and continuous decline in turgor of the plants is observed. The degree of loss in turgor varies with crop varieties and the stage of growth.

Water stress can affect photosynthesis directly, by affecting various biochemical processes involved in photosynthesis, and indirectly by reducing the intake of CO_2 through stomata as a result of their closure in response to water stress. The translocation of assimilates can also be affected by water stress, and the resulting assimilable saturation in the leaves may limit photosynthesis (Hartt, 1967). An advetse water regime also reduces the leaf area and hastens leaf senescence, thereby decreasing the productivity of the crop to a greater extent than the depressive effect resulting from a reduced net assimilation rate (Fisher and Hagan, 1965).

More severs drought lowers water content and respiration CKaul, 1965).

Periodical water stress leads to many anatomical changes. These include a decrease in the size of cells and of intercellular spaces, the growth of a thicker cell wall, and greater development of the mechanical tissues (May and Milthrope, 1962).

Severe water deficits affect metabolic reactions such as enzymic activity, the nucleic acid system and hormonal distribution in plants particularly in the contents of cytokinis and abscisic acid.

As tissues become desiccated, protoplasm becomes increasingly dense and its viscosity gradually increases; when dehydration is severe, the complete gelation of the protoplasm occurs and finally it may become rigid to the point of brittleness (Levitt, 1956).

Root development is affected less than that of the aerial parts under soil-moisture stress. Roots become finer, lighter and suberized. The activities of such roots are reduced. Root growth in each soil layer appears to be independent of the moisture content in other soil layers or in the shoot (Newman, 1966).

Growth is suspended during moisture stress and resumed upon its elimination. The extent of damage caused to the plants depends on their physiological age, the degree of water stress, and the species concerned (Gates, 1968).

Reproduction and grain development are seriously affected by moisture stress during this period. In crops with determinate flowering in which the period from pollination to seed-set is relatively short, the most critical period with respect to water stress begins with the appearance of pollen mother cells. In cotton, an indeterminate flowering plant, water stress during the period of early flowering causes shedding of new flower buds, but has no effect on current flowering or on boll retention. A similar stress during peak flowering causes shedding of flower buds and bolls.

When the yield consists of most or the aerial part of the crop (forage crop, tobacco) the effects of stress on yield are much the same as those on total growth. When the yield consists of storage organs other than seeds or fruits (stock beets, potatoes) it will generally be as sensitive to moisture stress as in the plants' total growth. When the yield consists of the seeds or fruits, the situation is very different and it has been shown for a number of crops that the dry matter stored in the seeds or grains is mainly the result of photosynthesis that occurs after flowering (Throne, 1966). The effect of water stress will depend on the stage of growth at which it occurs. At an early stage, the number of primordia formed may be reduced.

When the yield is a chemical constituent (sugar, alkaloid, fibre) the economically valuable part of the crop is only a small fraction of the total dry matter produced and moderate stress that affects growth may have no or even a beneficial effect on yield.

Slatyer (1967) has described the sequence of effects on plants of water deficits as they become progressively severe. At first metabolism is impaired only during the daily periods of maximum deficit; but these periods become progressively longer. Stomatal closure during these periods leads to reduced transpiration and also CO_2 absorption: photosynthesis is reduced and leaf temperatures are increased. Increased respiration further reduces apparent photosynthesis.

As the moisture stress in the soil approaches wilting point, turgor pressure in the plants approaches zero. Cell enlargement virtually ceases, and the rate of cell division is markedly reduced. The stomata remain closed for most of the day and transpiration is limited to cuticular transpiration. Leaf temperature increases markedly. Most metabolic processes, including respiration, slow down apparent photosynthesis and the production of dry matter becomes practically nil.

The disruption of normal cell metabolism is accompanied by the breakdown of proteins and carbohydrates, causing an increase in the concentration of sugars; leaf phosphorus and nitrogen migrate from older leaves to the stem. Finally, as protoplasmic dehydration continues, individual cells and tissues die. Root hairs die and the roots suberize. In some cases shoots die before roots, in others the opposite occurs.

Crop plants under dryland situations are subject to renewed moisture supply after a stress period. When crops at their different stages of growth suffer periods of stress of varying duration and severity, but do not desiccate completely, the stress period is terminated each time by a renewed supply of moisture. The recovery of photosynthetic efficiency occurs immediately on the reduction of moisture stress, sometimes the rate of photosynthesis exceeding the original level. If the moisture supply is renewed only shortly before death would have occurred, metabolic processes do not immediately return to normal. The death of the root hairs and the suberization of the roots delay recovery—because of reduced capacity of water uptake. The adverse effects of water stress are less severe at some stages of growth than at others. At certain stages, there may be irreversible damage which affects the final yield while at other stages, a renewed water supply may stimulate growth to such an extent that the plant may 'catch up' (Arnon, 1975).

11.10.6 Nutrient Management

Our dryland areas are hungry as well as thirsty. They are universally deficient in nitrogen and frequently limited in phosphorus. Potassium is adequate in many areas except in light textured soils or at continued high levels of production. Zinc and sulphur application shows good response in some crops.

The use of fertiliser helps to increase water use-efficiency [$WUE = Y/ET$ i.e., the yield of marketable crop produced (Y) per unit of water used in evapo-transpiration (ET)] by accelerating growth of both shoot and root. Adequate root growth helps to extract moisture from a greater depth and volume of soil. Balancing soil nutrient status helps the development of hardiness of plants against water stress and insect-pest hazards. If 'nutrient stress' is properly taken care of plants are capable of utilising a favourable soil-moisture period more efficiently.

It has been found that areas receiving more than 700 mm rainfall can be supplied with fertiliser with much less risk. In areas receiving rainfall less than 700 mm, it is necessary to use lower levels of fertiliser (Venkateswarlu, 1981). Regarding fertiliser management it is necessary that the initial fertiliser application on sowing should be such that it does not allow too much vegetative growth specially in the post-rainy crops (De, 1983). In general cereals respond to N and P, short duration pulses respond to P and deep rooted crops such as castor and mustard respond more to N. Therefore, in single cropped areas a need-based application is advocated. In cropping system fertility balancing may be adopted by applying P to legumes and N to cereals and growing these crops alternately.

The efficiency of applied fertilisers can be enhanced by the placement of fertisers near the rows/roots. Deeper placement in the *rabi* crops is essential. For a lower dose, fertiliser can be placed in the seed rows but for a higher dose the placement should be on the sides or in the broader rows. Organic manures and fertilisers may be mixed and placed in the seed rows to avoid salt injury. During *kharif* all P and K and one-third N as basal in the seed row and the remaining N in one or two splits may be applied. In double cropped areas phosphate should be used for the *rabi* crop and organic manure should be applied to the *kharif* crop. A nutrient should be applied to the most responsive crop and the crop succeeding a legume may be given less nitrogen. It is always advisable to use moderate levels of fertilisers to avoid risk. In intercropping systems a part of N along with P and K are to be applied as basal to each crop and the remaining N is to be side dressed to non-legume (cereals/millets/oil-seeds) components only.

Integrated nutrient management involving fertilisers, organic manures, green manures, biofertilisers, intercropping with legumes and legume crop/fodder cultivation reduces the cost and increases yield and the quality of the produce, if they are adopted after considering soil test values and soil and moisture conditions.

11.10.7 Pests and Diseases

A large number of pests, parasites, pathogens and weeds interfere in crop production. Some of them are soil-borne, seed-borne or air-borne/migratory in nature. Integrated pest management helps to maintain a better environment for crop plants with a poorer environment for the pests. Under unavoidable circumstances pesticides may be used. The cost may be reduced and the efficiency can be improved if compatible nutrient elements, other agri-chemicals (anti-transpirants and growth regulators) are combined together with pesticides for spray application. All pests should be controlled at their weakest stage of development before they cause economic injury and prior to the susceptible (critical) stage of crop plants. For this a regular surveillance service may be necessary considering the efficacy, ecology and economy of the practice of management to be followed.

11.10.8 Technological

At present a number of new technologies are available. Some of them are most appropriate for specific locations while others are of wider applicability even under changing circumstances. Some consist of the selection of ideotypes suitable for the soil, season and systems of cropping: the time of sowing with the least consumption of seasonal moisture; the method of sowing, the maintenance of an optimum plant population, the use of improved implements, and agri-chemicals; water harvesting and increasing water use efficiency, mitigating aberrant weather conditions, checking soil erosion, soil fertility build-up, reclamation of problem soils, land improvement and increasing total yield and stability of production. All these not only alleviate the economic conditions of the farmers for the current year but provide a sustainable response.

Drylands require a tillage practice which conserve the basic resources of soil and water. Tillage practices also aim at complementing basic soil and water conservation measures such as graded bunds and land levelling. The rapidity of tillage operations for keeping the field ready for seeding immediately after the rain sets in is an important key to the success of the crops. Proper tillage also facilitates the establishment of a crop and its support during the growth period. It gains further importance when a second crop is to be sown in sequence after the harvest of the first crop. Tillage practices also have a major role in providing drainage and/or *in situ* moisture conservation depending on the soil type, the crops grown and the season. It is better to apply mulch after the completion of tillage operation for improving moisture conservation either for the use of the crop and/or for extending the date of seeding.

Deep tillage has proved highly beneficial under alluvial and red soils. Deep tillage has been found useful in increasing crop yields. Other benefits of deep tillage are the retention of soil moisture, the breaking of hard pan, helping tap rooted crops, the incorporation of phosphate fertilisers deep into the seedbed, allowing the residual effect of manures and fertilisers to remain, retaining moisture and making mulching and weed managemant more effective. Imposing suitable land treatment prior to seeding helps in regulating run off for the use of crops. Ridge and furrow seeding have wide application. Ridge seeding provides drainage and furrow seeding allows *in situ* moisture conservation depending upon the location, season and crop. Set-row cultivation does not affect the yield. however it reduces tillage energy which can be used later for intercultivation. Reducing the peak tillage energy requirements and the efficient use of fertiliser are the added advantages.

Seed-cum-fertiliser drills provide the option of line sowing with metering of seed and fertiliser, covering and compacting the seeded furrow for quick and better emergence. Betides, line seeding with appropriate crop geometry and fertiliser placement, seeding principles other then flat sowing are important. These are ridge and furrow seeding, semi-deep furrow seeding and dry seeding. Dry seeding is forced under circumstances where the soil does not allow the operation of seeding on the monsoon in certain areas. Dry seeding in anticipation of rain helps to establish a crop at an early opportunity and provides the advantage of better utilisation of rainfall during crop growth. The tractor drawn lister and the bullock drawn ridger seeder have been found superior while undertaking dry seeding compared to other tools (Sriram *et al.* 1982).

For the efficient use of rain water to achieve high crop production, crop management such as plant density, crop geometry and weed control are important. Optimum plant density under the rain fed system is likely to be less than that of irrigated agriculture. Excessive foliage leads to a quicker loss of soil moisture through transpiration and higher plant density may be counter-productive if rainless periods become prolonged during the critical stages of crop growth and development.

The time of sowing is a crucial factor in crop production in rain fed farming. Early sowing of rainy season crops always results in higher yields. Early sown crops have also been observed to escape major insect and disease damage, and to mitigate drought effects in case of the early cessation of monsoon rains. In sequential cropping areas, the early sown crop releases land earlier so that the succeeding crop can be sown at the optimum time at optimum soil moisture. The harvesting of the fisrt crop at physiological maturity helps in the early sowing of succeeding crops (Subramanian *et al.* 1982).

In red soils graded bunds may be preferred over the contour bunds because of the scope for harvesting run off water for supplementing irrigation which provides a pay-off to the tune of 2.0 q/cm/ha. Bund spacing can be at 90 m on two per cent slope. The optimum furrow length should be 60 m for favourable performance.

In black soils with their low infiltration characteristic, cultivated fields suffer from the problem of water stagnation. Run off losses can be reduced by adopting the graded bunds. Besides bunding, creating corrugations across the slope on a mild gradient of 0.2 to 0.4 per cent provides an opportunity to infiltrate into the self-tilled soils (due to cracking) as the run off water flows along these corrugations at a safe and non-erosive velocity.

In red soils, flat sowing and ridging seems to have an edge over the 'bed and furrow' system

In black soils, the addition of gypsum has increased the intake rate by four to seven times. Vertical mulching, or filling sorghum stubble into the trenches made at regular intervals across the slope has proved to be beneficial in further raising the intake rate of these soils (Sharma *et al.* 1982).

Harvested water from the watershed that is stored in tanks is treated as a scarce commodity. It is used only for 'life saving' irrigation at the most critical stages of crop growth or for extending the cropping season. In principle, Alfisols which lead to droughts because they have a lower water holding capacity but produce a greater run off early in the monsoon season would benefit more from the application of water during the dry spells occurring during the monsoon. In Vertisols which have a higher water retention capacity and generate a run off more towards the latter part of the monsoon when the soil is already fully recharged or nearly saturated, collected run off water can be put to better use in establishing a sequential post-rainy season crop or in irrigating the same crop at the most critical stage, such as at flowering.

Yields of sorghum and maize on Alfisols were approximately doubled when five cm of irrigation was applied during a 30-day drought in late August and early September. The post-rainy sorghum in Vertisols responded to supplemental irrigation at the grain-filling stage with an increase in yield from 2570 kg to 3570 kg/ha. The post-monsoon chick pea in Vertisols produced increased yield from 817 to 1441 kg/ha with one irrigation at the flowering stage (Miranda *et al.* 1983).

Weed management systems in all crops should start with preventive methods (clean seed, sanitation in non-cropped areas and seasons). This must be followed by appropriate cropping and cultural practices that give the crop a competitive advantage (rotations, fertiliser placement, seeding methods). Mechanical methods may vary from crop to crop and region to region and must take into consideration pre-seeding, post-harvest, inter-row and intra-row weed removal by efficient and effective tillage or hand-weeding methods. Herbicides also have an important part to play in this package, especially for 'hard-to-kill' weeds (*Striga, Cynodon, Cyperus*) when tillage practices cannot be carried out due to close crop spacing or when it is difficult to enter fields due to wet soil conditions.

Sometimes one method enhances the effectiveness of another method. In maize (Dehradun) the efficiency of simazine on weed flora was increased by 23 per cent when combined with tillage. Such combination treatments increased maize yields by 30 per cent compared to hand weeding alone and reduced soil nutrient loss by 30 to 40 per cent. At Solapur, propanil applied pre-emergence plus hand weeding gave significantly higher groundnut yields than either treatment alone. Atrazine at 0.3 kg a.i./ha over the row plus two intercultivations have given the highest yields of sorghum at Hyderabad.

The year-round tillage (at Hyderabad) not only leads to increased sorghum production, but also reduces the infestations of weeds in the rainy season to such an extent that hand-weeding and inter-row cultivations are more efficient and less time consuming (Friesen, 1983).

Regarding the management of insect-pests and diseases, the use of resistant varieties, seed treatment, early planting, adoption of high seed rate and the removal of infested seedlings, the balanced application of fertiliser, crop residue destruction, weed control, the use of light traps, off-season tillage and the application of pesticides are the different ways.

It is essential to assess the reasons for the income gap that plagues small farmers who have not realised the full income potential of their farming holdings. The cultivation of remunerative substitute crop varieties helps to alleviate the stringent pecuniary conditions of the peasants. In Deccan areas safflower is found to be a superior substitute of the wheat crops. In the Bellary region, sorghum is a more productive substitute of dryland cotton (*G. arboreum/herbaceum*) crops. In the post-rainy season in Agra and Varanasi mustard and chick-pea are identified as more productive than the traditional wheat crop.

With respect to the stability of yield, pearl millet is more stable than sorghum in Anantapur. In Varanasi, the rice yield is more stable than maize and black gram is more stable than pigeon-pea in the rainy season. In the post-rainy season, barley and chick-pea are found to be the most stable crops. In Hissar, chick-pea yield had, as high as mustard in frost-free years. However, when frost occurred during the third week of January and second week of February, only chick-pea maintained its productivity while the yield of mustard was drastically reduced.

A higher grain yield of wheat was recorded with 100 ppm phenyl mercuric acetate (PMA). The number of spikelets/sqm and the number of fertile spikelets and grains/spike had contributed to the higher grain yield. Higher N, P and K uptake was recorded due to the application of PMA at 100 ppm, half at tillering and half at the boot leaf stage (Reddy and Misra, 1986). The spraying of transpiration suppressants (cetyl alcohol, PMA, Stearic acid and chlormequat chloride) at 1000 ppm with a low volume sprayer at 100 l/ha on wheat at the boot+ anthesis stages significantly increased the grain and straw yields. The additional yields were of the order of 22.7 q grain and 28.0 q straw/ha. Amongst the chemicals chlormequat chloride recorded the highest grain and straw yields of wheat of the order of 62.8 and 71.8 q/ha, respectively from a rain fed crop at Solan, Himachal Pradesh (Singh and Awasthi, 1984).

Seed treatments with two-and-a-half per cent KCl increased the dry wheight of the shoot and the yields of straw and grain to the tune of 15 per cent in wheat (Misra and Dwivedi, 1980). Seed treatments with Na_2HPO_4 12 H_2O solution (358 ppm for six hours) and water (for 24 hours) increased the grain yield of barley by 37 and 24 per cent respectively (Singh and Chatterjee, 1980).

11.11 CROPPING SYSTEMS

Cropping systems differ according to climate and soil types. In regions of low rainfall (400 mm), a short growing season (two to two-and-a-half months) and shallow soil as in western Rajasthan and parts of Haryana and Kutch, grass is the principal crop. In addition to grasses, short duration pulses or agroforestry based systems might hold some promise in such areas. With increasing rainfall and a longer growing season, grain crops such as pearl millet, sorghum, maize or upland rice are grown. The areas with 400 to 750 mm annual rainfall, monocropping with traditional long duration crops is common. Even in areas with an annual rainfall above 750 mm and soil storage capacity of 150 to 200 mm of available moisture, farmers have been growing either a rainy season crop or a post-rainy season crop on residual soil moisture after fallowing during the rainy season.

Generally adaptable crops of this region are cereals and millets such as rice, wheat, maize, barley, oats, sorghum, pearl millet and finger millet; oil-seeds such as groundnut, sesame, castor, safflower, sunflower, linseed, rape and mustard, niger and taramira; pulses such as pigeon-pea, black gram, green gram, chick-pea, cow-pea, dolichos, horse gram, cluster bean, soybean and lentil; fibre crops such as cotton and other crops such as colocasia, tobacco, chillies and vegetables.

Among fodder crops non-legume fodders such as sorghum, pearl millet, maize, *sanwa* (*Echinochloa* spp.), Dinanath grass and legume fodders such as cow-pea, cluster bean, velvet bean, field bean, horse gram, rice bean, tetrakalai and *moth* are important. Pasture legumes such as siratro, stylo and phasemy beans are also used as green fodder or as hay.

When the rainfall is between 500 and 700 mm with a distinct period of moisture surplus, the intercropping system can be adopted. In areas with more than 750 mm annual rainfall intercropping facilitates the growing of either cereal+legume or legume+legume in a system of different maturity patterns. Intercropping in such areas is helpful in quick drainage (ridge-furrow system of intercropping). Here the choice of varieties of component crops are important to utilise the space. The future intercropping systems in drylands should be cereal+legumes; the objectives are to minimise fertiliser use, pest and disease incidence on legumes, produce balanced food, provide balanced fodder for cattle and to take advantage of the extended growing season (Singh, 1983). With the recent emphasis on intercropping systems, it has now become possible to harvest almost the full yield of a cereal component and a bonus yield to the tune of 60 per cent of the associate legume crop. Examples are: sorghum+pigeon-pea, pearl millet+pigeon-pea, sorghum+green gram, sorghum+soybean, groundnut+pigeon-pea and oxtail millet+pigeon-pea.

In areas with more than 750 mm annual rainfall with a soil

storage capacity of 150 mm or more of available moisture sequential cropping is possible. In the Vertisols in the Vidarbha and Malwa plateau and the Gangetic alluvial belt besides the sub-humid red soils of northeast India, it is possible to grow pulses and oil-seeds. In the Gangetic alluvial belt, rice can be followed by chick-pea. In the Kandi belt, maize can be followed by chick-pea. In the Vidarbha region, sorghum or green gram can be followed by safflower. In the Malwa plateau, maize and safflower/sorghum—chick-pea and maize—chick-pea are the possible crop sequences. In the Bihar plateau, rice can be followed by lentil. In the Orissa sub-humid red soils, horse gram can be the second crop.

In sequence cropping the objective should be to maximise the returns from the system rather than from the individual crops. In developing such a system consideration should be given to soil fertility management, the growing season, integrated insect-pest and weed management and above all much needed food, particularly pulses and oil-seeds.

In addition to improved crops and cropping systems, adoption of timely sowing and suitable methods of sowing, contingent planning for weather aberrations, fertiliser use and adequate insect-pest management are essential prerequisites for making the cropping system a successful enterprise. The collection of run off water and its recycling adds a new dimension to the flexibility of cropping systems. During the winter season, pulses and particularly oil-seeds have more flexibility regarding dates of sowing compared to cereals (Singh, 1983).

In sequential cropping, the first rainy season crops should not be grasses and legumes as they leave the soil in a very dry condition and have a high water requirement which considerably limits their use in such conditions. On the other hand, row crops such as maize, do not utilise all of the available soil moisture.

The nitrogen response was linear upto 40 kg N/ha in barley grown after pearl millet but grown after any of the legume fodders, the optimum yields were reached with 20 kg N application to barley. Thus growing short season fodder legumes and harvesting them at least ten days earlier than the expected end of monsoon rains would provide nutritious fodder and at the same time raise the productivity of soils sufficiently to benefit the subsequent crop grown on the conserved moisture to the exten of 40 kg N/ha(De, 1983).

Cropping systems suitable for aberrant weather involve a choice of alternate crops, ratooning and thining, the use of urea spray for rapid regeneration, emergency nurseries in case of upland rice and the providing of life saving irrigation to moisture stressed crops. In case of delayed onset, cereals should be replaced by suitable pulses and oil-seeds, for instance green/black gram/sesamum in the place of upland rice (Singb, 1983).

The effect of green gram was found to be either more or equal to that of fallowing with safflower and wheat. Though the yields of barley, wheat and sorghum were marginally low after green gram in comparison to fallowing, the sequence cropping was highly remunerative. The post-rainy season crops in sequence of soybean—wheat/safflower/and black gram—barley were found to be benefited by the legume component. (Rao and Das, 1982).

11.12   MANAGEMENT TECHNIQUES

The dryland farming practices may be summarised in the following four steps to increase production from these areas. These are:

a) Reduction of moisture loss due to evaporation and transpiration;
b) Reduction of run off and increasing infiltration;
c) Improvement of under-use efficiency by crops; and
d) Maintenance of fertility and yield stability.

11.12.1   Redaction of Moisture Loss doe to Evaporation and Transpiration

Some measures reduce the loss of moisture received by the soil. The following measures were found to be useful.

a) growing early maturing adaptable crop varieties with a deep and ramified root system and with a reduced number, size and horizontally orientation of leaves; such crop varieties should have wider adaptability with respect to the ever-dynamic thermic and hydrologic conditions of the atmosphere and soil, the chemical, physical and biological capacities of the soil and the higher potential of yield with moderate stability,

b) maintaining optimum plant population per unit area,
c) sowing crops either in dry soil anticipating rainfall suitable for early crop establishment with the first shower, and subsequent growth and development with subsequent rainfall, or in optimum soil moisture but with a minimum expenditure of seasonal moisture for land preparation and sowing;
d) keeping the field free from weeds,
e) adopting mixed or intercropping to utilise the slow growth phase of wide spaced crops, to restore soil fertility and to check soil and water loss,
f) using mulches,
g) using agri-chemicals;

1. anti-transpirants or transpiration suppressants such as hydroxylamine hydrochloride, phenyl mercuric acetate (PMA), silicone, cetyl alcohol, stearic acid, chlormequat chloride and alachlor.

2. plant modifiers or growth retardants such as abscisic acid, a-NAA, phosphon, daminozide, chlormequat chloride, mepi quat chloride, ethephon, diphenyl, alkenyl succinic acid, 2, 4-dinitrophenol, 8-hydroxyquinoline, chlorflurenol, methyl ester, DMAS, TIBA, MH and CMH.

3. desiccants or defoliants such as 2,4-D, paraquat, H_2SO_4, Na-chlorate, diquat, dinoseb, metaxuron, ethephon, glyphosate, arsenic acid, Mg-chlorate, ametryn, endothal, Na-chloroacetate, Na-pentachlorophenate, TCA, CCC, chlorflurenol, N-dimethyl morpholine chloride and 2, 3-dichloroisothiazole-5-carboxylic acid.

4. crop ripeners such as picloram, ammonium isobultylate, glyphosine, bromacil, CCC, tetrahydrobenzoic acid, cacodylic acid, cycloheximide, mefluidide, cetyltrimethyl ammonium bromide, vanillin, carbamyl urea, isoaureomycin, penicillamine, disugran, isobutanol, paraquat, asulum, anisomycin, endothal, 2, 3-6 TBA, fluoridamide, bacitracin, mineral oil, ripenthal and desugran.

5. anti-evaporants.

a) for reducing evaporation from the soil surface by reducing albedo value i.e., as synthetic or chemical mulch, for instance Encap Esso, plastic emulsion, Kaolin water.

b)   for water reservoirs such as hexadecanol, octadecanol, cetyl alcohol and stearyl alcohol.

6.   antiseepage, such as bentonite clay, Na_2CO_3, cement and plastic sheets.

11.12.2   Reduction of Run Off and Increasing Infiltration

Whenever rainfall is received a large portion (30 to 50 per cent) is lost as run off because of poor infiltration due to the nature of the soil and the slope of the area. The following measures help to prevent this:

A. *Land management*

The following land management measures were found to be promising in several locations with varying degrees of response:

1.   bunding fields on contour lines,

2.   division of the land into compartments,

3.   land shaping, levelling, terracing according to slope and configuration of the land,

4.   graded bunds with grassed waterways and box-type masonry drainage outlets in arable fields,

5.   ploughing across the slope and growing low value crops in catchment areas; the row direction should be across the slope,

6.   growing high value crops in level, run off concentrated strips,

7.   the ploughing of deep soils should be done once in three to four years immediately after the harvest of *rabi* crops. The ploughing of light, shallow and medium soils should be omitted, they may instead be hoed. Frequent hoeing with rapid and light implements is advocated to receive and retain moisture,

8.   reclaiming problem soils,

9.   incorporating a liberal quantity of organic matter.

B. *Other technologies*

1.   growing both erosion permitting and erosion restricting crops with permanent grass strips with *Dichanthium annulatum*,

Cynodon dactylon, Brachiaria sp., *Eulaliopsis binate* and *Saccharum* sp. and growing cover and strip crops,

2. growing a shelter belt of forage, fuel and timber trees to protect the land from wind erosion,

3. using mulches particularly organic mulches and practising stubble mulch farming,

4. adopting water harvesting by inter-plot, inter-row, modified inter-row, broad-bed and furrow system (ICRISAT) for *in situ* conservation and by collecting the run off in tanks for future use as life saving irrigation for crops.

11.12.3 Improvement of Under-Use-Efficiency by Crops

The traditional crop varieties or land races or ecotypes are generally longer in duration, poor in yield, harvest index, and quality but moderate to high in tolerance against adverse biotic and abiotic conditions and stable in yield. Therefore, the use of such varieties generally does not permit double cropping and higher yield, exploiting the available moisture conditions. The adoption of scientific measures helps to alleviate quantitative and qualitative yields per unit area, time, water and other investments. Such measures are:

a) adopting a potential cropping system in relation to rainfall and soil type

Rainfal (mm)	Broad soil group	Growing season (week)	Cropping system
275-625	Red soils and shallow black soils	20	Single *kharif* cropping
375-625	Deep sierozems and alluviums	20	Either *kharif* or *rabi* cropping
375-625	Deep black soil	20	Single *rabi* cropping
625-800	Red, black and alluviums	20-30	Intercropping
800 and above	Alluviums, deep black soils, sub-montanous and deep red soils	30	Double cropping

The traditional cropping systems are characterised by low risk and low yield. Emphasis should be given to combine low risk with high yield. The following approaches may help in solving this problem:

1. Adopting cropping systems to meet the aberrant weather.

2. Combining low monetary inputs with a high level of management in the selection of suitable crop varieties, the choice of sowing dates, sowing methods, maintaining an adequate plant stand and weeding.

3. Supplementing the natural resources with monetary inputs, for instance fertilisers and life saving irrigation.

4. Flexibility in the choice of crops in the cropping systems. In many cropping sequences pulses and oil-seeds provide more return than cereal-cereal rotation.

5. Growing crops harvestable at any time and always usable as contingent crops.

    b) growing crops in quick succession or as overlapping crops, or *paira* crops with till planting, land preparation fertiliser application and sowing simultaneously with other operations.

    c) applying fertiliser considering the soil-test-value, the cropping system and nutrient balancing, in the recommended form and dose by band or localised placement and by broadcast or spray during monsoon days.

    d) practising line sowing after applying manures and fertilisers deep into the soil.

    e) treating seeds to induce.

    1) drought hardiness with $CaCl_2$, $NaCl$, Na_2SO_4, KCl, $MgSO_4$, Y-ray auxin, boron, agrosan, CCC, citric, fumaric, succinic, malic acids, inorganic and organic salts, purines, pyrimidines, caffein, uracil, xanthine, uridine diphosphate, $GA_4 + GA_7$.

    2) cold hardiness with chlormequat, daminozide, abscisic acid, 2-amino-6 methylbenzoic acid, polyamine compounds such as 5-chloro-4 quinoline carboxylic acid, 2-chloro-4-quinoline carboxylic acid and 2-trifluromethyl quinoline carboxylic acid.

    f) growing ideotypes considering the season, soil, cropping and cultural systems and environment.

    g) controlling insect-pests, parasites and disease in time.

h) providing drainage facilities particularly in low lying spots.

11.12.4. Maintenance of Fertility and Yield Stability

Dryland areas have low yields and high yield fluctuations. The adoption of adequate management practices of soil, water and crop reduces fluctuations in year to year production. The maintenance of soil fertility is a problem in such areas as for a considerable period of the year the soil remains uncropped and there is a loss of plant nutrients. The continuous loss of the fertile surface due to erosion leads to a decline in soil fertility. Some measures help to build up soil fertility and reduce the fluctuation of crop yield. They are:

a) the combined use of farmyard manure and green manure with inorganic fertiliser,

b) the incorporation of crop residues,

c) the inclusion of fodder-legumes/legumes in the cropping system/rotation,

d) the use of organic fertilisers such as *Azospirillum*,

e) the adoption of suitable methods of application of fertilisers, and mulching to improve use-efficiency of both fertiliser and available moisture to increase the yield,

f) checking loss of the surface soil by using soil stabiliser/sealant chemicals such as, $Na_2 CO_3$, polythene sheet, mulches, and plastic sheets,

g) developing small agricultural watersheds for run off collection and recycling for life saving irrigation to crops in moisture stress, conserving and upgrading crop lands for the stabilisation of crop yields,

h) allowing a portion of the holding as chemical or legume fallow which on cultivation provides a substantial yield during famine,

i) adopting alternate land use planning in conjunction with regular cropping to improve the income of the farming families from fibre, fuel, fruit, furniture-timber, fodder and farm animals along with food.

CHAPTER 12
Weed and Weed Management

12.1 WEEDS DEFINED

The word weed has been derived from the old English woed. According to Onions (1956) the term weed is defined as a herbaceous plant not valued for use or beauty, growing wild and rank and regarded as cumbering the ground or hindering the superior vegetation. According to Herbert (1962) in the Dictionary of Ecology weed is a general term for any troublesome or otherwise undesirable plant, usually introduced, grows without intentional cultivation. According to Peter (1967) in the Dictionary of Biological Sciences weed is used by gardeners in the sense of a plant growing without human encouragement; compound names of lower plants/animals. According to the Encyclopedia Americana (1962) a weed is a plant growing out of place, where it is not wanted, either because of its inherent disagreeable or poisonous character or because it is taking the place reserved for something else. Weeds are plants growing in places and at times when man wanted either some other plants to grow or no plants at all (NAS, 1971; Gupta and Lamba, 1978; Rao, 1983). A weed is a plant growing where it is not desired (ISA, 1987). Thus weeds are the uncultivated (volunteer) plants that are not desired by man with respect to place and time of occurrence. Weeds do not belong to a particular group of plants. Under certain circumstances, each plant species can be a weed. Hardly any plant species is a weed as such. All weeds are unwanted plants but all unwanted plants may not be weeds. If plants grow wild without interfering in man's affairs they are not regarded as weeds. For instance, *Achyranthes aspera* growing in waste lands is not regarded as a weed but when it grows in crop lands and interferes in man's activities it is considered a weed. *Cynodon dactylon* is a world's worst weed of crop fields but is a desirable plant in lawns, rangelands, and permanent bunds. Sometimes, the shattered crop seeds may grow voluntarily along with crops grown in succession i.e., they come up at the wrong time and in the wrong site and thus are regarded as weeds. Sometimes seeds of weeds or other crop plants are sown as an admixture with the crop seeds. This is either through ignorance or negligence of the farmer but is not intentional, as is the sowing of crop seeds. Thus the relative position of a plant is important to determine whether it can be designated as a weed or not. Volunteer crop plants such as rice in a wheat field is regarded as a relative weed while *Cyperus rotundus* is considered as an absolute weed.

Lantana camara, Argemone mexicana etc. that grow wild near industrial sites where no plants are desired for the possibility of fire hazards, are considered weeds. *Parthenium hysterophorus and Ipomoea carnea* which are poisonous to mammals or *Xanthium strumarium, Martynia annua,* and *Asphodelus tenuifolius* which affect the quantitative and qualitative products of livestock, are considered as weeds. Some allergenic, poisonous and spinus plants are directly noxious to man. *Pistia* sp. and *Eichhornia crassipes* hinder the economic use of water bodies and thus are regarded as weeds.

In a particular habitat there may be four groups of plants: crop plants, wild plants, rogues and weeds. Crop plants are those that are either grown or harvested for obtaining yield. Wild plants grow voluntarily in nature in an uncontrolled way and do not impair man's activities. Rogues are off-type plants, in crop fields. Weeds are those plants which are out of place, unwanted, non-useful, often prolific and persistent, competitive, harmful, even poisonous which interfere with agricultural operation, increase labour, add to costs, reduce yields and detract from the comforts of life (Crafts and Robbins, 1973).

Weeds interfere with the utilisation of land and water resources.

Man, the master of agriculture, wants to exploit certain habitats that are favourable for raising crops. In such an environment it is not always possible to exclude weeds completely and also it may not be desirable. Weed management is thus the reduction of the population and growth of weeds to a level where its usefulness is greater than the damage it could reasonably be expected to cause.

12.2   CLASSIFICATION OF WEEDS

There are over 30,000 species of weeds around the world of which about 18,000 cause serious losses. Each weed species does not require specific control measures/and their nature and mode of infestation and the possible harmful effects also differ. Several weed species need complete eradication from certain geographical area, some species may require prevention from reinfestation by adopting certain measures and still others require control or suppression for the time being for obtaining the maximum return from the habitat being utilized for production. Several weed species respond alike to a common weed management technique. This is because of the similarity in some aspects of their ontology, physiology, morphology, anatomy and overall biology and also the requirements of the agro-climate. These enable us to generalise about the effectiveness or failure of a weed management practice against weeds as a class instead of against individual weed species that have already come up or may come up in different flushes regularly or irregularly. Of course, while considering weeds as a group even within the effective weed management methods some weed species may escape which may result in the shifting of flora to a particular site, the dominated species may become dominant due to the repeated use or adoption of same method of weed management.

Weeds are classified in many ways: some of them are mentioned below:

12.2.1   According to the Origin

a)   Alien (foreign in origin), for instance, *Argemone mexicana, Parthenium hysterophorus, Eichhornia, crassipes, Lolium temulentum.*

b)   Apophytes (indigenous to a country: India), for in-stance, *Saccharum spontaneum, Melilotus indlca, Acalypha indica, Croton bonplandianum.*

c)   Anthrophytes (introduced by man), for instance, *Phalarts minor, Corchorus aeutangulus, Avena ludoviciana.*

12.2.2 According to Ontogeny (Life Cycle)

A. *Annuals*: Those which complete their life cycle within a season or year and propagate by seeds. Annuals are subdivided according to the season of prevalence:

1. *Kharif seasonal annuals*: Those which come up and tend to complete their life cycle during the warm-wet season (June-July to September-October). Examples are *Ammania baccifera, Aeschynomene aspera, Dopatrium junceum, Sphenoclea zeylanica, Sagittaria sagittifolia, Ludwigia parviflora, Cyperus dubius, C. difformis, Fimbristylis miliacea, Scirpus* sp.

2. *Rabi seasonal annuals*: Those which are prevalent and complete their life cycle during the cool-dry (October-November to January-February) season. Examples are *Chenopodium album, Spergula arvensis, Vicia hirsuta, Anagallis arvensis, Phalaris minor, Asphodelus tenuifolius, Gnaphalium purpureum, Avena fatua.*

3. *Summer seasonal annuals*: Those which grow and complete their life cycle during the hot-dry season (February to June). Examples are *Solanum nigrum, Physalis minima, Portulaca oleracea, Heliotropium indicum, Argemone mexicana, Tephrosia purpurea, Trianthema portulacastrum.*

4. *Multi-seasonal annuals*: Those which are capable of growing and completing their life cycle almost at any time of the year i.e., they do not have a strict preference for any season. In general they are thermo- and photo-nonsensitive or weakly sensitive. They may complete their life cycle more than once in a season or year. Therefore, they are multi-cyclic and multi-seasonal annuals. Examples are *Echinochloa colonum, Eclipta alba, Eleusine indica, Digitaria sanguinalis, Phyllanthus niruri.* (De and Mukhopadhyay 1984).

B. *Biennials*: Those which complete their life cycle within two years. They may propagate either by seeds or vegetative parts or by both. Biennials generally do not come up in annual crop fields but they infest perennial crop fields, pastures, lawns and orchards. Examples are *Daucus carrota* and *Zingiber casumunar.* Under unfavourable weather conditions biennials may behave as annuals by bolting in the first year itself.

C. *Perennials*: Those which live for three or more years and produce seeds more than once in their life cycle. They may propagate by seeds and vegetative parts or both. Perennials may be of the following types:

1. *Simple perennials*: These may reproduce solely by seeds. They do not have a quick regenerative capacity from root stocks on crowns. Examples are *Ipomoea carnea* and *Lantana camara*.

2. *Bulbous perennials*: These propagate by bulbs, bulblets as well as by seeds such as *Allium vineale*.

3. *Creeping perennials*: These spread by lateral extension of stems along the soil surface (such as *Paspalum distichum, Cynodon dactylon*), beneath the soil surface (such as *Convolvulus arvensis, Sorghum halepense*), by the roots (such as *Ambrosia trifida, Sonchus arvensis*) or by seeds. They may be (1) shallow rooted creeping perennials such as *Cynodon dactylon, Agropyron repens, Dichanthium annulatum* and *Chrysopogon aciculatus* and (2) deep rooted creeping perennials such as *Sorghum halepense* and *Imperata cylindrica*.

12.2.3 According to Stem Character

A. *Aerial stem*: According to aerial stem character weeds may be herbs (such as *Chenopodium album, Eclipta alba*), shrubs (for instance, *Abutilon indicum, Sida rhombifolia*), brushes (for instance. *Zizyphus nummularia, Flacourtia ramontchi*), trees (for instance, *Ficus religiosa, F. bengalensis*) and filamentous (for instance *Chara* sp. *Nitella* sp.).

B. *Subaerial stems with storage organs*: A number of weeds have underground storage tissues which help them to propagate and perennate besides other modes of reproduction. Such organs may be nuts (such as *Cyperus rotundus*), rhizomes (such as *Inula indica*).

C. *Subaerial stems without storage organs*, such as runner (*Oxalis repens*), stolon (*Colocasia antiquorum*), offset (*Pistia stratiotes*).

12.2.4 According to Forms of Stem

A. *Erect or strong*: They may be (1) herbaceous, such as *Sphenoclea zeylanica* or (2) woody, such as *Lantana camara*.

B. *Weak*: They may be (1) trailing—this may be prostrate, such as *Oxalis repens*, procumbent. such as *Tridax procumbens*, diffuse such as *Boerhavia diffusa* (2) creeping, such as *Sorghum helepense* and (3) climbing, such as *Convolvulus arvensis*, *Vicia* sp.

12.2.5 According to Plant Family

A. Poaceae (Gramineae), such as *Eleusine indica, Imperata cylindrica.*

B. Asteraceae (Compositae), such as *Tridax procumbens, Eclipta alba.*

C. Solanaceae, such as *Solanum nigrum, S. xanthocarpum.*

D. Euphorbiaceae, such as *Phyllanthus niruri, Euphorbia hirta.*

E. Tiliaceae, such as *Corchorus fascicularis, C. acutangulus.*

F. Leguminosae, such as *Melilolus indica, Lathyrus aphaea.*

G. Liliaceae such as *Asphodelus tenuifolius.*

H. Chenopodiaceae, such as *Chenopodium album, C. murale.*

I. Amaranthaceae, such as *Amaranthus viridis, A. spinosus.*

12.2.6 According to the Plant Morphology

A. Dicot or broad-leaved, such as *Cleome viscosa, Eclipta alba.*

B. Grasses, such as *Echinochloa colonum, Cynodon dactylon.*

C. Sedges, such as *Cyperus rotundas, Fimbristylis miliaceas.*

D. Filamentous, such as *Chara zeylanica, Nitella hyalina.*

12.2.7 According to the Type of Habitat

A. Terrestrial, such as *Chenopodium album, Solanum nigrum.*

B. Aquatic: They may be (a) Floating (1) Free floating, such as *Eichhornia crassipes, Pistia stratiotes*, (2) Rooted floating, such as *Trapa bispinosa, Ludwigia adscendens,*

(b) Submerged, such as *Hydrilta verticillata, Vallisnaria spiralis.*

(c) Emerged, such as *Typha elephantina, Sagittaria sogittifolia.*

C. Amphibious, such as *Ranunculus aquatilis, Scirpus supinus.*

12.2.8   According to Soil Reactions

A. Saline soils: such as *Salsola* sp., *Suaeda maritime.*

B. Alkaline soils: such as *Cressa erecta, Sporobolus diander.*

C. Acid soils: such as *Rumex acetosella, Pteridium aquilinum.*

12 2.9   According to the Site of Predominance

A. *Obligate weeds:* Those which grow only in association with man and bis agriculture and never in wild form such as *Chenopodium album, Anagallls arvensis.*

B. *Fecultative* & faculative *Weeds:* Those which grow both wild and in cultivated habitats such as *Argemone mexicana, Euphorbia hirta.*

Both obligate and fecultative & faculative weeds may predominate specific habitats often disturbed by man:

1. *Crop lands:* (a) dry land: *Cyperus rofundus, Cynodon dactylon* (b) irrigated land: *Echinochloa colonum, Chenopodium album* (c) wet land: *Ammonia muldflora, Cyperus difformis, Sagittaria sagittifolia, Sphenoclea zeylanica.*

2. *Pasture lands:* *Argemone mexicana, Tephwia purpurea, Celosia argentea, Vernonla cinerea, Blumea lacera.*

3. *Waste and abandoned lands:* *Abutilon indica, Achyranthes aspera, Calotropis gigantea, Zizyphus rotundifolia.*

4. *Rail and road sides:* *Lantana camara, Datura stramonium, Ageratum conyzoides, Parthenlum hysterophorus.*

5. *Playgrounds and airfields:* *Chrysopogoh aciculatus, Dactyloctenium aegyptium, Euphorbia hirta, Kyllinga erecta.*

6. *Lawns and public gardens:* *Cyperus rotundus, Gomphrena celosioides, Phyllanthus simplex, Zornia diphylla.*

7. *Nurseries and flower gardens:* *Cyperus rotundus, Melilotus Indica, Eleusine indica, Echinochloa colonum, Cleome viscosa, Amaranthus viridls, Solanum nigrum, Chenopodium album.*

8. *Buildings and walls* (Chasmophytes): *Ficus religiosa, F. bengalensis, Andrographis paniculata, Cyperus rotundus.*

9. *Along the water courses:* *Ipomoea carnea, Xanthium strumarium, Panicum sp., Centella asiatlca, Ruellia prostrata.*

12.2.10 According to Dependence on Other Hosts

A. Total parasitic: *Cuscuta reflexa.*
B. Semi-parasitic: *Orobanche indica.*
C. Independent: *Cyperus rotundus.*

12.3 CHARACTERISTICS OF WEEDS

Weeds being nourished by nature are more tolerant of adverse edaphic, climatic and biotic factors compared to domesticated crop plants. Weeds bear certain special characteristics which help in their perpetuation, multiplication, dissemination, stabilisation and overall adaptation.

12.3.1 Perpetuation

Weeds perpetuate from generation to generation through seeds and/or vegetative propagules. When adult mother weed plants experience unfavourable environmental conditions they are capable of producing some propagules before their death. Propagules are resistant to extreme adverse conditions and thus maintain their survival in nature.

Seeds may be produced and ripened at one time or synchronously (as in *Phalaris minor*) or over an extended period of time (as in *Echinochloa colonum, Eclipta alba*) when they continue to shed or detach from the mother plant as they become mature. In a particular habitat all weed plants of a single species do not germinate, grow or bear seeds synchronously. Thus some seeds escape even though adequate measures of control are adopted. In general, vegetative propagules of weeds start to form from comparatively an early age and many of them are capable of producing new plants within a short period. Some weeds can survive even without any specialised structure of reproduction but only with fragments of stems or roots or both. Some weeds have a strong potential to regenerate repeatedly from their root stocks (*Imperata cylindrica*). Some weeds persist in nature because they can reproduce by different means.

The seeds or vegetative propagules once formed may remain dormant but viable for years when buried underneath the soil.

Such age-old propagules produce new plants when there is a favourable environment. However, some of these buried seeds die because of age. Such dormant propagules not only persist in the locality but are distributed both spatially and temporally.

12.3.2 Multiplication

The number of seeds or propagules produced per weed plant may be at par with the crop plants of the same genera but the number of seeds produced per unit area sometimes exceeds crop plants one hundred times. Weeds are in general prolific with high fecundity. In a crop field or in a bare field habitat, a large number of weed species co-exist and produce seeds and thus in a season a few million weed seeds may be produced per m^2 of field and enrich the weed seed population in the soil, although a considerable fraction of seeds may be non-viable.

Out of the total number of viable seeds produced at a particular time and site a large number are lost as they become the food of birds, insects and microbes, a fraction is washed or blown away to unfavourable habitats and a fraction is buried alive in the soil. The rest of the viable seeds may produce plants. A fraction of these plants may be lost during competition, due to management and/or due to insect attack, grazing or trampling. The remaining plants may produce seeds.

The fate of vegetative propagative materials may be similar to the seeds but those weeds which have both seed and vegetative propagules multiply enormously (De *et al.* 1986).

All management practices are not equally effective in destroying all the weeds that grow at different times. Those which escape, multiply profusely by bearing seeds or propagules.

12.3.3 Dissemination

The dispersal of seeds or propagules of weeds takes place by mobile agents: man and his activities, animals including birds, wind and water. Man is most important for the dissemination of weeds over some distance and in a particular direction. Weed seeds disseminate by man through handling of contaminated foodgrains, feed grains, crop seeds, sheaves, organic mulch, trashes, chaffs, manures, tillage implements, soil and water. The movement of armies and conveyances help to disseminate weed seeds. Most weed seeds escape digestion by animals and birds and thus remain viable and disseminate easily.

Fruits and seeds of some weed species have appendages which enable them to be carried easily by wind and water. Seeds of some weed species are very light per unit volume and small and thus they are easily blown by the wind or they float with water.

Most of the weed seeds have the characteristic of dormancy which helps them to estivate and escape an unfavourable environment during transit. Weed seeds can migrate from one continent to another and establish themselves in a new area even after a few years.

Some weeds have explosive mechanisms for seed dispersal (such as *Ruellia prostrata*).

12.3.4 Stabilisation

Weeds find their suitable sites and times for establishment by their intrinsic nature which breaks dormancy through the triggering action of edaphic, climatic and biotic factors. Some weed species germinate periodically at regular intervals. Some have a less regular pattern while a few species germinate throughout the year under conditions of moderate soil and climate. Some weeds of the desert area act as a 'rain gauge' while some weeds of an aquatic environment require a 'dry period' for germination.

The seedling phase is a very sensitive time for weeds. During this period the reserved food in the seed or other propagules is exhausted for metabolic and morphogenetic activities but the synthetic activities are not fully developed so they have a poor source and sink and also poor root growth for extracting water and nutrient from a greater volume of soil. At this stage they are also sensitive to chemicals as they are very active in growth and thus receptive to the absorption and translocation of chemicals. In monocot weeds (grasses and sedges) the growing terminal tissue remains covered with leaf sheaths and buried in the soil. In this phase they have a very poor regenerative capacity if the aerial parts are removed. They can easily be uprooted by pulling or hoeing. Once the weeds are well established the removal of the aerial parts provides for rank growth because of the higher root: shoot ratio.

Seedlings of certain perennials resemble annuals. The young soft and tender seedlings with temporary or seminal roots are very vulnerable to soil disturbances. However, some species re-establish more quickly at this stage than at maturity.

Weeds grow profusely after establishment. Weeds establish themselves without deliberate action by man and when present they are difficult to eradicate. They are capable of exploiting habitats created or modified by man and of forming an extensive population that interferes with agriculture.

Established weeds can easily endure competition through their physiological and morphological regulation and compensation capacities in a mixed flora with zonation (over-storey, under-storey, transgressive and ground strata taking crop canopy as a standard) and stratification (as floating, emergent and immersed under an aquatic environment for photic and thermic layers particularly in lentic water bodies). Some of them are more aggressive (the relative biomass increases of a species in a mixed community) and competitive than crop plants particularly under stress situations when crops suffer a lot while weeds grow and dominate over crop plants. Some weeds may outgrow crop plants and enjoy the benefits of a canopy. Some weeds have a tolerance to more crowding per unit area. They have a higher relative frequency and density compared to other plants in the community.

The occurrence of all the weed species are not synchronous even in a particular habitat. Therefore, their critical phases do not coincide. They exert competition against crop plants which have definite phenologic responses. Weeds compete with crop plants for light, nutrient, water, air and space within the microenvironment and some weeds release toxins to exclude others from the space near them (allelopathy). Weeds also provide feeding, breeding, heeding and hiding sites for insect pests, pathogens and parasites because of their staggered physiological stages. These pests infect the crop plants in their vulnerable/susceptible stages in swarms and weaken crop plants as weeds enjoy the opportunity in their favour.

12.3.5 Adaptation

Weeds are euryoecious (wide range of tolerance) compared to crop plants which are more stenoecious in nature. Such a character is responsible for their widespread and troublesome occurrence.

Weed seeds have a wide range (very low to very high) of moisture requirement for seed germination. Therefore, whenever and whatever soil moisture is available is utilised by some weed species conveniently. Some weeds can modify their morphology by reducing (he leaf area, sending roots deeper and wider as well as by rain roots and physiology which maintains a higher osmotic potential and by translocation and storage and photosynthetic pathways to mitigate drought. They have quick regenerating, restoring, rejuvenating and resurging capacities of both roots and shoots. They are quickly responsive to favourable environments after the removal of stress.

A large number of weeds can spread by vegetative means even though they are prevented from seed formation. Some weeds resemble crop plants and seeds. Thus it becomes difficult to recognise and separate them. Some weeds imitate the general appearance, colour, shape or particular feature of another plant or animal which has developed a special weapon of defence. Some weed species have special devices of armature such as thorns, spines, prickles, bristles, stinging hair and glandular hair or pois-onous substances such as latex, alkaloids, irritating substances or substances that are bitter in taste, repulsive in smell or disagreeable in odour which help to protect them from natural enemies. Some weeds develop a thick cuticle, cork and bark as a defense mechanism against fungi, insects and adverse climate.

Some weed species either endure, escape or evade unfavourable periodic as well as non-periodic changes in the environment by their special characteristics. They may adopt some dormancy forms such as hemicryptophyte, chamaephyte, microphaenophyte, therophyte, geophyte and monophyte and radicoid forms, such as rhizomatous, clonal and non-clonal forms.

Some weeds of an aquatic environment bear aeranchymatic cells, active epidermis, epidermis containing chloroplast, no or functionless stomata on the lower surface of leaves, breathing roots, no root hair and root cap instead of root pocket and even no root.

Some weeds bear very minute or inconspicuous flowers (*Trianthema portulacastrum*) and these flowers often produce mature seeds much earlier before they are recognised to flower. In the nursery bed some weeds (such as *Echinochloa colonum*) produce mature seeds before the crop seedlings are ready for uprooting for transplanting. Some weeds (such as *Phalaris minor*) can produce a number of mature seeds even though they are uprooted and heaped at the flowering stage. The time of maturity of some weeds (such as *Avena ludoviciana, Cleome viscosa*) coincides with the maturity of the crop and shed their seeds before the harvest of the crop or the weeds containing seeds are harvested with the crop plants. Some weeds of the ground stratum escape the reaping and' uprooting type of harvesting and produce large number of seeds after the harvest of the crops. Some weed species (such as *Eriocaulon* sp., *Cyperus* sp. and *Sphaeranthus indicus*) come up during the post-harvest period and produce seeds vigorously.

Some weeds have adventitious roots from aerial nodes (such as *Cynodon dactylon*) which help them to remain firmly in the soil. Even if such weeds are pulled or grazed a considerable part remains intact on the soil and regenerates quickly.

Weeds with radicoid forms easily escape drought, fire, soil erosion and man-made modifications or disruptive forces such as repeated cultivation and irrigation.

12.4   HARMFUL EFFECTS OF WEEDS

Weeds interfere with man's utilisation of land and water in various ways. The direct harmful effects of weeds may not be apparent in many cases while other pests and pathogens produce noticeable symptoms of infestation or infection. Weeds extend the harmful effects slowly, steadily and inconspicuously and such effects are realised almost at an irrevocable stage. The nature and extent of damage varies from one habitat to another. The habitats may be grouped into crop fields, non-crop fields and aquatic environments for easy understanding.

12.4.1 Harmful Effects of Weeds in Crop Fields

a) *Reduce the yield of crops*: A crop field is a favourable eco-system where a large number of weed species greminate in flushes and grow profusely. If no restriction is imposed they compete with crop plants for nutrients, light, air, water, space and other factors in the micro-environment. The competition is more acute under a stress situation. As weeds comprise a variety of plant types they are capable of exploiting the environment more intensely than the crop plants. Weeds not only affect the standing crop but the crops in sequence and also have an effect during the uncropped season or fallow period. Weeds may produce a higher biomass than crop plants per unit area and time. Some species of weeds are capable of absorbing more amount of N, P and K and other major and minor plant nutrients than crop plants. Parasitic weeds may even kill the crop plants. In dry lands, weeds exhaust the soil moisture and put the crop under water stress earlier and thus reduce the yield. The extent of yield reduction of individual crops varies from zero to hundred per cent while in different cropping systems it varies from five to fifty per cent. In India an estimated annual loss of agricultural production is Rs. 1,650 crores due to weeds.

b) *Increase the cost of agriculture*: The cost of tillage operations comes to be around ten per cent of the cost of crop cultivation. More than 50 per cent of tillage cost is incurred for controlling weeds. Besides this, tillage operations during uncropped seasons for weed control adds to the total cost. The presence of weeds in crop fields hinders the progress of work. Intercultivation, harvesting, threshing and cleaning take time in the presence of weeds and thus add to cost.

c) *Increase the irrigation requirements*: Weeds transpire irrigated water and thus increase the frequency and depth of irrigation water which is very often expensive and diffiicult to obtain. In weedy plots the movement of water is restricted and this reduces the efficiency of irrigation.

d) *Reduce the value of produce*: The quality of produce in weed infested plots becomes poor and fetches a lower price. Agricultural produce mixed with seeds and trashes of weeds are poorly accepted and offered lower prices. If the seeds of objectionable weeds are mixed with crop seeds/grains such produce is discouraged or banned for sale.

e) *Reduce the quality and quantity of products from live-stock and birds*: Some weeds when eaten by milch animals produce an undesirable flavour in their milk (such as *Cleome viscosa, Paederia foetida*). At times, death, disorder, or permanent deformity/disability may occur among animals and birds after the consumption of poisonous plants (such as *Datura stramonium*) or their seeds. The fruits and seeds of some weeds (such as *Xanthium strumarium* and *Achyranthus aspera*) entangle with the wool and hooks of *Martynia annua* cause injury to hides which fetch lower prices.

f) *Harbour insect-pests, pathogens and parasites*: Some weeds provide shelter to pests. They act as alternate hosts of a number of fungi, bacteria, virus and nematodes. During uncropped seasons these weeds harbour pests and during the crop season they aid in their reproduction. Weeds also provide food and shelter to field rats which infest the crop plants [subsequently.

g) *Reduce the value of land*: Severe infestation of perennial, problematic, pernicious and poisonous weeds reduce the value of land as they incur heavy expenditure to clear or to eradicate.

h) *Impair the purity of varieties*: Cross pollination with compatible weed plants, impairs the genetic purity of the crop seeds.

i) *Cause health hazards to animals and men*: Some poisonous weeds cause irritation, allergy and eczema to animals and labourers when they come in contact with them. Such weeds are *Parthenium hysterophorus, Chenopodium album, Argemone mexicana* and *Heliotropium indicum*. Some weeds such as *Mimosa pudica, Tribulus terrestris, Centipeda minima*, and *Datura stramonium* cause wounds, headache, sneezing, giddiness and vomiting.

j) *Cause allelopathic effects*: Some weeds (such as *Parthenium hysterophorus*) release certain toxic substances in the environment which impair the germination and growth of crop plants.

12.4.2 Harmful Effects of Weeds in NOB Cropped Areas

a) *Detracts from the beauty of public places:* Road sides, rail yards, air fields, parade grounds, picnic gardens, lawns and parks become ugly due to the presence of some weeds such as *Cyperus rotundus, Chrysopogon aciculatus, Zizyphus oenopolla,* and *Mimosa pudica.*

b) *Causes fire hazards:* Some annual weeds deposit pyres after death or some perennial weeds shed their dry leaves which when they come in contact with sparks of fire from factory or rail engines, cause serious fire hazards. Weeds around factories, electric power houses and rail lines cause such hazards.

c) *Causes accidents:* The unsightly growth of weeds obstructs the view on roads. Hinders the speedy movement of vehicles and causes accidents. At sharp turns, weed growth seems to be more dangerous.

d) *Causes nuisance and pollution:* Weeds tend to occupy the spaces unoccupied by other plants. No plant is desired in certain locations such as courtyards, rock gardens, manure pits, threshing and drying yards. The occurrence and growth of weeds at various heights and of various colours and forms detracts from the beauty of the area and it becomes a nuisance to eradicate them and to maintain weed free conditions at such sites. These weeds also attract a number of insects, rodents, birds and reptiles which pollute the environment.

12.4.3 Harmful Effects of Weeds in Aquatic Environments

a) *Restrict the movement of water flow.* Weeds that grow at the bottom or that float on the moving water of channels, canals and rivers restrict the quick movement of water. Weeds can restrict up to 95 per cent of flow and cause (1) loss of water by seepage and percolation, breaking or overflowing dykes resulting in the onrush of water, flood, siltation, loss of crops and lives, (2) delay in irrigation resulting in poor irrigation efficiency, (3) rise of the water-table which may result in the upward movement of injurious salts or leaching loss of soluble salts or the creation of inundated fallow lands, (4) siltation in the beds of earth channels resulting in them being shallow but wider and thus needing renovation, (5) reduction in the storage capacity of water.

b) *Restrict drainage*: Profuse weed growth in catchment areas restricts drainage and thus impedes water harvesting. Weeds also cause a delay in draining excess water resulting in damage to crops.

c) *Choke the closed water passages*: The mat of weeds choke the sluice gates, culverts and intake pipes of the pumps. Filamentous algae choke the strainers, siphons and nozzles. All these require repeated cleaning to provide easy passage through such closed waterways.

d) *Increase wastage of water*: Weeds transpire more water than merely evaporation from the open water surface.

e) *Interfere in pisciculture*: Weed infested water bodies have poor photic and thermic layers which affect the growth of phytoplanktons which are the major food constituents of fishes. Weed roots liberate CO_2 which when dissolved in water affects the aeration of fishes. The weed residues decompose anaerobically, which result in the production of toxic gases such as CO, H_2S, CH_4 and other toxic substances and also increase the acidity of water which is harmful to fish. Weed infested water bodies encourage fish-predators and fish-weeds which affect the cultivation of fishes. Weeds, such as *Spirogyra* sp., and *Cladophora* sp. stick to the gills of fishes and seriously affect respiration and sometimes cause death. A weed infested fishery affects fish breeding and netting. Fishes reared in weed infested water become discoloured and distasteful.

Rearing of ducks in such ponds are also affected.

f) *Impair the cultivation of aquatic crops*: The cultivation of flooded rice, deep water rice, lotus, water chestnut, cork plants, reeds and *Ipomoea* sp. becomes difficult in weed infested water bodies.

g) *Impair recreation in water bodies*: Weed growth in an ugly fashion or in mats in the water impair its beauty as well as the view. Weeds impair angling, boating and water-sports.

h) *Pollute the environment*: Weeds harbour mosquitoes, leaches, snakes, otters, frogs and other animals which may impair the lives of animals, birds and humans. Weeds impair sanctuaries and aquariums and the seasonal pursuits of migratory animals and birds.

Weeds on decomposition pollute the water and make the

water unfit for irrigation, washing and drinking.

i) *Increase the infestation of weeds in crop fields*: Irrigation water from weed infested water bodies carries seeds or weeds of different species to the crop fields where they grow profusely. Weeds such as *Chara* sp., *Nitella* sp., *Pistia* sp., *Eichhornia crassipes, Ludwigia adscendens, Aeschynomene indica, Hydrolea zeylanica* and *Cyperus difformis* are carried by irrigation water.

j) *Impair navigation in waterways*: Emergent weeds and mats or chunks of floating weeds do not permit the easy movement of boats, launches, steamers and ships. Weeds also induce siltation by reducing the movement of water thus impair the navigability of waterways (De, 1980 b).

12.5 BENEFICIAL EFFECTS OF WEEDS

No plant in this earth is completely worthless. Each one has some contribution to make which justifies its existence. The magnitude may vary from species to species and habitat to habitat. If weeds are handled or managed properly they prove their value. Such values may not always be worthwhile to human being. Some of the beneficial effects of weeds are listed below:

12.5.1 Valued for Increasing Organic Matter Content in Soil

a) Weeds when incorporated into the soil, increase the organic matter content which is an essential constituent of arable soil. Weeds such as *Echinochloa colonum, Cynodon dactylon* and *Medicago denticulatea* are used for green manuring. Aquatic weeds, such as *Eichhornia crassipes and Pistia stratoites* are used for composting.

b) *Valued for increasing soil fertility*: Weeds belonging to the legume family, algae and fern (*Azolla* sp.) fix atmospheric N biochemically in their bodies. The incorporation of such weeds increases soil fertility. Most of the weeds absorb nutrients from the deeper layers of soil. On incorporation such weeds enrich fertility in the plough layer from where crop plants absorb much of their nutrient requirements. Weeds are composed of all essential plant nutrients in various proportions. The incorporation in the soil of such mixed flora of weeds helps to balance the nutrient proportions in the soil.

c) *Valued for checking soil erosion* : Weeds tend to form a living mulch which resists the beating action of rain drops and the sweeping action of wind. Weeds form turfs and sods with a compact canopy and matted roots intertwine the soil and resist surface flow at non-erosive velocity and tnus reduce soil erosion. Such living mulch induces the infiltration of water and the siltation of suspended soil particles. It also resists saltation, suspension and surface creep types of wind erosion. Thus weeds prove their values as guardians of the soil.

d) *Valued for inducing soil formation:* Weeds are capable of sending their roots deeper through rocks and compressed layers. Penetrating roots make them porous. Deposits of weed residues accelerate other activities of weathering of rocks and gradually induce soil formation.

e) *Valued for supplying food, feed, medicine:* Weeds, such as *Chenopodium album*, *Trianthema portulacastrum* and *Amaranthus* sp. are used as greens; *Cynodon dactylon, Echinochloa colonum, Digitaria sanguinalis, Dichanthium annulatum, Eclipta alba, Avena fatua, Phalaris minor* and *Melilotus indica* are valued as palatable succulent feed; *Andrographis paniculata* and *Centella asiatica* are valued for therapeutic purposes.

f) *Valued for economic utilization:* A large number of weed species are valued for different economic utilisation: *Imperata cylindrica, Saccharum munjo* and *Typha elephantina* are used for thatching roofs of huts and hermitages; *Eulaliopsis binate* and *Sansevieria roxburghiana* are valued for making ropes and strings, Cyperus rotundus and Andropogon squarrosus are used for making joss-sticks, Phragmites karka and A. squarrosus are used for making mats and screens; *Cymbopogon citratus, C. nardus* and *C. martini* are valued for essential oils. Weeds are also used as raw materials for paper making, basket making and solid fuel. Some weeds are used for extracting plant growth hormones and pesticides.

g) *Valued as indicator plants:* Investigators get an approximate indication of agro-ecological characteristics and possible fiora to be grown in any location from the study of the floristic composition and nature of growth of weeds in that locality. Weeds also indicate the incidence of disease, deficiency, disorder and frost that may cause a poor yield of crop plants.

h) *Valued as the source of genetic material for crop improvement*: Weeds are the sources of genes responsible for various agronomic and/or physiological characteristics that are desirable for incorporation in crop improvement.

i) *Valued for reclamation of the soil*: Incorporation in the soil of weeds such as *Argemone mexicana, Cynodon dactylon* etc. help to reclaim the alkalinity of soils.

j) *Valued as a host for beneficial organisms*: Some weeds act as alternate hosts of pollinating insects, such as the honey bee and butterfly, predators, such as spiders, symbiotic bacteria, such as *Rhizobium* sp. when there is no crop plant or regular host in the field.

k) *Valued for protecting bunds*: Sod, turf and brush types of weeds (such as *Cynodon dactylon, Dichanthium annulatum* and *Saccharum squarrosus*) protect the different types of bunds, such as boundary bunds, field bunds, dykes of the dams, irrigation and drainage channels, embankments of rivers, streams, estuaries, confluences, mounds of tanks, ponds and pools as well as sides of rail and roadways.

l) *Valued for cleaning and purifying water*:
Suspended materials in water adhere to the surface of the aquatic weeds and thus weeds keep the water clean. Some weeds, such as *Eichhornia crassipes* absorb heavy metals and other toxic substances dissolved in water.

m) *Valued as ornamental plant*: Some weeds bear attractive flower, foliage, fruits, fluids, flavours and forms which favour them for use in furnishing drawing rooms, balconies, courtyards and beautifying limbs and braids of hair particularly of maids.

n) *Valued for religious and ritual purposes*: A number of weeds, such as *Cynodon dactylon, Eragrostis cyanosuroides, Ocimum sanctum* and *Andropogon squarrosus* are used for various religious and ritual purposes.

o) *Valued for maintaining biological equilibrium* : Weeds as components of the biosphere play an important role in maintaining the equilibrium between animals and plants by providing food, shelter and oxygen and by providing a green ward in denuded and deserted areas as well as purifying the polluted, highly populated and poverty stricken environment (De, 1986).

12.6 CROP-WEED ASSOCIATION

There are a number of weed species which do not have any specificity with respect to season and site of occurrence Such weeds may be perennial (*Cynodon dactylon*) or annual *Echinochloa colonum*). Some weeds prefer to grow in crop fields only and some of them have a distinct association with some specific crops.

The climatic, edaphic and biotic factors of environment determine the distribution of species, their competitive ability, their prevalence and their intricate relationships. Apart from natural factors and their seasonal effects, crop fields are subjected to man-made modifications by the management of soil, water, nutrient, crops and pests in individual crops and crops in sequence. Since, weeds have unique characteristics for adaptation they try to shape themselves under the changed situations through acclimatisation.

Often man is indifferent towards weeds, ignoring the introduction and building up of a population of weed seeds in field soils. Field investigations show that different crop varieties that are grown under specific sets of agro-ecological situations (such as season, land situation, cropping systems and cultural practices) have their characteristic weeds. The important factors contributing to a common occurrence of a phytosociological association of crops and weeds are similarity of seed size, time of ripening and germination, and various tillage, cropping and harvest practices (Crafts and Robbins, 1973). They also requiring similar agroecological conditions.

Once a certain weed species is introduced, its abundance or scarcity in a crop field microclimate is determined largely by the degree and duration of co-operation or competition offered or endured by the crop plants and the nature and extent of management practices adopted in favour of crop plants. On the other hand, the competitive ability of weeds depend on behaviour, for instance readiness of seed germination, seedling growth, nature and type of vegetative habits and growth including root and shoot growth, the time and extent of the life cycle, the regenerative capacity and reproduction. These are modified by the management practices. As weeds do not belong to a particular plant species, at least some weed species enjoy the favour of a crop field habitat and they flourish as well as crop plants. Some weeds that are almost always associated with respective crop plants are:

Rice	—*Echinochloa colonum, E. crus-galli.*
Wheat	—*Chenopodium album, Phalaris minor.*
Maize	—*Eclipta alba, E. colonum.*
Barley	—*P. minor, Avena fatua, C. album.*
Sorghum	—*Striga lutea, Taidax procumbens.*
Pigeon-pea	—*Cyperus rotundus, E. colonum.*
Chick-pea	—*Croton sparsiflorus, Cleome viscosa.*
Peas	—*Lathyrus aphaca, Gnaphalium indicum.*
Black gram	—*Solanum nigrum, Physalis minima.*
Rape and mustard	—*Cleome viscosa, Spergula arvensis, C. album.*
Groundnut	—*Phyllanthus niruri, Ageratum conyzoides.*
Potato	—*C. album, S. arvensis, Anagallis arvensis.*
Cotton	—*C. rotundus, A. conyzoides, Abutilon indicum.*
Jute	—*Corchorus acutangulus, C. rotundus.*
Tobacco	—*C. rotundus, Orobanche indica.*
Berseem	—*Cichorium intybus, Cuscuta reflexa.*
Sugar-cane	—*Sorghum halepense, Imperata cylindrica etc.*

Plant competition is a powerful natural force responsible for the suppression or extinction of weaker plants There may be intra-plant (as among crop plants or weed plants) or inter-plant (between crop and weed plants) competitions. A principle of plant competition is that the first plants to occupy an area have an advantage over late comers (NAS, 1971) Rapidly growing plants with rapid coverage of both above and below ground areas have an advantage over slow growing ones. Density, geometry and architecture of plants have contributing roles in competition. Competition is usually most severe when competing plants are alike in vegetative and reproductive habits, morphology and physiology and depend upon common resources. Competition is most acute when and where available resources become limiting to plant growth.

The extent of competition is influenced by the season at which the crop is sown, the weed species present and by the habit, growth rate and density of both crop and weeds. The time of weed emergence relative to that of the crop is particularly important and may affect the outcome of competition between weeds and crop. Crops that are slow to establish and cover the ground are particularly susceptible to weed competition and a setback in the early stages may have a lasting effect on crop growth and development. Weeds that emerge once the crop has become established are likely to be suppressed, although the extent to which this occurs depends on crop habit and density (Frayer and Makepeace, 1977). Some weed species come up in flushes after rain or irrigation and/or top dressing of nitrogen fertiliser or after prolonged submergence or alternate wetting and drying and turning of the soil at different depths. These weeds also compete with crop plants for current resources such as light, water, nutrient, air, space and the micro-environment meant for standing crop plants and also resources that are to be used by crops in sequence. Some late coming weeds complete their life cycle within a very short period and produce effective propagules which may come up and generate competition in the following crops or in crops grown during the same seasons of the following years. In some cases volunteer seeding of crop plants or sowing of crops early in the season causes an infestation of short-lived weeds which die after exhausting the moisture and nutrients thus creating stress for the standing crop plants.

A broadcast seeded crop which comes up almost uniformly may shade the ground quickly and uniformly and suppress the growth of weeds effectively. Such suppression is not possible in wide spaced row crops where weeds grow luxuriantly in the interspaces as well as in the crop rows. If the crop stand is thin or lacks vigour, weeds flourish. The conditions or methods unfavourable to the growth of useful plants permit the invasion and development of weed population (Crafts and Robbins, 1973) and growth.

Factors in weed-crop competition

Weeds compete with crop plants for water, light, nutrients, space, air and the micro-environment. A specific situation dictates the factor or factors of competition. In wet land rice, water is not a limiting factor and competition may be for other factors such as nutrients or air. Competition begins when the supply of any one of these factors falls below the requirements of both weeds and crop. An abundance of any factor may induce competition for other factors. For example, abundant nutrient supply may generate competition for water, light and space.

Effect of weed competition upon crops

The chief effect of weeds is to decrease yields of crops. The extent to which yield is depressed by weed competition varies greatly. It depends partly on the adequacy of the resources available, but mainly on the competitive ability of the crop relative to that of the weed population and growth, including biomass accumulation and dominance. Besides the reduction of yield up to the point of complete failure, weed competition can reduce the quality of produce, particularly of root crops by altering the size and distribution or even piercing the fleshy roots or tubers by the stolons of *C. rotundus*. Parasitic weeds affect crop plants directly and may kill the host plant completely.

Weeds impair tillering, branching, fruiting, the development of grains or seeds, stock elongation, leaf enlargement and biomass production in different crops and ultimately reduces the quantitative, qualitative and generative yields.

Critical periods of competition

The early growth phase of the plants are found to be the most critical with respect to competition with weeds. With the progress of crop growth they secure a relatively higher competitive ability to suppress weeds or to escape competition generated by weeds.

Competing ability of different crop plants

Crop plants and their varieties differ in their competing ability. The characteristics that enable a species to be successful in competition are rapid and uniform germination, rapid development of foliage and root, tolerance to high density and the formation of a close-in canopy at the early growth phase. The crops vary in their ability to suppress various weed species. Crops such as barley sown at about 100 plants per m² emerge early and take up rapid seedling growth with somewhat decumbent type of tufted tillers with planophyllic leaves that suppress a variety of weed species. Crops, such as mustard sown at about 100 plants per m² emerge early and rapidly develop a close-in canopy by spreading planophyllic leaves and thus suppress weeds. Crops, such as cow-pea smother the weeds through their thick foliage and twining or climbing habits. On the contrary, onion and garlic which have very poor coverage over ground are the least competitors. It has been well established that some crop varieties can suppress weeds only by obscuring light to a greater extent (smother crops) while for other factors there are vary few crops which can compete with weeds. Some weed species can sustain themselves even at ten per cent of incidental light (Sciophytes) and therefore the curtailment of light by a crop with a single layer of leaf (leaf area index of 1.0) may not always have a suppressing effect on weeds.

Some plant species excrete some phytotoxins in the environment to evict their neighbours. Such allelopathy has a bearing on competition. On the contrary, some plants secrete stimulatory substances that induce the germination and growth of weeds such as *Orobanche* and *Striga*.

12.7 FACTORS THAT DETERMINE VIGOROUS AND UNIFORM STANDS OF CROP PLANTS

It is apparent that in weed management, consideration must be given to the factors that determine vigorous and uniform stands of crop plants and that ignore or discourage the invasion and growth of weeds. These factors are variety of crop, soil-water relations, soil fertility, soil reaction, tillage, date, rate and method of seeding, crop rotation, cropping and cultural systems, use of herbicides, insect pests and disease management. Any single factor is not capable of providing a vigorous crop stand while a combination of a number of such factors with respect to the crop concerned helps to suppress weeds, induce use-efficiency of inputs and cultural practices and thus increase the quality of produce of crop plants per unit area, time and input.

12.8 PRINCIPLES OF WEED MANAGEMENT

Weeds have both harmful and beneficial effects. When and where the harmful effects of weeds are greater than the usefulness there is a need to reduce their population and growth through management practices to such an extent that the nature and extent of damage they cause are within permissible limits. On the other hand, the cost of weed management must be less than that of the expected value of produce that will be reduced by the infestation of weeds. However, the value of the expected reduction in every situation cannot be judged in terms of money.

On many occasions weed growth and biomass development are more important than only the density of weeds. The floristic composition of weeds is more important than that of the abundance of a single species in a habitat. A specific weed management practice may be adopted to kill a predominant species without much or no effect on another species which may then become a predominant one (shifting flora). Thus weed management should be aimed at reducing both the population and the growth of all the species together. This is most difficult and tedious. All the species of weeds and even individuals of the same species do not emerge, grow and reproduce, simultaneously. Thus the method employed to manage weeds must have some residual effects.

To deal with problems posed by weeds some methods of management are to be adopted. Such methods are based on some basic principles. These principles are related to (1) life cycle of weeds, (2) characteristics of weeds, (3) mode of reproduction of weeds, (4) habitat, location and season, (5) soil and weather conditions, (6) area, (7) farming and cultural practices,

(8) availability of resources, and (9) economics of the method.

12.8.1 Life Cycle of Weeds and their Management

The understanding of the life cycle of weeds determines the effective method and optimum time of weed management.

Annuals complete their life cycle within a season or year. If the introduction and seed production of annuals are prevented, they can be managed better than biennials and perennials. Annuals should be killed before flowering. It is better to adopt stale seed-bed technique (allowing germinable weed seeds in soil to germinate by preparing a weed seed-bed and destroying them thereafter before sowing the crops), pre-sowing irrigation pre- or post-weed emergence but pre-crop emergence application of broad spectrum translocated herbicides or post-weed and crop emergence application of selective translocated herbicides or intercultivation to destroy annual weeds successfully.

Annuals in the early growth stages are easily and cheaply destroyed. Once the shoots are killed, the roots are unable to regenerate and rob water and nutrients, meant for crop growth and production.

The best time to destroy biennials is during the seedling stage of the first year when they behave like annuals. Once they form underground storage organs they escape a number of management practices, and they may even multiply due to the shearing and tearing action of tillage implements. The application of translocated herbicides at the beginning of the aerial growth in the second year or digging out of storage organs helps to destroy them. They can also be killed by creating extreme adverse conditions, such as impounding water for a fortnight at their resumption of growth in the second year, particularly for biennials of aerobic soils. Heating soil by burn. ing trashes (rabbing) after exposing the storage organs also kills the weeds successfully.

Perennials may be grouped according to forms of vegetative reproduction: simple, bulbous or rhizomatous and creeping perennials. In combating perennial weeds two problems are to be encountered (a) to check the spread or reinfestation by seeds or vegetative propagules, and (b) to destroy the established plants. They should be prevented from seeding and forming vegetative propagules by the repeated destruction of aerial shoots and underground roots by cutting, ploughing, digging, drying, and flooding during their regeneration period when food reserves tend to be exhausted.

Some rhizomatous weeds (such as *Imperata cylindrica* and *Sorghum halepense*) hide their storage organs well below the plough layer. Such organs are to be dug out and destroyed or translocated herbicides should be applied during their regeneration period. The burning of grown up brushes destroys the shoot and then digging or the application of translocated herbicides after the beginning of regeneration makes it easy to kill them.

If required, perennial weeds may be controlled by leaving the land uncropped for the season when several measures, such as repeated ploughing, puddling, flooding, green manuring, digging and uprooting of storage organs, stocks and stumps, burning, grazing, and shifting of soils may be adopted. The cultivation of cleaning crops, fodder crops for repeated cuttings at shorter intervals and rotational cropping reduces the infestation of perennial weeds. A combination of several methods accelerates the destruction of such weeds and thus gets rid of the perennial problems posed by them.

12.8.2 Characteristics of Weeds

Predominance of a particular group of weeds (such as broad-leaved/grasses/sedges) determine the method of management. Individual species of the groups may also be important in a particular habitat because of their noxious, problematic, poisonous or predominant nature. Such individual weed species may have special characteristics by which they severely react to the environmental conditions. For example, *Cyperus rotundus* and *Alhagi camelorum* cannot tolerate waterlogged conditions. They may be killed by flooding or growing wet land rice. Some weeds (such as *E. colonum*, *E. indica*) can re-establish even if a single root is left undisturbed. Complete uprooting and feeding to cattle or composting and the application of herbicides are the possible measures to destroy them. Some weeds particularly during their grown up stages (such as *Trianthema portulacastrum*) take a few days to desiccate even under hot dry conditions.

There should be a considerable gap between mechanical uprooting and irrigation to dry them up completely.

12.8.3 Mode of Reproduction of Weeds

While selecting a suitable method of weed management detailed information regarding the reproduction of a prevalent weed species is essential. Information pertaining to the propagation by seeds or vegetative means or by both, the number and time of production of propagules, the methods for their dispersal, their dormancy, viability, time of germination or sprouting, association with crop plants, mode of perennation, stabilization and adaptation under adverse environmental conditions dictate the methods and the scheduling of them for their control.

12.8.4 Habitat, Location and Season

In the crop field habitat, targetted plants (weeds) are to be killed or suppressed withont affecting crop plants, beneficial organisms and their activities. Such selective treatments are essential in crop fields considering the standing crop and crops in sequence and also the surrounding environment. The relative age and tolerance of crop plants and weeds are to be considered. Such considerations may not be required for non-cropped areas where other factors, such as soil erosion and degradation may be considered.

The crops and weeds of uplands are quite different from those of low lands. Cultural practices and the workability of such lands also differ, thus weed management practices vary.

The occurrence, composition and abundance of weeds vary from season to season and therefore suitable weed management practices also differ. In the *kharif* season a large number of terrestrial weeds can be suppressed by continuous submergence in water of a depth of 5 cm to 10 cms.

12 8.5 Soil and Weather Conditions

Soil texture nad moisture are the important factors in selecting a suitable method of weed management. Light soils have a poor water holding capacity and therefore flooding cannot be adopted. In heavy soils with poor drainage mechanical weed management is impeded. Dry soils resist mechanical operations while soils with optimum moisture content are suitable for mechanical weed management. Types of weed flora differ with soil reaction and therefore weed management practices vary.

The prevalent weather conditions determine the workability of the soil, the effectivity of operation and the efficiency of treatments. Continuous rain, drought, cold and a high velocity of wind affect field operations. Under such conditions the management operations are either pre- or postponed or an effective method is sought for combating the infestation of weeds.

12.8.6 Area of Weed Management

In localised spots or small areas (such as nurseries) labour intensive but most effective methods, such as hand pulling may be adopted. In greater areas (say up to two to five ha), mechanical methods and in larger areas, chemical methods of weed management may be adopted. In many cases an integrated approach may be more effective and remunerative than that of individual methods.

12.8.7 Farming and Cultural Practices

Methods of weed management in farming with seasonal field crops differ from farming with fodder or seed production and multiplication. Similarly, weed management practices in rain fed or dryland farming differ from that of irrigated farming. Methods of weed management in mixed or intercropping differ from that of pure cropping. Weed management methods in direct seeded upland crops differ from transplanted low land crops. Management practices such as good land preparation, sowing at the optimum time and by the best method, proper irrigation and drainage, the application of a balanced dose of fertilisers at the right time and by the right method, intercultural operations, such as earthing up, insect-pests and disease management at the appropriate time and by the correct method favour crop growth and development and at the same time suppress the population and growth of weeds.

Weed management in ratoons and repeatedly cut crops, differs from that in plant crops. The best time for weed management in ratoon crops is the pre-regenerative stage of the crops.

In green manure crops weed population and growth increase the green matter and weed management in such crops is an unnecessary expenditure and rather a loss.

In potato, groundnut and sugarbeet, earthing up and the preparation of irrigation and drainage channels are the essential intercultural operations when weed management becomes simultaneous while in cotton and jute interculture is directed mainly for weed management. Therefore, the method and time of weed management differ in these types of crops.

The method and time of weed management varies with different methods of harvesting crops. Potato, sweet potato and groundnut are harvested by digging when all the weeds are uprooted and destroyed. This is not possible with the reaping, picking or uprooting methods of harvesting.

In mechanised farming, weed infestation increases as they are not identified separately by the machines. Weeds that come up in the crop rows, grow and produce seeds and are harvested with crop plants and remain as an admixture with crop seeds for the next sowing. Under certain conditions mechanical weed management becomes impossible. In such cases, chemical weed management becomes indispensable, effective and cheap.

The same crop species may be grown for different purposes, such as maize, oat, *jowar* and cow-pea, for grains, fodder or seeds. Some species of weeds may be allowed to grow for fodder, still less for grain and no weed in seed crops. Therefore, weed management practices differ with the purpose of the crop grown.

12.8.8 Availability of Resources

Farm labour, implements, power and herbicides are the major inputs in weed management. The availability and cheapness of these inputs determine the method of weed management. Although other factors such as the availability of time, the effectiveness and workability are also contributing factors in selecting the suitable method.

Agriculture in our country is very dependent on the whims

of nature. Therefore, the demand for labour reaches its peak during sowing, interculture and harvesting as every farmer wants to complete such operations in a short span of time. During such periods skilled labourers are in short supply and thus wage rates are also high.

Apart from some hand tools there are a few improved speedy and effective implements for weed management in our country. These also have limitations of use under the various conditions of soil and crop.

The availability of appropriate herbicides with the desirable formulation is rare in our country. Sometimes they are costly. The knowledge and experience of the farmers are also lacking regarding herbicide use.

The availability of resources therefore determine the method of weed management.

12.8.9 Economics of the Methods

Most of the farmers of our country are poor and live at sub-sistence level. Even though some methods are found to be pro-fitable, they do not care for them. The level of production is, in general, low and therefore the use of costly inputs for weed management becomes risky. Except in areas with commercial crops, farmers prefer multi-purpose traditional operations such as hoeing which improves the physical conditions of the soil, earthing up and the incorporation of top dressed manures and fertilisers and at the same time weed management which on apportioning the cost is found to be the least expensive and most effective though time consuming and tedious. In the recent past the introduction of integrated weed management involving non-monetary, less monetary and monetary inputs becomes popular, effective and less risky.

12.9 METHODS OF WEED MANAGEMENT

Weed control and weed management are two terms used in weed science. Weed control is the process of limiting infestations so that crops can be grown profitably or other operations can be conducted efficiently. Only the controlling of weeds is not always sufficient. Some pernicious, problematic, poisonous, perennial and objectionable weeds need eradication or extermination that is the complete elimination of the species including its different parts and propagules from an area for ever. Some weeds grow on the bunds, along the irrigation channels and road sides where they are desirable to protect the soil from erosion. However, weeds from such areas encroach or intrude on the crop field or other valuable areas. This is not desirable and needs to be prevented.

Weed management includes (!) prevention, (2) eradication, and (3) control which includes (a) regulated use, (b) restricting invasion, (c) suppression of growth, (d) prevention of seed production, vegetative propagules and their dispersal, (e) complete destruction. Thus weed control is one of the aspects of weed management.

Weed management methods are also divided into two: (1) preventive method, and (2) curative or remedial method. Curative method is subdivided into (a) eradication and (b) control.

12.9.1 Preventive Methods

There is an axiom: "Prevention is better than cure." Prevention has two dimensions with respect to time and space: the measures taken to prevent the infestation prior to weed germination and to prevent the introduction or spread to new areas. Now-a-days in respect to ecology, environment and economy, preventive measures are not always desirable whereas, under conditions of certainty of infestation, the proverb "prevention is better than cure" is absolutely true. With respect to weeds 'once they are introduced they take several years to eradicate'.

The following methods can be adopted to prevent the introduction and infestation of weeds in crop fields:

a) Use pure seeds and seedlings;

b) Do not use whole and viable seeds of weeds with other materials for feeding animals and birds. Such feeding materials may be cooked, ground or ensiloed;

c) Do not use fresh or partially decomposed organic manures containing viable weed seeds;

d) Do not use organic mulches containing weed seeds;

e) Do not permit the livestock to move from an infested

area to a clean area as they carry weed propagules through fur, fleece, hooves and droppings;

f) Farm implements and machinery should be properly cleaned before moving from an infested to a clean area;

g) Avoid shifting of soil from infested to clean areas;

h) Keep the banks of irrigation and drainage channels free from weeds; at least weeds of such areas may be prevented from seeding;

i) Avoid irrigation water contaminated with weeds, seeds and other propagules of weeds. In the canal system of irrigation it is difficult to avoid contaminated water. In such cases strainers may be used. Sometimes water collected from the lower depth of courses carries a fewer number of weeds and weed propagules than water on the surface;

j) Keep the field bunds, field channels, boundary bunds, fence lines and head lands free from weeds by scrapping, scything, burning, mowing and grazing at least once in each cropping;

k) Keep the threshing yard, granary and compost pits free from weeds or prevent seed production in these areas and introduction to crop fields;

l) Follow legal and quarantine measures.

12.9.2 Eradication Methods

When and where a new weed species is found it must be destroyed immediately before its multiplication, dispersion and acclimatisation. Some of the absolute weeds are undesirable under a variety of situations. If such weeds are left without extermination, they produce seeds and the axiom "One year's seeding seven years' weeding" becomes absolutely true. Eradication can be done by (a) destroying the species at the initial stage of introduction and before it produces any propagule or enforces its regenerative capacity i.e., at an early growth stage, and (b) degenerating the buried dormant but viable seeds by fumigation, flooding, heating and other methods.

12.9.3 Control Methods

Weed control methods are grouped into (a) cultural, (b) biological, and (c) chemical methods.

a) Cultural methods are subdivided into (1) mechanical and (2) cropping and competitive methods.

(1) *Mechanical methods:* These include the use of different implements and power to uproot, cut, incorporate, burn and decompose the weeds to reduce infestation and growth. Important mechanical methods are hand pulling, hoeing, spudding, tillage (preparatory tillage, intercultivation, fallow tillage), clipping, mowing, flooding, burning (heating), and mulching with non-living materials.

(2) *Cropping and competitive methods:* These include cultural practices (also known as non-monetary or less monetary practices) in favour of crops and suppressing of weeds. These practices are good land preparation, growing smother, competitive (such as sunnhemp, cow-pea and rice bean), and cleaning crops in rotation, the use of land rotation, stale seed-bed technique, pre-sowing irrigation, green manuring, puddling, selection of adaptable variety, optimum date, rate and method of sowing, sowing pre-germinated or sprouted seeds or transplanting, intercropping in wide spaced and slow-growing crops, skipping the basal dose of fertiliser nitrogen in direct-seeded crops, the incorporation of bulky organic manures sufficiently ahead of land preparation, correction of soil reaction, exact soil, crop, water, nutrient, insect pests and disease management, appropriate time and method of harvesting and adoption of multiple cropping.

b) *Biological methods*: In biological methods of control, the natural enemies belonging to plants or animals are used for the destruction or suppression of weeds. Before the release of such agents 'starvation tests' need to be conducted regarding the safety of non-target flora in and around the area under a specific habitat.

Among microbes, fungi, such as *Alternaria eichhorniae* and *Uredo eichhorniae* on *Eichhornia crassipes; Colletotrichum gloeosperioides* on *Aeschynomene virginica, Fusarium* sp. on *Opuntia* spp. are found to be useful. Among insects *Dactylopius tementosus* on *Opuntia dillenii, Chrysolina gemellata* on *Hypercium perforatum; Gasonula punctifrons* on *Eichhornia crassipes, Crocidosema lantana, Agromyza lantanae, Theelea echion* and *T. bazochi* on *Lantana camara* are found to be effctive. The use of fish such as *Tilapia mossambica, Ctenopharyngodon idella* to control aquatic herbs, *Ospharonemus olfax* on *Hydrilla verticillata* are some of the examples of the biological control of weeds.

Cattle, camel, equine, sheep, goat, deer, rabbits and pigs, feed on different weeds selectively. Most of these domestic animals are allowed to graze or browse freely or by tethering, on weeds that grow in crop fields and field bunds during uncropped seasons or that grow on the banks of irrigation channels and other uncropped areas. Repeated grazing or browsing removes the regenerated organs and tops of weeds and thus reduces their reproduction and multiplication. Trampling also help to destroy a number of weeds. Pigs in particular devour underground storage tissues of aquatic and semi-aquatic weeds and control them effectively.

A number of birds, such as ducks, hens and turkeys feed on weed seeds and nipple on the growing twigs of tender weeds. The release of such birds in cotton helps to control a number of weed species. Ducks help to destroy floating aquatics, such as *Pistia stratiotes, Lemna minor* and *L. trisulka.*

c) *Chemical methods*: In this method chemical energy is used in controlling weeds. Some of the weed controlling chemicals are able to kill a group of plant species without affecting others, that is, their action is selective which is a unique property among pesticides. Although the selectivity of herbicides (the chemicals used for killing or inhibiting the growth of unwanted plants) depend on the differences in physiological absorption, translocation and bio-transformation between weeds and crop plants and also on the time and dose of application, the same chemicals are also used for growth regulation, dehaulming, defoliation and ripening of crop plants at different doses and times of application.

According to use (methods of application and mechanism of action) herbicides are classified as:

1. Selective:
   A. Foliage applicant:
      1) Contact: DNBP, H_2SO_4, PCP

    2)   Translocated: 2, 4-D, propanial, MCPA.
- B.   Soil applicant (residual): siroazine, butacblor.
2. Non-selective:
- A.   Foliage applicant:
  - 1)   Contact: Na-arsenitc, Na-chlorate. ·
  - 2)   Translocated: dalapon, amitrole.
- B.   Soil applicant (residual):
  - 1)   Fumigant: Vapum, EPTC, CS_2
  - 2)   Sterilant: diuron, monuron, glyphosate.

According to the time of application herbicides are classified as:

(1) pre-emergence to crops and weeds: fluchloralin, metbabenzthiazuron, alacblor;

(2) pre-emergence to crops but post-emergence to weeds: paraquat; and

(3) post-emergence to both crops and weeds: 2, 4-D Na-salt, propanil.

According to the chemical structure and mode of action herbicides are classified as:

1. Phenoxy carboxylic acids: 2, 4-D, 2, 4-DB, MCPA, MCPB;
2. Aromatic acids: 2, 3, 6-TBA, dicamba, chloramben;
3. Amides and Nitrites: benzadox, propyzamide, diphenamid, dichlobenil, bromoxynil, ioxynil;
4. Anilides: propanil, benzoylprop-ethyl, alachlor, buta-xnior, propachlor;
5. Nitrophenols: dinoseb;
6. Nitrophenyl ethers: nitrofen, fluorodifen;
7. Nitroanilines: trifluralin, benfluralin, nitralin,
8. Carbamates: barban, propham, asulum;
9. Thiocarbamates: EPTC, butylate, molinate, diallate, triallate, benthiocarb;
10. Ureas: fenuron, monuron, diuron, metoxuron, isoproturon, linuron, mcthabenzthiazuron;
11. Heterocyclic nitrogen compounds:
   - A) Triazines: simazine, atrazine, ametryne;
   - B) Pyridines: picloram, cyperquat, paraquat, diquat;
   - C) Pyrimidines (uracils): terbacil, bromacil, lenocil·
   - D) Pyridaztnes: pyrazone, norflurazon:

E) Unclassified: difenzoquat, aminotriazole, metribuzin, metamitron, isocarbamid;

12. Heterocyclic compounds: benazolin, bentazone, methazole, dazomet, oxadiazon;
13. Organo-arsenic compounds: MSMA, DSMA;
14. Inorganic salts and acids: Na-chlorate, $FeSO_4$, $CuSO_4$, borax;
15. Unclassified: dimexan, glyphosate.

Herbicides are available in different formulations: (a) solid (dust, wettable powder, granular), (b) liquid (emulsion, emulsifiable concentrate, solutions of water or oil), (c) semisolid (paste or slurry), and (d) others (gelatin capsule, glomule, cellulose ester, wax bar). Herbicides are applied by various methods. The important methods are broadcast or blanket, band, directed spray, spot, basal bark/frill or stump application, sub-surface and liquid application (both over head and soil application with high, medium, low and ultra low volume sprayer or by drip, trickle or sprinkling method). Some herbicides are used as a fumigant.

Each method has its effectiveness under specific conditions but if all suitable methods or techniques are utilised in as compatible a manner as possible to lower the population and growth of weeds so that their infestation remains below the level of economic injury, it is said to be integrated weed management. It is multi-disciplinary in approach and unidirectional in nature.

CHAPTER 13

Methods of Cropping

The way in which the cropping system in a particular field or farm or operational holding or locality is practised in an agricultural year (July-June) is said to be the cropping pattern of that site. The cropping system is the crop production activity of the farm or holding. It comprises all cropping patterns adopted on the farm or holding and their interactions with farm resources, other household enterprises and the physical, biological, technological and socio-economic factors or environments. The cropping pattern is the yearly sequence and spatial arrangement of crops on a given land area. The method of utilising the land resources by the cropping pattern is said to be the method of cropping. There are a large number of cropping methods that are followed in different agro-climatic conditions.

13.1 SHIFTING CULTIVATION

This consists of the slash and burn type of shifting cultivation or natural vegetation or sylva of a particular area. Under this system crops are grown for years until the land becomes either poorer in fertility or infested by bio-agents such as weeds, pests, pathogens and parasites particularly in the pedosphere and then the site for cultivation is shifted leaving the land for degradation. It is a method of cultivation in which several crop years are followed by several fallow years with the land not under management during the fallow years and involves shifts around a permanent homestead to the location where the fields for cultivation are moved.

13.2 SETTLED OR PERMANENT AGRICULTURE

In contrast to shifting cultivation, settled agriculture is practised in an area either owned or taken on lease by the peasants where cultivation goes on permanently. Therefore, the tiller is interested in improving, maintaining, and protecting the land for his agriculture for generations and to maintain the livelihood as a member of the nation or the domain.

In this type of agriculture both horizontal and vertical land-use have been taking place by adopting an intensive cropping pattern, growing modern plant types, using inputs for high production potential and improved management practices and thus increasing land capacity (land area needed to feed a human or an animal). Farming practices may be (a) specialised (more than 50 per cent of farm income comes from a single enterprise such as sugar-cane, cotton, jute, rice, vegetable, fish, livestock, poultry or goat), (b) diversified or mixed (no single enterprise contributes 5 per cent of farm income such as jute+rice+vegetable, rice+potato+vegetable, dairy+poultry+goat keeping+piggery) and (c) integrated (two or more types of enterprises are taken together to have an increased use-efficiency of by-products of each enterprise in a cyclic order for instance, field crops+livestock+fish, field crops+forage crops+vegetable+livestock+poultry, dairy+duckery+fish, field and forage crops + dairy +duckery+apiary+sericulture+agariculture +fernery) farmings.

13.2.1 Methods of Cropping in Settled Agriculture

Several methods of cropping are followed in such settled and stable agriculture. They are as follows:

a) *Monoculture*: The cultivation of a single crop in a given field in a crop season such as rice, jute, sugar-cane, cotton, tobacco, potato or green gram.

b) *Dual culture* : The cultivation of two types of crops grown together or two types of enterprises on a piece of land in a crop season such as rice+fish, rice+Azolla.

c) *Polyculture*: The cultivation of more than two types of crops grown together on a piece of land in a crop season for instance, subabool + papaya + pigeon-pea + Dinanath grass, mango+pineapple+turmeric, litchi+papaya+tuberose+soyabean, guava+banana+cow-pea, or banana+marigold+bersem, under agri-horti-sylvipastoral systems.

d) *Monocropping*: The cultivation of a single crop, season

after season or year after year in a given field such as rice, sugar-cane, *jowar*, cotton or wheat.

e) *Double cropping*: The cultivation of two crops in succession on a piece of land in an agricultural year for instance, rice followed by wheat or potato or mustard, maize followed by groundnut, *jowar* followed by gram or wheat followed by cotton. Under this situation cropping intensity is 200 per cent.

f) *Multiple or polycropping*: The cultivation of more than two crops in succession in a field in a year for instance, rice-potato-green gram, rice-mustard-maize, jute-rice-potato, jute-rice-wheat, rice-radish-rice, rice-potato-sesame, rice-potato-rice, sunnhemp-cauliflower-potato-black gram, or *jowar* (fodder)-rice-berseem-cow-pea. Cropping intensity is more than 200 per cent under this situation.

When the farm or operational holding as a whole is considered the multiple cropping index (MCI) is determined by the total area planted divided by the total arable area. When the value is three or more it is said to be a most promising farm. This is also called intensive cropping.

g) *Triple cropping*: The cultivation of three crops in succession in a field in a year so that cropping intensity is 300 per cent, for instance, rice-potato-rice, rice-potato-cow-pea or cow-pea-mustard-jute.

h) *Relay cropping*: The cultivation of crops in succession. Here the crop in succession is sown or planted either in the field of the standing crop which is going to be harvested soon or in the nursery so that immediately after the harvest of the standing crop the subsequent crop can be transplanted in the same field without any allowance for keeping the field uncropped or fallow even for the turn around period (the period between the harvesting of the preceding crop and the planting of the succeeding crop in a specific field). It is analogous to a relay rice. This is done by *paira* cropping, till planting, and raising seedlings in nurseries and possible combinations are jute-rice-mustard-onion, rice-cauliflower-onion-summergourds. Here before the harvesting of jute, the rice seedlings are raised in the nursery and before the harvesting of mustard, onion seedlings are made ready in the nursery for transplanting. Similarly, before the harvesting of rice, cauliflower seedlings are made ready and before the harvesting of cauliflower, onion seedlings are kept

:eady for transplanting and before the harvesting of onion, summer gourd plants are ready to occupy the field.

i) *Continuous cropping*: The cultivation of crops without breaking the continuity. In this system fields are kept continuously under crops for instance, rice-rice-rice, sugar-cane-sugar-cane-sugar-cane, napier-napier-napier, colocasia-rice-peas-maize (fodder) or sunnhemp-radish-potato-cow-pea (fodder).

j) *Overlapping cropping*: The cultivation of crops in such a way that one crop overlaps the other crop. This is practised mostly for fodder crops. For feeding animals a portion of the standing crop may be harvested and before the harvest of the whole field some other succeeding crop may be sown in the portions from where the preceding crop has been harvested, for instance maize-berseem-*jowar*-cow-pea-oats-rice bean-stylo.

k) *Rotational cropping*: The cultivation of crops in recurring succession in the same field which consists of growing the same crop or crops in the same seasons or years rotationally such as rice (*kharif*)—potato (rabi)—green gram (summer) in the first year and a similar sequence will continue after one or more years. Similarly, two, three, or four years' rotation may be adopted.

There may also be rotational use of land (land rotation), growing crops followed by fallowing either for a part or for the whole of the field.

l) *Sequential cropping*: The cultivation of crops in sequence or raising a crop after harvesting one. In general, the sequence of crops is maintained season after season for one or more years, for instance (*kharif*) rice—potato (*rabi*)—summer maize, rice-potato-jute, rice-rape-onion, sugar-cane (spring planted)—rice nursery, rice-black gram/Lathyrus/linseed/lentil/gram/safflower/niger, rice-wheat-rice nursury, rice-onion nursery-black gram/green gram/sesame, or rice nursery-rice-onion.

m) *Multi-storeyed cropping*: The cultivation of different crops of different canopy heights in a field simultaneously such as arecanut + betel vine + papaya + pigeon-pea+pineapple+ginger, pigeon-pea + sesame +groundnut, maize+green gram +groundnut, pigeon-pea+ upland rice+ black gram, amaranth +lady's finger+colocasia, spinach+radish+onion and brinjal +lady's finger+basella+colocasia.

n) *Crop cafetaria*: The cultivation of different crops of

various uses in the same field at the same time. This may or may not follow a definite pattern of planting. Examples are potato+radish+coriander+barley + peas, colocasia + lady's finger+basella+brinjal+chillies+turmeric+ pigeon-pea+sunflower+ridged gourd+ bitter gourd+palwal+soybean, brinjal +spinach+peas+marigold+cauliflower+cabbage+beet+ carrot+radish+fenugreek+coriander+onion+garlic.

o) *Mixed cropping*: Cultivating two or more crops simultaneously in the same field without keeping their identity with respect to field area such as maize+cow-pea or wheat+ mustard.

p) *Intercropping*: Cultivating two or more crops simultaneously in alternate rows in the same field such as sesame +green gram or pigeon-pea+groundnut.

Crops may be grown single or mixed or as intercrops with or without a definite sequence. The field or farm may be occupied by one crop in a year (monocropping) or by two crops (double cropping) or more crops (multiple or intensive cropping) in sequence in a year.

The climatic, edaphic and socio-economic diversity of the Indian crop production scene is ever dynamic with many cropping patterns. With a geographical area of 328.048 million hectares, stretching between 8°N and 36°N latitude and between 68°E and 98°E longitude, its altitude varies from mean sea level to the highest mountain ranges of the world. India has the station with the highest mean annual rainfall in the world and simultaneously locations with dry, semi-desert areas. It presents a range of diversity of climate, soil, flora and fauna including human being, the masters of agriculture. The major area under cultivation is dependent on monsoonal rains which very often differ with respect to onset, intensity, duration, distribution and retreat and therefore, Indian agriculture is a gamble with the weather making the farmers wait, watch and worry for the favours of the monsoon and/or financing agencies or face famine with folded hands, frequently.

Population growth, social instability and political insecurity and the influx of unskilled labourers into the agricultural sector are reducing the efficiency and lowering the per capita farm income. New technologies are dwindling the traditional, stable and inherited skills of the peasantry. The jobless, poor, ill-fed ill-clad, poverty stricken farming community including labourers are continuing to blur the rural perspective. Most of the time they have to depend on emaciated animals for draft power. Most of the members of this community are poor in their capability and capacity to absorb, adapt and adopt new knowledge and technologies suitable to the specific agro-ecological situations and conditions even though such technologies are proved to be profitable to agribusiness, suitable and adaptable by the means of the farming community.

The one commodity which will be in the greatest demand throughout the globe is food grain which plays a pivotal position in the development of mankind. As food grains are not only consumed by humans but by animals, birds and fishes, and they convert carbohydrate to animal proteins, fats, and other energies, the demand on food grain will continue as long as mankind exists as a component of the biosphere in nature. Surprisingly man is a competitor with his animals for food grain

The average size of an operational holding is tending to get smaller and smaller land provision (land area that may be provided per person) is decreasing rapidly. Land is an inelastic site for agriculture and good arable soil is continuously encroached upon for other uses such as the construction of houses, factories, roads, canals and for brick making. The biological potential of the land is reduced by erosion, desertification and other faulty uses of the land. The productivity levels of most of our cropping systems are low and consequently the cost of production is high. Unless we can maximise agricultural production through intensive cropping with increased productivity we will not be able to be self-sufficient, self-reliant and will not sustain ourselves in the world picture.

The total number of operational holdings (land which is wholly or partly used for agricultural production and is operated as one technical unit by one person alone or with others without regard to title, legal form, size or location; it includes both cultivated and uncultivated areas) in the country in 1981-82 was 900 lakhs including about 500 lakhs marginal holdings (in which production equals the cost) of less than one hectare each. Such marginal and uneconomical operational holdings are the major bottlenecks in adopting modern technologies which are mostly costly and risky to the farmer at the beginning but become more remunerative subsequently than the traditional ones. The piece-meal distribution of fragmented and odd-configurated fields with ownership titles may potentially make land reforms a success but will perturb the production potential of the peasantry on the possessed land. This particularly true to those who do not or cannot wholly depend on what they produce and thus cannot defend their landed property from the lucrative price offered by the well-to-do.

13.2.2 Multiple Cropping

The only scope in improving agricultural productivity in the same land is the adoption of intensive (multiple) cropping with intercropping wherever and whenever possible.

A. PREREQUISITES OF MULTIPLE CROPPING

In adopting a multiple cropping programme some basic needs are to be fulfilled. They may be:

1. considerably level fields with highly productive soil in which the climatic conditions of the region are not limiting to crop production;

2. the field should not be prone to be affected by seasonal or occasional flood or drought or severely infested by weeds, pests, pathogens and parasites;

3. the use of short duration, fertiliser responsive modern type crop varieties or hybrids that fit into an annual cycle of operations for at least three crops in a year. It is better to use determinate and synchronously ripening crop varieties;

4. adequate and assured irrigation and drainage facilities throughout the year;

5. adequate and timely availability of inputs such as seed, fertiliser, implement, labour, draft power, plant protection chemicals and funds;

6. appropriate facilities to remove or utilise crop residue from the field immediately after harvest;

7. the surrounding fields should be cultivated with crops requiring similar agronomic practices. Each crop should be grown in a cluster or in a compact way to avoid hazards of

approaching the field for various operations with bullocks or machinery, water management and pest management as these difficulties are associated with isolated plots distributed haphazardly;

8. there should be suitable post-harvest opportunities so that crops may be harvested when they are physiologically mature rather than field mature (at the ripe stage) and there should be or a lower requirement of a field drying period which causes delay and probably difficulty in land preparation and sowing of the succeeding crops;

9. there should be appropriate technical know-how to plan, programme and execute exact technology considering the soil-crop-environment and the ultimate goal;

10. there should be immediate marketing facilities even from the field or door of the farmers with remunerative prices. Most of the time agricultural commodities become the victims of distress or credit sales particularly during the harvest season when almost all the farmers are compelled to dispose of at least a part of the produce to pay the cost and the interest of the capital spent on the cultivation of the crop, as well as to meet the cost of cultivation of the succeeding crops or to pay the premiums for the insurance of the crop and also to meet the recurrent expenditure on capital assets.

Melested (1954) has mentioned that multiple cropping is a philosophy of maximum crop production per acre of land with a minimum of soil deterioration. The philosophy is based on the concept that:

a) the high production is good for the soil;

b) the minimum tillage promotes soil tilth and conserves soil organic matter thus resulting in good physical conditions;

c) the cover in the form of living mulch is a protection against all forms of erosion and weeds;

d) there is a proper utilisation of residual effects of manures and fertilisers, moistures and management practices;

e) the successive crops provide an opportunity for the efficient utilisation of land, labour, capital and other available resources;

    f)   there is scope for higher production with a higher rate of turn out.

Pal and Shiks (1969) demonstrated the production of 45 q wheat grain (Jan—15th April), 4.5 q seeds of green gram (20th April—30th June), 60 q maize grain (7th July—30th Sept.) and 15 q toria seed (Oct-Dec.) totalling 124.5 q/ha/year economic yield of staple food, pulse and oil-seed crops grown in sequence. Mitra *et al* (1972) reported a profit of Rs 10,000/ ha/ yr with the cropping system consisting of green gram+jute—rice—potato with an annual production of 241 q of food and fibre. Potato-jute-rice sequence under different fertiliser management showed that the maximum production of potato (173 q/ha) jute (35 q/ha) and rice (48 q/ha) could be obtained with nitrogen, phosphorous (P_2O_5) and potassium (K_2O) each at 100 kg/ha (50 per cent nitrogen as organic and the rest as inorganic) to potato with no fertiliser to jute (with leaves incorporated into the soil) and 60 kg N/ha to rice (Biswas *et al*, 1985) i.e. a total production of 256 q/ha/yr. There is still ample scope to increase production and profit per ha/yr as the field crops have the potential to produce more than 30 t dry matter per ha/yr.

B. POTENTIAL OF MULTIPLE CROPPING

Modern intensive cropping systems have created the following potential:

    1. improved stability of food and feed supply throughout the year;

    2. increased productivity per unit area, time, input and total production accompanied by an increase in the total income of the farmer;

    3 improved distribution of income throughout the year with quick out-turns and thus an increased possibility of recycling working capital;

    4. increased total employment and distribution of labour and other capital use throughout the year and opportunities for on-farm seed production, preservation, processing and marketing.

    5. minimised the scope of soil erosion and degradation;

    6. maximised the possible utilisation of land, residual effects

of manures, fertilisers, moisture and management practices;

7. minimised the rental value, irrigation charge and other imputed costs per unit of production;

8. broadened the scope to select and substitute crop varieties matching the agro-ecological situation, the cropping pattern and programmes based on home requirements and market competitions;

9. extended the possibilities of the complete removal of weeds as reduced fallow periods minimise the reproduction of weeds;

10. improved the nutrition for the farm family from crop diversification;

There are some drawbacks of multiple cropping systems particularly when they are not adopted properly. They may be:

1. a very short time is available for land preparation;

2. it requires more intercultivation to compensate for inadequate preparatory tillage;

3. the cleaning of stubble and stumps becomes a great problem;

4. crops with residual toxicity (allelopathic effect) affect crops in quick succession;

5. it increases weeds, pests and disease hazards if not handled properly;

6. it causes deficiency and disorder due to an inadequate and imbalanced supply of plant nutrients;

7. the lack of appropriate post-harvest technology becomes a limiting factor;

8. inclement weather and unworkable soil conditions hinder quick succession;

9. it restricts the scope of soil and land improvement;

10. long duration crops cannot be accommodated.

Even though there are some shortcomings yet this approach is having a tremendous impact on the total productivity, stability, profitability and managerial capacity in the present day integrated farming and intensive cropping systems of agribusiness. Modern, photoperiod insensitive and weakly sensitive, very short duration crop varieties with improved plant types; improved input supply system coupled with improved cultural practices have widened the scope of intensive cropping utilising even pre- and post-monsoon rains and residual soil moisture under rain fed conditions. Soil and water conservation, the increasing use-efficiency of inputs, the choice of efficient crop varieties matching the soil and water availability extend the scope of intensive cropping even under dry land agriculture.

13.2.3 Intercropping

Intercropping is a method of cropping in which there is a greater utilisation of the interspaced area, the spatial distribution of light, nutrients, moisture, air and micro-environment of both the rhizosphere and phyllosphere and the temporal use of resources during the slow growth phase of a crop by the subsidiary crop or crops or vice-versa either in monocropping or intensive cropping systems. A cropping system is a combination of crops in space and time and the objective of any given system should be to provide the farmer with a stable level of returns. In agronomic terms, the systems that best meet this objective are those that make efficient use of the basic resources necessary for plant growth, especially any resources that are limiting. This depends partly on the inherent efficiency of the individual crops that make up the system, and partly on complementary effects between those crops that are grown in association (Willey *et al*, 1980). Intercropping can often produce higher yields than sole crops (Mead and Willey, 1980; Reddy and Willey, 1981; De, 1980 a).

Intercropping has long been recognised as a common practice throughout the developing tropics. In India, its importance was highlighted by Aiyer (1949). With the introduction of line sowing equipment and modern varieties requiring advanced and exact methods of cultural practices such as weeding, hoeing and earthing up and requiring more fertiliser application in spiits and the incorporation of fertiliser into the soil, the need-based application of pesticides as preventive and curative measures as well as the harvesting of harvestable economic parts when the crop is still in the productive stage for the subsequent harvest by picking requiring frequent entry inside the crop field, the practice of growing more than one simultaneous crop in rows in general, the mixed cropping term becomes inappropriate. The mixed cropping was an age-old practice in agriculture right from the primitive age and till today it is in practice in different forms. More recently, it has been realised that intercropping remains an extremely widespread practice and is likely to continue so far at least as the forseeable future (Arnon, 1972; Francis *et al*, 1975).

A 'component crop' is used to refer to either of the individual crops making up the intercropping situation. 'Intercrop yield' is the yield of a component crop when grown in an intercropping system and expressed over the total intercropped area (the area occupied by both crops). A simple addition of both intercrop yields thus gives a 'combined intercrop yield'. A sole crop refers to a component crop being grown alone at optimum population and spacing. 'Combined sole crop yield' is the combined yield achieved when a unit area is divided between the two sole crops in some given proportion (Willey 1979). All intercrops are therefore, a simple replacement to a certain per cent of one crop or the other.

13.2.3. A. Advantages of intercropping

1. Intercropping provides yield advantages compared to sole cropping. These are not by means of costly inputs but by the simple expedient of growing crops together.

2. It economises the space and time of cultivating two or more component crops of comparable agronomic practices, grown separately.

3. It provides greater surety and stability of higher yield over different seasons. This is particularly important to subsistence farmers as when one crop fails or grows poorly, the other component crop or crops compensate this loss to a great extent.

4. It helps to restore soil fertility when legumes are included as component crop.

5. It utilises a greater total volume of soil efficiently by the differential root system with varying cation exchange capacities of different component crops (Drake, 1964). Thus there is a more intimate use of the below ground environment consisting of soil nutrients, moisture, air, heat, micro- and macro-organisms.

6. It utilises a greater total volume of above ground environment embracing incidental light, heat, CO_2 and rain

water by different crop geometry, canopy structure, leaf volume, leaf area density, leaf thickness and photosynthetic pathways (C_3 & C_4 types).

7. It utilises the slow growth phase of a long duration crop. It provides for the temporal use of resources as different crops demand resources at different times of their growth patterns and for the spatial use of resources by the combined phyllosphere and rhizosphere especially in wider spaced crops.

8. It helps to grow more balanced food or feed with a higher biological value as cereals are deficient in lysine, threonine and tryptophan whereas pulses are deficient in methionine and tryptophan and their 67 : 33 proportion provides a somewhat balanced diet and satisfies different dietary requirements without additional cost.

9. It covers the ground more continuously thus providing a better protection of the soil from erosion. It also helps in siltation by choking an easy run off from bare spaces. It reduces weed population and growth by its smothering effect.

10. It provides physical support, shelter and nourishment to the component crops. The component crop or crops may act as a barrier to both soil and wind-borne pests and pathogens.

11. It helps to avoid intra-crop competition and thus a higher number of crop plants can be grown per unit area.

12. It ensures an ideal farming particularly in uneven topography with moderate to high slopes where pastures are raised by intercropping legumes for rotational grazing and in this way the marginal lands are gradually introduced to more productive lands.

13. It helps to shape lands by growing strip crops (erosion permitting and erosion restricting crops alternately).

14. It provides more employment and distribution of labour and thus spreads labour peaks with more human appreciation value by growing and harvesting different crops at different intervals in the same piece of land.

15. It provides for income in instalments although with a smaller amount and reduces marketing risks.

13.2.3.B. Disadvantages associated with intercropping

There are difficulties in the practical management of inter-cropping, especially where there is high degree of mechanisation or where the component crops have different requirements of fertiliser, water, depth of sowing, pesticides, growth regulators and post-planting operations. There may be a yield decrease not because of adverse competitive effects, although it is rare. Harvesting is difficult particularly when intercrops do not need to be harvested simultaneously. Intercropping may reduce qualitative and generative yields though it may provide a higher quantitative yield. These difficulties are typically associated with better developed agriculture; the poorly developed farmer not only seems able to handle intercropping but often seems to have a strong inherent preference for it. It is the small farmer of limited means who is most likely to benefit much from inter-cropping (Willey, 1979).

13.2.3.C. Criteria for assessing yield advantages

Different intercropping situations may have to satisfy differ-ent requirements. Therefore, yield advantages are not always satisfied. Based on sound objectives three different situations can be distinguished from different intercropping systems. They are:

1. Where intercropping must provide a *full* yield of the main crop (as much as sole cropping) and some *additional* (bonus) yield of a second crop. In this situation the primary requirement is for a full yield of some staple food crop with the aim of maximising the yield of the second crop without re-ducing the yield of the main crop.

2. Where the combined intercrop yield must exceed the higher sole crop yield: This is the criterion which has tradi-tionally been used for assessing yield advantages in grassland or grass + legume (*jowur* + cow-pea as fodder) mixtures. It is based on the assumption that the unit yield of each compo-nent crop is equally acceptable and therefore, the requirement is simply for maximising yield regardless of the crop from which it comes. Growing only the higher yielding sole crop is a valid alternative to growing both except when some special

advantages are not sought for.

3. Where the combined intercrop yield must exceed a combined sole crop yield: This criterion is based on the assumption that a farmer usually needs to grow more than one crop, for instance to satisfy dietary requirements, to even out labour peaks and to guard against market risks etc. In this situation, a yield advantage occurs if intercropping provides higher yields than growing both the component crops separately: in fact, the combined intercrop yield does not now have to out-yield the higher yielding sole crop, since by definition, growing only the latter is not an acceptable alternative to growing both crops (Willey, 1979).

13.2.3.D. Competitive relationships

Three categories of competitive relationships between intercrops are recognised. They are:

1. When the actual yield of each species is less than expected [where expected yields are those that would be obtained if each species experienced the same degree of competition in mixture as in pure stand i.e., if inter-specific competition was equal to intra-specific competition]; *mutual inhibition*, though it is rare in practice, has been observed by some workers.

2. Where the yield of each species is greater than expected— *mutual co-operation*, or synergistic, is not unusual.

3. Where one species yields less than expected and the other more—*compensation*—the commonest situation. In this situation the competitive abilities of the two species obviously differ. According to their more or less competitiveness they are called the dominant and dominated species.

The mutual inhibition situations (may be due to allelopathic or shading effect of a component crop) cannot give a yield advantage whereas, the mutual co-operation situation must do so. Where compensation occurs, the possible advantage of intercropping is not so clear.

13.2.3.E. Land equivalent ratio (*LER*)

In an ideal intercropping system there is a yield advantage. To determine which of the component crops (dominant and dominated species) contributes how much to this yield advantage, *LER* is calculated; *LER* is the relative land area under sole crops that is required to produce the yields achieved in intercropping: it is usually stipulated that the 'level of management' must be the same for intercropping as for sole cropping. *LER* is analogous to 'relative yield total' (RYT).

$$LER = RYT = La + Lb = \frac{Ya}{Sa} + \frac{Yb}{Sb}$$

where *La* & *Lb* are the *LERs* for the individual crops;

Ya & *Yb* are the individual crop yields in intercropping; and *Sa* & *Sb* are their yields as sole crops.

Table 13.1 :
Yield and *LERs* of pigeon-pea intercropped with sorghum

Crops	Yield (kg/ha)	LERs	Total LER	Intercrop benefit
Sole sorghum	3952 (Sa)			
Sole pigeon-pea	1699 (Sb)			
Intercrop sorghum	3804 (Ya)	0.96	1.47	0.47
Intercrop pigeon-pea	850 (Yb)	0.50		

Now it is misleading to argue that intercropping 1 : 1 (Table 13.2) is better than intercropping 2 : 1 proportion on the basis of higher *LER*. Intercropping 2 : 1 may be preferred if the farmer is desirous to obtain a higher yield of maize.

Table 13.2 :
Yield and *LERs* of pigeon-pea intercropped with maize at 1 : 1 and 2 : 1 proportions

Crops	Yield (kg/ha)	LERs	Total LER	Yield proportion of maize $[Lm/(Lm+Lp)]$
Sole maize	3398 (Sa)			
Sole pigeon-pea	1035 (Sb)			
1 : 1 Intercrop maize	2234 (Ya)	0.66 (Lm¹)	1.53	0.43
1 : 1 Intercrop pigeon-pea	896 (Yb)	0.87 (Lp¹)		
2 : 1 Intercrop maize	3130 (Yaa)	0.92 (Lm²)	1.47	0.63
2 : 1 Intercrop pigeon-pea	571 (Ybb)	0.55 (Ly²)		

Reproduced from Mead and Willey (1980).

If a farmer wants to obtain the same amount of maize yield that has been obtained under the intercropping 2 : 1 situation how much sole maize area would have to be grown in addition to one hectare of intercropping 1 : 1 may be calculated as:

Assume E = additional sole maize area (ha) required;

the required proportion $\dfrac{\text{maize}}{\text{maize}+\text{pigeon-pea}} = \dfrac{LM+E}{LER+E}$

where LM and LER refer to the intercrop actually being grown in intercropping 1 : 1 proportion.

Thus $0.63 = \dfrac{0.66+E}{1.53+E}$ or, $0.66E = 0.63 \times 1.53 + 0.63E$

$$= 0.9639 + 0.63E$$
$$\text{or, } 0.9639 + 0.37E = 0.66$$
$$\text{or, } 0.37E = 0.3039$$
$$\text{or, } E = 0.82 \text{ ha.}$$

Therefore, to achieve an equal amount of maize yield obtained in intercropping 2 : 1 proportion, intercropping 1 : 1+0.82 ha of sole maize is to be grown (Mead and Willey, 1980).

If a farmer wants to quantify his benefits (advantages) in terms of money considering prices of maize and pigeon-pea at Rs. 4/- and Rs. 6/- per kg respectively, then the Price Equivalent Ratio (PER which refers to the relative monetary advantage per unit area in intercropping—may be of different proportions of component crops as it is with LER) of each intercropping situation is to be determined (Table 13.3) and com-

Table 13.3 :

Price equivalent ratios ($PERs$) of different intercropping systems

Crops	Yield (kg/ha)	Price (Rs/kg)	Total value (Rs.)	Return (Rs/ha)	PER
Sole maize	3952	4 /-	15,808/-	} —13,001/-	1.00
Sole pigeon-pea	1699	6/-	10,194/-		
1 : 1 Intercrop maize	2234	4/-	8,936/-	} —14,312/-	1.10
1 : 1 Intercrop pigeon-pea	896	6/-	5,376/-		
2 : 1 Intercrop maize	3130	4/-	12,520/-	} —15,946-/	1.22
2 : 1 Intercrop pigeon-pea	571	6/-	3,426/-		

parisons are to be made. While the monetary advantage for a seasonal cropping was considered, the 2:1 proportion gave 1.22 times (PER = 1.22) more advantage per hectare than when the same area was grown with the component species separately, as sole crops and where each sole crop occupied 50 percent of land area which was followed by 1 : 1 proportion of intercropping whereas, the sole ciopping of pigeon-pea provided the least return. However, this situation is true only when the cost of cultivation remains the same for ail these cropping systems and other benefits such as residual values of pigeon-pea cultivation as a restorative crop are ignored. Therefore, it is being found that LER values do not always indicate the superiority of different intercropping systems .and thus valid comparisons cannot be made.

In another intercropping situation with different proportions of sesame and black gram sown in different row directions (Table 13.4) the highest LER value (1.26) was obtained in 1 : 2 proportion of intercropping with north-south row orientation and this treatment had the second best PER value (1.30) whereas, the highest PER value (1.56) was obtained in 2 : 1 proportion of inter-cropping with north-south row orientation which showed the second best LER value (1.22); this intercropping situation provided 1.56 times more return than when the same area would have been grown with the component crops separately as sole crops occupying 50 : 50 (1 : 1) proportion of land area. In other situations of intercropping LER and PER values varied greatly and in general, an intercropping situation other than 1 : 1 proportion gave proportionately higher monetary advantages. Therefore, from the farmers' point of view PER value seemed to be a more appropriate determination in assessing the advantages from intercropping systems than the LER values (De, 1986). Recently, area-time-equivalency-ratio ($ATER$) which is the ratio of number of hectare-days used in monoculture to the number of hectare-days used in intercropping to produce identical quantities of each of the component crop, is considered where area-tirae-price-equi valency-ratio ($ATPER$) seemed to be more appropriate.

In a standard 2 sorghum (82 days): I pigeon-pea (173 days) row arrangement sorghum growth was not affected by the presence of pigeon-pea and the farmers' primary objective of maintaining a full sorghum yield was achieved if the density of the intercropped sorghum was equivalent to the sole crop optimum. The initial growth of pigeon-pea was suppressed by the presence of sorghum, but some compensation in growth after the sorghum harvest, and a much higher ratio of seed yield to total above ground dry matter, resulted in seed yields of upto 73 per cent of a sole pigeon-pea yield. The optimum density for intercrop pigeon-pea was considerably higher than the sole crop optimum.

Table 13.4 :
PER and LER of different intercropping situations

Direction of sowing	Crop	Yield (kg/ha)	LER	Value of produce (Rs./ha)*	PER
E-W **	Sole S	972.5		2,917.50/-	
,,	Sole BG	270.8		1,354.00/-	
,,	2 : 1 S+BG	752.5+36.6	0.90	2,440.50/-	1.14a
,,	1 : 1 S+BG	562.5+87.0	0.90	2,122.50/-	0.99
,,	1 : 2 S+BG	747.5+104.4	1.15	2,764.50/-	1.29
N-S***	Sole S	912.5		2,737.50/-	
,,	Sole BG	231.2		1,156.00/-	
,,	2 : 1 S+BG	942.5+43.4	1.22	3,044.50/-	1.56b
,,	1 : 1 S+BG	705.0+68.5	1.07	2,457 50/-	1.26
,,	1 : 2 S+BG	622.5+134.6	1.26	2,540.50/-	1.30

S=Sesame; BG=Black gram.
* @ Rs. 3/- and Rs. 5/- per kg of sesame and blackgram respectively.
** E-W=East-west row orientation; ***N—S=North-south row-orientation.
a=2440.50÷[(2917.50+1354.00)÷2]
b=3044.50÷[(273 7.50+1156.90)÷2]

Reproduced from De, 1980.

The spatial distribution of roots after 30 and 60 days of growth did not appear to be altered by intercropping, and roots of the two crops were found to intermingle freely (Natarajan and Willey, 1980).

In a sorghum+pigeon-pea intercropping the response to five pigeon-pea populations in a 150 cm-bed furrow system at three row arrangements per bed: (a) 1 row sorghum: 1 row pigeon-pea: 1 row sorghum at 45 cm between rows (SPS); (b) 1 row sorghum: 2 row pigeon-pea: 1 row sorghum at 30 cm between rows (SPPS); (c) 1 row pigeon-pea: 2 row sorghum: 1 row pigeon-pea at 30 cm between rows (*PSSP*) was studied. The distance between the outer rows of adjacent beds was 60 cm. Pigeon-pea seed yield in the intercropping system responded to plant populations above the sole crop optimum of 40,000 plants /ha but the response for the combined yield of both crops was less because of decreasing sorghum yield. Maximum land equivalent ratio and gross monetary returns were at 70,000 plants/ha for the *SPS* arrangement and at 40,000 plants/ ha for the *SPPS* and *PSSP* arrangements. The greater number of pigeon-pea rows in *SPPS* and *PSSP* produced more pigeon-pea yield but less sorghum yield; this resulted in a net benefit for the *SPPS* arrangement though not for the *PSSP* arrangement. The sorghum intercrop reduced the total branch number in pigeon-pea but had little effect on the number of pod-bearing branches. Intercropping also increased the harvest index of pigeon-pea because sorghum suppressed the early vegetative growth but was harvested before the reproductive phase (Rao and Willey, 1983).

In an experiment with four plant populations of chick-pea in combination with 15 systematically arranged populations of safflower in 1 : 1 and 2 : 1 row arrangements, for populations of the sole crops were included. Safflower was usually dominant and increasing the total population (both crops combined) made it more so. Safflower yield was little affected by changes in its own population and was independent of changes in the chick-pea population. Chick-pea yield increased with the increase in its own population and it was the dominant crop at high chick-pea+low safflower populations. An initial increase in the safflower population caused an increase in chick-pea yield at the 1: 1 row arrangement; otherwise increasing safflower population decreased chick-pea yield. *LERs* at 2 : 1 indicated no real evidence of yield advantages for intercropping. At 1: 1 advantages ranged up to 19 per cent, with a maximum where the highest chick-pea population was combined with a low safflower one (Willey and Rao, 1981). Intercrop systems of maize+pigeon-pea or sorghum+pigeon-pea and a three crop system of maize+pigeon-pea+chick-pea appeared very promising. The gross return was usually much higher for the improved systems that utilised both the rainy and post-rainy seasons (Reddy and Willey, 1982).

The reason for higher yields in intercropping compared with sole cropping is that the component crops make complementary use of resources and therefore, achieve better overall resource use when growing together. In an intercropping combination of early sorghum (82 days) and later maturing pigeon-pea (173 days) in a row arrangement of 2 sorghum: pigeon-pea, light interception by the intercrop combination was almost as high as in sole sorghum before the harvest of sorghum. After the sorghum harvest, light interception by the remaining pigeon-pea was very poor and it is suggested that pigeon-pea yield could be increased with higher plant population density and better plant distribution. Soil water measurements indicated that this would increase the amount of water being transpired through the crop but would not increase the total evapo-transpiration demand. Higher nutrient concentrations in the intercrop pigeon-pea compared with sole pigeon-pea was not limited by nutrient stress, though the total uptake of nutrients by both crops was much greater from intercropping than from sole cropping (Natarajan and Willey, 1980). In an intercropping combination of 1 millet: 3 groundnut at 30 cm row spacing with 50 kg P_2O_5/ha as basal for all treatments and N at 80 kg/ha for sole crop and at 20 kg/ha in intercropping as expressed over the area occupied by both crops were top-dressed, measurements of light interception showed that intercropping did not intercept more light energy than sole cropping but this energy was more efficiently converted into dry matter. Intercropping made slightly greater demands on soil moisture but the higher yields were mainly associated with a higher production of the water demand being channelled through the crop as transpiration rather than being lost as evaporation from the soil surface. The intercrop provided a yield advantage for total dry matter ($LER=1.32$) and higher total root length ($LER=1.18$). Nutrient uptake was greater and commensurate with the higher yields ($LERs$ for N, P and K were 1.26, 1.28 and 1.26 respectively) when the same intercropping treatments were repeated under a stress situation (under reduced water supply) where below-ground resources were more limiting. The stress experienced crop gave a comparable yield advantage ($LER=1.25$). The reduction in dry matter yield of sole millet was 13 per cent and sole groundnut was eight per cent. Intercrop millet produced a yield slightly lower than expected and groundnut compensated more after millet harvest. The reproductive yield advantage was slightly higher but there was less leaf area advantage ($LER=1.15$ to 1.20) (Reddy and Willey, 1980).

In the study of root growth of monocropped and inter-cropped pearl millet and groundnut it was found that mono-cropped millet produced a longer root length per unit ground area than monocropped groundnut and also rooted deeper. The distribution of root length also differed for the two monocrops while the intercrop appeared intermediate both in total length and in its distribution. Roots in adjacent rows of millet and groundnut in the intercrop were mixing midway through the growing season. This suggests that root interaction between crops may occur during intercropping; it was a major factor contributing to the increased yields. Intercropping resulted in additional root growth and during the later stages of growth it produced ten to fifteen per cent more root length as compared to the monocrop (Gregory and Reddy, 1982).

In an experiment the intercrop row arrangement was 1 millet (82 days): 3 groundnut (105 days) and the intra-row spacing of each species was the same in the sole crop and the intercrop. The dry weight of sole cropped millet increased line-arly with intercepted photosynthetically active radiation (PAR) during the vegetative and much of the reproductive phases. In contrast, the dry weight of sole cropped groundnut only increas-ed linearly in the vegetative phase. During the first half of pod filling, there was no increase in dry weight despite a substantial quantity of PAR interception. In the second half, the dry weight of groundnut increased by a further 30 per cent. Similar relations were observed for the two components of the inter-crop.

Intercropping gave 28 per cent more total dry matter (LER $=1.28$) that growing the two crops separately. The processes producing the intercropping advantage are separated by defining two ratios: the Resources Capture Ratio (RCR) and the Con-version Efficiency Ratio (CER). These ratios compare, on a per plant basis, the performance of the component species in the intercrop relative to their respective sole crops in terms of the interception of radiation and the production of dry matter/unit of radiation intercepted, respectively. **Per row, the** millet intercepted 2.1 (RCR=2.1) times more PAR in the inter: crop than in the sole crop and used it with a similar efficiency (CER=0.97) to produce twice as much dry matter. Per row in the intercrop, the groundnut intercepted with 46 per cent (CER =1.46) greater efficiency to yield the same (Marshall and Willey, 1983).

In an intercropping situation with 1 millet : 3 groundnut, in groundnut, yield/plant and yield components were similar in intercropping and sole cropping. In millet, on a per plant basis the dry matter accumulation, leaf area development and tiller production were all substantially greater in intercropping compared with sole cropping: the final seed yield per plant was just over twice as high in intercropping, this being achieved by an increase in heads/plant and seeds/head (Reddy and Willey, 1981).

13.2.3.F. Criteria for selecting component crops for intercropping

Maximising intercropping advantages is a matter of maximising the degree of complementary between the components and minimising intercrop competition. On this basis, intercropping advantages are more likely to occur where the component crops are very different. Although, it has been found that when any sole crop is grown with skipped rows after a regular interval of solid rows, the border rows (rows by the skipped rows) produce more yield per unit length than in the middle rows under solid sole cropping. These skipped rows may be economically utilised with any component crop which does not exert competition for resources but rather utilises more of the left over resources or the resources that are in excess to the main crop for the time being.

The main way that complementarity may occur is when the growth patterns of the component crops differ in time so that the crops make their major demands on resources at different times—thus with a better temporal use of resources. Its importance is indicated by some very substantial yield advantages that have occurred when there have been marked differences in the maturity periods of component crops. Advantages have also occurred where the only difference between component crops has been one of time rather than crop type such as the inter-cropping of early and late potatoes. In line sown radish (which was harvestable from 15 to 75 days after sowing) harvesting alternate plants at one week intervals yielded far more than harvesting once 75 days after sowing. This 'high density and progressive thining' cropping practice widened the resource use more than sowing the crop in the recommended spacing. Three-crop combinations have given greater advantages than two-crop ones and they also emphasise the importance of this type of complementarity because the effects have largely been attributed to a better temporal distribution of crops. In crop cafetaria with vegetable crops, for instance, onion, spinach and radish at 1: 1: 1 proportion in rows or in the same row sown with high density and the progressive thining of spinach and radish, leaves the single crop of onion to grow and utilise full resources for the last one and half month period. During the earlier part of this period (sole onion period), the harvesting of spikes at the green stage induces bulking at a higher rate. Under good management conditions the onion crop gives a full yield and the yields of spinach, radish and onion spikes are the additional or bonus yields.

Baker (1974) suggested that yield advantages were unlikely unless there was at least a 25 per cent difference between two shorter growing periods less than 1.75 of the longest. Baker and Yusuf (1976) estimated that no advantage would occur unless there was approximately a 30-to-40-day maturity difference. The crop plants with indeterminate growth habits such as pigeon-pea, green gram and mustard may produce such advantages because their economic yield mostly depends on current photo-synthesis whereas in determinate plants such as sunflower, economic yield depends mostly on the photosynthesis prior to anthesis (where economic yield depends on the generative organs) and thus only a 30-to-40-day maturity difference will not result in much. Again, when the economic yield is derived from organs other than seeds, for instance, tubers in potato, roots in beets, sweet potato and carrot, leaves in tobacco and tea and bark in jute and sunnhemp only a 30-to-40-day maturity difference will not be sufficient to produce the full yield of the crop, A mature crop of potato requires about a 120-day field duration. The crop starts stolonization at about 30 to 40 days after planting and thereafter tuberization takes place and at about 60 days it becomes ready for harvesting as a truck crop. Though the yield is poor the higher price compensates for the quantitative yield loss. After tuberization the tuber bulking rate (*TBR*) is faster up to about 90 days and thereafter maturity takes place. In the case of rice, panicle initiation (*PI*) takes place about 65 days prior to maturity for harvesting. Prior to panicle initiation the growth and reserves in the plant body determine the panicle density (number of florets per panicle) and even the effectivity of tillers per hill. In jute which is harvestable 100 to 120 days after sowing (*DAS*) the rapid elongation period coincides with about 40 to 80 days when abundant sunshine, warmth, water and nutrient supply are greatly needed. In tobacco, leaf expansion takes place much before maturity.

The shifting of dates of sowing of intercrop components may lead to an escape from such a situation. This is most prevalent under the still planting system (sowing the succeeding crop in the standing crop field much before harvest but it is different from the *paira* crop where only for a fortnight or so, for a brief period they simultaneously form a pair). The growing period of the base crop is suitable for germination and the very slow growth of the intercrop component does not offer any competition and starts vigorous growth practically after the maturity of the base crop (Mazumdar, 1984). Under the avenue cropping system where perennial crops are lopped leaving the stumps intact and lopped materials are used for green manuring during seasonal cropping, there is a full use of resources for crop production both directly and indirectly. Ratooning, the planting of aged seedlings, closer planting, dual culture, high nitrogen application as basal dressing, the selection of crops with large leaves, their rapid expansion rate, a marked increase in the size of successive leaves and early branching or tillering (with sod type or procumbent type in tufted tillers) utilise resources more intensively from the very beginning of the crop duration in the field.

Whilst the emphasis is given on temporal complementarity, the use of *LER* may be misleading because it does not consider the differences in the growing periods of the cropping situations but these differences may have a very important effect on the overall cropping systems. Intercropping must be compared with realistic practical alternatives. Temporal differences are most important. It is better to use a long duration crop as a base crop where the effective growing season is longer than that of an early maturing crop but not long enough for two sequential crops or the agro-climatic situation is not suitable for the growing of the succeeding crop. This is very common in semi-arid and drought-prone zones of India.

A comparison with other crops under relatively good growing conditions does little to indicate the real value of the pigeon-pea (duration exceeding 120 days) crop to a farmer. Because of its hardiness, one of its main advantages is its ability to produce some yield under conditions too harsh for many other crops. This is particularly true where its relatively long growing period enables it to survive and yield on residual soil moisture long after the end of the rainy season (Gooding, 1962; Rachie and Roberts, 1974). Of course other crops such as castor, cotton or cassava, can do this to some extent, but pigeon-pea, being a legume, can also make a valuable contribution to the nitrogen economy of the systems in which it occurs (Willey *et al*, 1980). Pigeon-pea being a predominantly rain fed crop in India, is becoming an important component of various multiple cropping systems in irrigated areas owing to the introduction of short duration and high yielding varieties. The concept of ratoon cropping of early pigeon-pea has been introduced (Ahlawat *et al*, 1985). In the north-eastern part of India *kharif* pigeon-pea has a longer duration than *autumn* sown pigeon-pea, which produces a good yield with a higher harvest index and closer spacing and a higher plant population can be maintained.

Other short duration leguminous intercrops are black gram, green gram and groundnut in long duration crops such as sugar-cane, cotton, castor and chillies. Where early maturing component crops can ensure efficient early use of resources and in the later maturing components of Andhra Pradesh and Tamil Nadu, *kharif* groundnut is normally intercropped with pulses and millets. In Karnataka, intercropping with cotton and chillies is commonly practised besides millets and pulses. In Maharashtra, in addition to these component crops, intercropping with sugar-cane and banana is gaining popularity. Groundnut+castor (2 : 1) for Gujarat, groundnut+hydrid sorghum (3 : 1) for Karnataka, groundnut+sunflower (4 : 2) for Maharashtra, and groundnut+pigeon-pea/black gram (3 : 1) for Tamil Nadu were found to be most profitable. Possibilities are enormous to grow groundnut as an intercrop in the new (non-traditional) areas with cotton, sorghum, chillies, sugar-cane, banana and cassava (Reddy *et al*, 1985). At the All India Coordinated Maize Improvement Project, Sabour Centre (Progress Report 1975-84) during *kharif* (1) maize+groundnut gave the maximum return (of Rs. 6,535/- per ha) fallowed by maize+black gram as compared to maize+soybean or maize alone; (2) intercropping of groundnut or black gram with maize at the inter-row spacing of 90 cm and 70 cm did not show a significant difference; (3) the increasing level of fertiliser had increased the profitability of the above 100 per cent of the recommended dose of maize+100 per cent of the recommended dose of legume was found superior to maize alone; (4) in the case of intercropping with maize, increasing the level of nitrogen — 120 kg/ha) had increased the combined yield of grain and of both grain and groundnut significantly, but the higher level of nitrogen had no impact on legume production; (5) the optimal time for interseeding legume after earthing up in maize, i.e., at 30 DAS increased the yield of maize by 200 kg/ha; (6) legumes sown at 30 DAS of maize and turned as green manure produced 318 kg more yield of wheat when grown in succession; (7) among two legumes the kidney bean was better suited than black gram as a companion crop with maize and as a green manure crop of succeeding wheat; (8) a maximum profit of Rs. 4,011.22 was obtained in maize (GS-2)+Arhar (Bihar) sown in the month of July under rain fed conditions; (9) under irrigated upland maize (GS-2)—potato (kufri Chandramukhi)—onion (Pusa red)—moong (Pusa Baisakhi) gave a maximum profit of Rs. 9,323.01 per ha/yr and this was followed by maize-berseem-maize-moong (Rs. 8,848.04/ha/yr); (10) during the *rabi* season maize alone was found to be superior to radish, Bengal gram, *tori*, linseed and lentil intercrops. The companion cropping of sugar-cane with early potato gave more yield of cane (autumn planted) than the late variety of potato. The ommission of nitrogen at planting or incorporating only one-third of the total recommended dose at planting was found to be the appropriate fertiliser application for the companion cropping of sugar-cane and wheat. The balance may be applied at the harvest of the companion crop (IISR Res. Report, 1982).

In addition to temporal complementarity between component crops, spatial complementarity is possible. In practice it is very difficult to distinguish between temporal and spatial effects as they are often inseparable.

In crop production land, light and carbon dioxide are not available in plenty. Wherever and whenever they are available we are to strive to utilise them to the highest extent to convert them into the primary product of carbohydrate by the interaction of the crop with water, nutrient and pest problems. Donald (1961) emphasised that light could not be regarded as a reservoir from which demands could be made as required: light is instantaneously available and has to be instantaneously intercepted if it is to be used for photosynthesis. Considering spatial light use Willey and Roberts (1976) stressed that given optimum plant populations, sole crops are themselves usually capable of achieving a peak value of light interception which leaves little scope for greater spatial interception by intercrops. Osiru (1974) found that some yield advantages resulted from intercropping sorghum genotypes of different heights, but the maximum increase was only nine per cent. Canopy architecture is an important consideration in spatial distribution and the interception of light which is strongly influenced by the shape, size and angle of leaves and their orientation (planophile or erectophile) (De, 1984). There is also the possibility of combining crops which have different inherent responses to light [C_3, C_4 and CAM type of photosynthesis though plants are not exactly homogenous with respect to C_3 into C_4 pathway. In maize, the plant initially exhibits C_3 photosynthesis which later changes into the C_4 type. In sorghum it is the reverse. In rice the main shoot does not exhibit photorespiration but the subsequent tillers do (Sen, 1985)]. The top of the canopy could consist of a component with a high light requirement (C_4) and the bottom with a component having a low light requirement (C_3). It may also be possible to have components adapted to the qualitative changes in light which occur down the canopy (Allen *et al*, 1975). Growing sciophytes in the interspace of mesophytes such as turmeric or ginger, in pigeon-pea, banana, or papaya utilise the transmitted light passing through the canopy of the tall plants due to the azimuthal movement of sun. A good example of efficient spatial use of light is multi-storeyed cropping (Nelliat *et al*, 1974; Chatterjee and Maiti, 1984) or with agri-horti-sylvi-pastoral system where crops ranging from tall perennial trees to lower growing annuals form different canopy layers. These systems consider stratification (a vertical layering of organisms or environmental conditions within a biotic community) and zonation (the horizontal arrangement of biotic and abiotic factors) comprising crop plants of over-storey, under-storey, transgressive, seedling and herbaceous strata on plains or on moderate slopes. However, in hilly areas with steep slopes, a 'tiered farming system' in which an area towards the foot hills is used for bench terracing for agricultural crops, an area of the middle slopes for horticulture and top for agro-forestry, is an ideal multi-disciplinary approach.

The non-arable areas adjacent to farming areas, forests, confluences, common lands of villages, embankments of rivers, canals and channels, deltas, estuaries, stream banks and steep slopes prone to land slides that have been denuded and/or degraded but are capable of sustaining trees, may be covered with vegetation adopting agro-forestry, social forestry or rural forestry. In these systems, annual crops are grown, mostly in intensive, mixed or intercropping methods, under the perennial forest trees or fruit-cum-timber trees. In some areas agro-forestry is called forest gardening. The multi-storeyed production of different species of economic plants with suitable plant geometry is the main feature of agro-forestry It is a 6 F programme that provides food for man, feed for livestock, fibre for clothing, fuel for village and urban homes, furniture timber and funds for the poor forest villagers and government development programmes. Growing annual crops under the perennial trees reduces weed growth, encourages the growth of perennial crops through periodic soil care and reduces the production cost of the plantation crop. Agro-forestry also decreases soil desiccation during the dry season and minimises soil erosion during the rainy season. The introduction and extension of agro-forestry in the forest belt helps to improve the quality of wild life and the life of the forest villagers and prevents the pilferage of national forest resources and deforestation by unemployed villagers. The development of year-round mixed farming systems in the agro-forestry areas can generate enough employment and energy in the form of fuel and food supplies for the farmers and forest labourers (Hoque, 1984).

Among underground resources plant nutrients and water are the most important factors specially in the rhizosphere. The mutual avoidance of different root systems of different component crops involves the exploitation of a greater total volume of soil. There is evidence of the intermingling of roots resulting in the intensive use of the soil and its components. There is also evidence that a deeper rooting component crop may be forced even deeper by the presence of a shallow rooting component. The water and nutrient status of different soil layers also determines the depths and distribution of roots. It is also evident that crop plants with subterranean fleshy roots or rhizomes or tubers when intercropped with component crops having stronger fibrous roots or suckers that may pierce through these soft organs, reduces the quality of produce.

There is evidence of greater nutrient uptake by intercrops. In general crop plants belonging to the monocot group have a preference for monovalent ions there and those of the dicot group for divalent ions in the soil solution. Therefore, a combination of monocot (cereals) with dicot (pulses, oil-seeds, fibre crops) utilises more nutrients of various types more intimately and simultaneously. However, all crop plants are in need of essential elements with different consumption rates. Each crop component has different peaks for absorption rates of different plant nutrients during its total duration. There may be intercrop competition for nutrients when they are in short supply. Crops such as potato and cauliflower grow with higher levels of plant nutrients and leave a considerable amount unutilised. Intercropping, still planting, *paira* cropping or growing crops in quick succession adopting rotational cropping utilises these residual nutrients. There is evidence of the complementary use of nitrogen in a legume + non-legume combination due to nitrogen fixation by the legume. Nitrogen fixation in legumes is wholly dependent on the activity of co-enzyme nitrogenase which is located within the bacteriodes of the *Rhizobium* microsymbiont with the energy for the reduction being totally derived from certain substances which pass into bacteriodes from the host cell of the nodule. The provision for energy by the host in return for reduced nitrogen from the bacteria links photosynthesis and the nitrogenase activity (symbiosis). There is evidence of nitrogen fixation in the rhizosphere of non-legume crop plants for instance, in *bajra* by *Azospirillum*, sugar-cane by *Azospirillum* and *Bejierinckia* sp. in acid soils and *Azotobactor* sp. in neutral and alkaline soils. Some factor other than the roots associated nitrogen fixation was responsible for the growth increase in inoculated plants— may be due to growth hormones such as indole acetic acid and gibberellins synthesised by the bacteria (Rao and Venkateswarlu, 1982). The main benefits of the *Rhizobium* sp. to soil and the succeeding or component crops occur when old nodules degenerate. Under certain conditions nitrogen, probably in the form of aspartic acid or B-alanin, are secreted into the soil from the actively functioning nodules and there is reason to believe that under a mixed or intercropping system of cereal and legume, the cereal benefits from the association. Microbial nitrogen fixation also stimulates other micro-organisms and exerts an antagonism towards pathogens.

A different temporal effect could occur where nutrients released from one crop, as a result of senescence of plant parts (both roots and shoots), are then made more readily available to another crop.

Better water use may be a common cause of yield advantages in semi-arid and drought-prone areas because such soils are more thirsty than hungry. In such areas the emphasis on matching the crop is given to the soil and water availability and not vice versa. Kassam (1973) reported that in the early stages of growth of a maize crop there was a water surplus. An early intercrop such as millet, would help to utilise this and so increase water use efficiency.

13.2 3.G. Plant population and spatial arrangement

Plant population defines the number of plants per unit area, which determines the size of the area available to the individual plant. The spatial arrangement defines the pattern of distribution of plants over the ground, which determines the shape of the area available to the individual plant. For crops regularly arranged in rows, spatial arrangement can be concisely defined by the rectangularity, which is the ratio of the inter-row spacing to the intra-row spacing (Holliday, 1963). In any cropping method, the geometry of individual plants, hills, stations and clumps and their population per running metre or square metre (centiare) have an important role on the growth and yield of the crops, as they have a regulatory or compensation capacity over the resource availability. With regard to plant number, both the total population (all components) and the component population (each component) have to be dealt with separately. When the population pressure on resources is considered, a single plant of a crop is seldom comparable to a single plant of another crop. This can be overcome by considering the optimum populations of sole crops as comparable. If they are taken as 100, component populations can then be expressed on a simple relative basis, for instance, a simple intercrop having half the sole crop optimum of each of two components is expressed as a 50 : 50 component population. If there is the full sole crop population of one component plus half the sole crop population of another, then it becomes 100 : 50 proportion.

With regard to the spatial arrangement of intercrops, rectangularity will have effects similar to those on sole crops. Then comes the proportional areas allocated to each crop at sowing. The proportional areas may be directly related to component populations. A crop plant in a community shares both the above and the below ground environment. Appropriate spacing between crop plants should be so determined that crop plants do not face an acute competition for light, nutrients, water, air, space and the micro-environment at any time of their duration. Plant spacing depends on the geometry of individual plants in a community and that forms a close in canopy at the maximum spatial coverage in their life (De, 1984). If a 50 : 50 component population is achieved by having equidistant alternate rows, the proportional areas will also be 50 : 50. The other factor of practical consideration is how the proportional areas are arranged with respect to each other. An intercrop which has proportional areas of 50 : 50 can be arranged as (1) alternate plants within the row; (2) alternate rows; (3) alternate double rows. There are different crop combinations and their ratios are also different. A large number of combinations are with two crops, some are with three crops and still others are with more than three crops. The proportion ratios vary from 1 : 1 (50: 50) to 1 : 30 ; 1 : 1: 1 to 2 : 4: 5, 2 : 3 : 5, 2A : 2B : 5C, 1A: 1-5B : 1-5C: 1-5B : 1-5C : 1A etc. of the main and subsidiary crops and vice versa. All these component crops appreciably increase the optimum total population and yield as compared to sole crops. Under situations where there are large temporal differences in growth patterns of the components, population increases are most likely and advantageous. Proportion ratios vary with the magnitude of spatial differences, growth patterns and the nature of component crops and their dealings. For example, during the establishment of a mango orchard when mango saplings are planted at 30×30 feet apart, the interplant space can efficiently be utilised with fruit plants requiring proportionately closer spacing such as litchi, guava, ber and custard apple. The interspace of these fruit plants can be utilised by planting fruit plants requiring further closer spacing and shorter in duration such as, lime, banana, papaya and pineapple and their interspace can be utilised by even closer spaced seasonal crops such as pigeon-pea, cotton, groundnut, brinjal, chillies, chrysanthemum, marigold, tube rose and gladeolus or biennials such as, turmeric and ginger or perennials such as, sugar-cane, napier and Guinea grass.

In other situations the space is allocated to the component crops by altering row arrangements without changing the population of the component crops. Under such a situation a full yield of the main crop with an additional yield of the subsidiary crop are achieved by pairing rows of the main crop. For example, the main crop is paired with 30 cm interrow spacing by reducing the recommended spacing of 45 cm and leaving 60 cm space between two pairs. In this space a subsidiary crop is grown without reducing the yield and population of the main crop. The subsidiary crop yield becomes a bonus over the same total area and with the same principle crop plants may be paired within the row and the interplant space is utilised with some subsidiary crop. For example, after such a pair of colocasia stations a lady's finger or brinjal or chilli plant is grown without reducing the population of the principal crop of colocasia in the row. Additional yield is derived from the subsidiary crop without depriving the yield of the principal crop.

In crops which need a border strip or check basin method of irrigation such as wheat, mustard, pulses and spices ridges are to be made after certain intervals. The crop plants that grow well in the intervening strip or basin, do not grow as well on the ridges. Some other crops tolerant to poor input supply such as mustard or safflower are grown on the ridges of the wheat field. This type of cropping increases the total productive area per unit of cultivable area in a season.

There is a long standing belief that the advantages of inter-cropping may occur only in low fertility, uncertain water supply and poor management conditions probably because of the fact that intercropping has emerged from poorly developed agriculture. Jodha (1976) showed this predominance in his survey of the semi-arid tropics of India and even indicated that as farming inputs become available, farmers tended to move out of inter-cropping, especially where irrigation was one of the inputs. Now farmers are supplied with recently introduced very short duration adoptable cultigens and cultivars, water for potential irrigation, implements, equipment and machinery, chemicals as fertilisers, pesticides, growth regulators and modifiers, biofertilisers and appropriate technologies which incline the farmer's preference towards more intensive cropping under integrated farming with sole cropping than that of mixed or inter-cropping.

The longer time and higher cost of sowing and harvesting of intercrops may be the major reasons for not adopting inter-cropping in advanced agriculture. Low literacy and fabracy are the other foctors. In mechanised and chemicalised farming, intercropping does not find its fair place. However, in kitchen garden, crop cafetarias and in crops which need straight rows for irrigation and earthing up such as potato or brinjal inter-cropping with crops such as, radish and garlic, which are needed in small quantities for the family, is adopted in a small scale. Whereas, in dryland farming and drought-prone areas and in conditions of poor and uncertain management practices farmers prefer mixed cropping to intercropping where row arrangement is a prerequisite, and which is more time consuming. In pre-carious weather conditions, with pre-monsoon short rains covering as much area as possible by seeding determines the fortunes of the farmers. Sometimes, to accomplish the objectives dry sowing of crops such as, millet or sesame is done, which does not allow other crops to be sown together.

Regarding fertiliser application and response component crops may vary greatly. In general, non-legumes respond to a higher dose of nitrogen whereas, legumes respond more to phosphates. For a legume/non-legume intercropping, separate rates of fertilisers may be used in pocket or pit or side-placement as basal dressing which may not be sufficient for the entire crop duration. Such an application method is more time consuming, costly and cumbersome and use-efficiency may be low because of the higher rate of absorption at an early stage of crop growth resulting in vigorous growth and scarcity .during the later period. This results in poor economic yield even though the crop produces a greater biological yield. Even if the response of applied fertiliser is high the net return will be narrowed down because of the higher cost of application. Again, with a full dose of fertilisers for each crop there will be higher total nutrient capacity of the soil and this will generate competition for other factors due to the luxurient growth resulting into dominant and dominated effects and there may be the lodging or higher infestation of insect-pests and pathogens and therefore, the reduction of both quantitative and qualitative yields.

In some areas the sowing times of the component crops differ widely depending on the availability of moisture in the soil. It is possible to obtain a yield advantage provided the component crops have the same field duration and are non-photosensitive. The delayed sowing of the component crop delays its harvesting and thus the total duration is extended. If the delayed sown component crop matures with the early sown component, yield advantage may not occur. Under conditions where the potential growing period is longer than that of either component crop but not long enough for two sequential crops such as in the nursery duration of rice or the same or other component crops, staggered sowing may be a very valuable way of ensuring that some crop is present on the land for the full period of the season of possible growth.

There may be time-of-sowing effects for instance, an earlier sown crop becomes more competitive (winner takes all) than when both are sown simultaneously. Temporal differences are increased by progressively delaying the sowing of one component and the yield advantages are diminished because of very poor yields of the later component and therefore, this condition is better dealt with relay cropping or *paira* cropping than simple intercropping.

Intercropping may be divided into the following four groups (Singh, 1985):

1. *Parallel cropping.* Under this cropping system, two crops are selected which have different growth habits and zero competition between each other and both of them express their full potential, such as black gram or green gram with maize, black gram, green gram or soybean with cotton.

2. *Companion cropping.* In companion cropping the yield of one crop is not affected by the other. In other words, the yield of both the crops is equal to their pure crops. Thus, the standard plant population of both crops is maintained such as mustard, wheat or potato with sugar-cane.

3. *Multi-storeyed cropping.* Growing plants of different heights in the same field at the same time is termed multistoreyed cropping. It is mostly practised in orchards and plantation crops such as Eucalyptus+papaya+berseem, sugarcane+potato+onion (seed crop) or sugar-cane+mustard+potato.

4. *Synergetic cropping.* Here the yields of both crops, grown together are found to be higher than the yields of their pure crops on unit area basis such as sugar-cane+potato.

13.2.3.H. Intercropping versus mixed cropping

The intercrops differ from mixed crops in the following ways:

1. The main object of intercropping is to utilise the space left between two rows of main crops especially during the early growth period of the main crop whereas, the main object of mixed cropping is to get at least one crop under any climatic hazard such as flood, drought or frost conditions.

2. In intercropping, the main emphasis is given to the main crop and the subsidiary crops are not grown at the cost of the main crop. Thus there is no competition between the main and

subsidiary crops whereas in mixed cropping, all crops are given
equal care and there is no main or subsidiary crop. Almost
all the crops compete with one another.

3. Subsidiary crops are of a short duration and in general,
they are harvested much earlier than the main crop whereas,
the crops under mixed cropping are almost of the same duration.

4. Under intercropping both the crops are sown in rows.
The sowing time may be the same or the main crop may be
sown earlier than the subsidiary crop whereas, under mixed
cropping the crops may be broadcast or sown in rows but the
sowing time for all the crops is the same.

13.2.3.1. Intercrop combinations

There may be several combinations between or among the
various component crops. They may be grouped as follows:

1. In relation to the life period:
   a) perennial+annual or permanent+temporary crops;
   b) permanent+perennial+annual crops;
   c) temporary+seasonal crops;
   d) annual+annual crops;
      1) long+short duration crops;
      2) both long or both short duration crops.

2. In relation to the type of crops:
   a) Cereal+pulse;
   b) Cereal+oil-seeds;
   c) Cereal+pulse+oil-seeds;
   d) Cereal+pulse+others;
   e) Cereal (fodder)+pulse (fodder);
   f) Cereal+pulse (fodder);
   g) Pulse (fodder)+oil-seeds or others (fodder);
   h) Cash crop+cereal or pulse or oil-seeds or fodder

3. In relation to morpho-agronomic characters:
   a) high+low crops;
   b) erect+creeping or climbing or twining crops;
   c) deep+shallow rooted crops;
   d) thin+thick foliage crops;
   e) fouling+smothering crops;
   f) multicut+multicut crops;

g) multicut + single cut crops;

h) reaping+picking crops;

i) hardier+ tender crops;

j) tall with sparse+low with dense canopy crops;

k) dicot+monocot crops;

l) tolerant +susceptible crops.

4. In relation to light energy utilisation:

a) C_3+C_4 type of crops;

b) C_4+CAM type of crops;

c) mesophyte+sciophyte;

d) xerophyte+mesophyte;

e) erectophyle+planophyle;

f) deeply lobed leaved+dorsiventral/isobilateral leaved crops.

5. In relation to organic matter and soil tilth:

a) leaf shedding+non-shedding type of crops;

b) avenue+cereal crops;

c) exhaustive+green manuring crops;

d) crops with digging+reaping method of harvesting;

e) crops with higher+lower quantity of decomposable residues;

f) crops with good preparatory tillage+zero or minimal tillage;

g) crops with repeated+zero or minimal interculture.

6. In relation to plant nutrient supply:

a) restorative+exhaustive crops;

b) crops requiring equal dose of manures and fertilisers

7. In relation to water intake and soil erosion:

a) crops requiring equal soil moisture status;

b) crops requiring high+How soil moisture;

c) crops absorbing soil water from surface+deep layers of soil;

d) crops of donor+receiver areas with respect to water harvesting;

e) erosion permitting-f-restricting crops;

f) both the crops are erosion resisting.

8. In relation to incidence of weeds, insect pests, pathogens and parasites:

a) fouling crop+ fcleaning/smother/competitive crops;

b) susceptible+toierant crops;

   c) both susceptible or tolerant crops;

   d) susceptible+trap/brake/guard crops.

  9. In relation to difference in height;

10. In relation to crop rotation;

11. In relation to relative reproductive rate

$$(RRR = \frac{A \text{ harvest}/A \text{ sown}}{B \text{ harvest}/B \text{ sown}} = 1);$$

12. In relation to price equivalent ratio (*PER*);

13. In relation to endurance to climatic hazards;

14. In relation to efficient utilisation of space and time;

15. In relation to synergetic effects;

16. In relation to economy of cropping.

13.2.3.J. Precautions in adopting intercropping

In selecting intercrop components some important points are to be kept in mind. They are:

1. Component crops should not be competitive to each other for factors of growth such as light, nutrients, water, air, space, microbes and the micro-environment;

2. the component crops should not have allelopathic effect;

3. crops with different management requirements should not be grown together as intercrops;

4. crops with similar agro-botanical characters snould not be regarded as intercrop components whereas, crops with similar agro-ecological situations should be taken together;

5. crops which are collateral hosts for important pests and parasites should not be intercropped;

6. crops which have a regeneration capacity or non-dormant seeds and voluntary seeding habit should not be taken as intercrops;

7. crops for seed production (generative yield) should not be intercropped especially with crops having the same maturity period;

8. crops with rough surfaces (with hairs or bristles, such as sunflower) should not be intercropped with tender or soft surface plants (such as jute);

9. crops grown for hybridisation should not be intercropped with crop plants which have genetic combining abilities;

10. crops with repeated and simultaneous pickings for har-

vesting should not be grown together as intercrops;

11. crops that are susceptible to lodging at any stage of growth may be used as intercrops with great care;

12. crops which have climbing and twining habits should be so used that they do not affect the growth and production of other crops of erect habits by their dominating effect.

Under garden and plantation crops some short statured, short duration seasonal crops of food, feed, fibre, fuel, flower and vegetable crops are cultivated during the early years of establishment of permanent trees or shrubs. Under field and fodder crops the most important and common component crops are legumes and non-legumes during the *kharif* and *rabi* seasons. Crops such as *jowar*+pigeon-pea/pea/cow-pea, cotton, groundnut, *bajra*+green gram/black gram/rice bean/cow-pea/ *guar*, pigeon-pea+groundnut, wheat+mustard, wheat+gram/ safflower, pea+*rai*, barley+gram are the most common intercrop components.

Suitable grass-legume mixtures are always desirable because of their complementary functions in providing nutritive, succulent, palatable forage or fodder, fresh or preserved feed for stall feeding or grazing animals. In addition, they are capable of producing much greater quantities of digestible dry matter and protein throughout the growing season than either component. Legumes usually maintain their quality better than grass even at maturity and being rich in protein, enhance the forage value and also enrich the soil with much needed nitrogen. These mixtures also improve the physical, chemical and biological conditions of the soil, check soil erosion, resist the encroachment of weeds and withstand the vagaries of the weather better than pure stands. They also help to check the spread of certain pests and diseases. Some of the important grass-legume intercrop combinations are: maize+cow-pea/green gram/black gram/*kudzu*, *jowar*+ cow-pea/velvet bean/*guar*/field bean/*kudzu*, *bajra*+cow-pea/ velvet bean/field bean/rice bean/tetrakalai/*kudzu*/*guar*, teosinte+ cow-pea/rice bean/field bean/*guar*, oats+*senji*/lucerne/peas, mustard+berseem/lucerne/Stylo, napier+cow-pea/rice bean/ tetrakalai/berseem/lucerne/*kudzu*/*gaur* + mustard/oats/Chinese cabbage/lettuce/maize, para grass+rice bean/Atylosia/Stylo/ Dolichos/Centrosema/lucerne/berseem/cow-pea, Guinea grass+

cow-pea/rice bean/velvet bean/Centrosema/Stylo/berseem+mustard, Setaria+cow-pea/rice bean/field bean/*guar*, Rhodes grass +Centrosema/Atylosia/cow-pea, Sudan grass+cow-pea, marvel grass + rice bean/Stylo, Dinanath grass + cow-pea/rice bean/ tetrakalai/soybean.

Suitable fodder trees for the sylvi-pastoral system are subabool/babul/Albizia/Cassia + anjan grass/marvel grass/Guinea grass/Dinanath grass/napier/Stylo/Atylosia/para grass/rice bean.

13.2.4 Rotational Cropping

Rotational cropping is the growing of different crops in recurring succession on the same land, in distinction to a monocrop system or a haphazard change of crops determined by opportunism or lacking a definite plan. It is one of the means of diversification in cropping in different seasons of the year or in different years.

Few soils are so rich that they can sustain continuous cropping with a single crop with relatively higher yields. To maintain yields from soils, productivity must be maintained. With the application of manures and fertilisers in balanced proportions and required amounts, soil fertility may be maintained but a number of plant nutrients and growth promoting substances would be unutilised or exhausted. In addition to these, there is the possibility that such a programme may encourage certain diseases and insects, and in such an event the continuity may be interrupted and other cropping practices initiated (Tisdale and Nelson, 1963).

There are six stages of development of farming: (1) land used to exhaustion; (2) naked fallow rotation; (3) legume rotation; (4) field grass husbandry; (5) scientific rotation and (6) modern intensive and integrated farming. The first stage is typically subsistence farming, i.e., natural husbandry and shifting cultivation or assertage which tended to use land to exhaustion. Natural husbandry refers to the situation where a harvest is taken from wild or semi-wild plants with a minimum of tending and without ploughing. Similarly, shifting cultivation is practised for a few years in a cleared and burnt forest area and shifted to other places when the productivity of the soil is depleted to a certain level. All these have caused soil exhaustion as the fields were used for cropping and grazing.

In the naked fallow system, a portion of the land is cropped and the other is clean-fallowed. Over the next year or years such use is alternated. Although the productive life of the land is increased, the work involved is enormous. Subsequently, the available land is divided into three fields with only one-third fallowed each year.

A legume rotation is used in which a legume crop is inserted in a sequence of crops in rotation. When a double cropping system is introduced the legume rotation becomes easy and grain-fallow/legume to restore fertility is introduced. In this system a field is divided into numerous small sub-fields (plots). Some fields are grown with grain, some with legume and some are kept clean-fallow for the season or the year.

In the field-grass-husbandry system, long rotation sequences and the grazing (free, regulated, tethered or rotational) of animals is integrated. Crop fields resting between crops are pastured, thus adding manure and sometimes planted to leguminous forages. This is a practice somewhat prevalent in mixed farming areas.

The primary object of growing crops in scientific rotation is the maintenance of fertility of the land and the production of heavy crops without recourse to manuring in an expensive manner. This should not be taken to mean that manuring can be dispensed with or that rotations would replace manuring. Manuring is necessary for maintaining the fertility and raising the productivity, and rotation is an aid in achieving this objective. The crops included in the rotation should be adaptable to the soil, climate, season, inputs available in the locality and the prevalent market conditions. Within broad limits, the choice of crops a farmer will grow are set for him by the principles of ecology and the laws of economics (Pearson, 1973). They should provide food for the family and the farm animals and funds from commercial crops for meeting family expenditure and commitments. The rotation should assist in maintaining the fertility of the land through non-monetary inputs such as, bio-fertilisers, for instance. by including leguminous crops, which add nitrogen to the soil and provide vegetable proteinaceous food for man and beast. The inclusion of cleaning and cover crops such as, sweet potato, potato and barley in rotation is helpful in checking soil erosion and both annual and perennial weed growth. Multicut seasonal fodder crops such as berseem and oats impede weed growth.

The modern intensive and integrated farming is designed to (1) obtain the maximum number of crops per year; (2) conserve available moisture and ensure its maximum utilisation; (3) improve soil fertility for a persistent supply of plant food throughout the duration of the crop; (4) increase physical, chemical, biological, mechanical and managerial capacities of the soil which may be expanded as "Soul of Infinite Life" (Goswami, 1986); (5) check soil erosion and degradation by loss of nutrients; (6) integrate crop growing with animal raising and recycling of their by-products for the production of usable economic yields; (7) check pests and diseases especially by manipulating cultural practices.

When different crops are grown on the land one after another, the yields obtained are invariably greater than when the same crops are grown season after season or year after year. For example, with the use of modern varieties, it is possible to grow *kharif*, *rabi* and summer maizes, groundnuts; *aman*, *boro* and *aus* rices but after a few cycles, the production level depletes considerably and reaches a stable threshold that may be the marginal level and thus places the agriculture at a subsistence level. If some other crop is included in the sequence of continuous cropping with the same crop, the yield of the introduced crop becomes more than the existing level and the crop in immediate succession also yields more than normal if the choice of crop variety is appropriate. Similarly, when sugarcane is grown year after year on the same land, there is a gradual decline in yield even though the crop is supplied with manures and fertilisers, sometimes by warping or allowing the siltation of suspended particles of the flood water. If some other crop, quite different in agro-botanical character and/or agro-ecological requirement such as, pigeon-pea, is inserted in the sequence, the succeeding crop of sugar-cane produces a proportionately higher yield.

The crop rotation may be single-course, two-course, three-course or many course and a rotational cycle may be of one, two, three or four years. A course represents a cropping in the yearly sequence forming a cycle. The rotational intensity of a crop is the number of times the crop recurs in a total rotational cycle. For example, in a three-course rotation in a two years' cycle with the following crops:

maize—potato—green gram   —1st year

maize—mustard—summer vegetable   —2nd year

the rotational intensity of maize is $\frac{2}{1} \times 100 = 200$ per cent whereas, rotational intensity of potato, green gram, mustard and summer vegetables is $\frac{1}{1} \times 100$ per cent each.

The rotational cropping intensity is the total number of crops grown on a land in a rotation cycle and is multiplied by 100 to express it in per cent. Under the above situation 3+3 crops are grown in one cycle (within two years), thus the rotational cropping intensity is $\frac{3+3}{1} \times 100 = 600$ per cent.

The cropping intensity is the total cropped area divided by the net area available for cultivation in a year and multiplied by 100. Under the above rotational cropping if the net land area is one hectare, the cropping intensity is $\frac{3}{1} \times 100 = 300$ per cent. Again, the multiple cropping index (MCI) is the sum of areas planted to different crops and harvested during the year, divided by the total cultivated area. In the above example $MCI = \frac{3}{1} = 3.0$.

In the examples (Table 13.5 : A-C) crops are rotated keeping the lands (blocks) fixed. Again, keeping the crops fixed, cropping in lands (blocks) may be rotated. This situation may be regarded as land rotational cropping (Table 13.5 : D).

Under the above situation all the blocks or fields must have equal productivity and other physical facilities such as irrigation and drainage and approaching facilities, soil type, topographical situation as well as shape and size.

13.2.4.A. Types of rotational cropping

Rotational cropping may be major or minor. In large areas, with fertile, well drained, relatively level lands, major rotation may be practised. Under these situatoins all the crops are given equal importance and are grown rotationally. Examples may

Table 13.5 A:
Single-course rotation (two years' sequence)

Years	Blocks or fields				
	A	B	C	D	E
1st	rice	jute	jowar (fodder)	vegetable	sugar-cane
2nd	jute	jowar (fodder)	vegetable	sugar-cane	rice
3rd	rice	jute	jowar (fodder)	vegetable	sugar-cane
4th	jute	jowar (fodder)	vegetable	sugar-cane	rice

or

Years	Blocks or fields				
	A	B	C	D	E
1st	rice	rice (MV)	rice (MV)	jute	vegetable
2nd	rice (MV)	rice (LIV)	jute	vegetable	rice (MV)
3rd	rice (LIV)	rice (MV)	rice (MV)	jute	vegetable
4th	rice (MV)	rice (LIV)	jute	vegetable	rice (MV)

LIV = Local improved variety; MV = Modern variety.

Table 13.5 B:
Two-course rotation (three years' sequence)

Years and seasons	Blocks or fields				
	A	B	C	D	E
1st* Kharif:	aus rice	aman rice	jute	vegetable	maize+cow-pea (fodder)
Rabi:	Bengal gram	Lathyrus	mustard	potato	vegetable
2nd Kharif:	jute	aman rice	vegetable	maize+cow-pea (fodder)	aus rice
Rabi:	vegetable	Lathyrus	potato	mustard	Bengal gram
3rd Kharif:	maize+cow-pea (fodder)	aman rice	aus rice	jute	vegetable
Rabi:	mustard	Lathyrus	Bengal gram	vegetable	Potato

*Crops in 1st. 2nd and 3rd years will be repeated in respective blocks or fields in 4th, 5th and 6th years

Table 13.5 C:
Three-course rotation (three years' sequence)

Years and seasons		A	B	C	D
			Blocks or fields		
1st*	Kharif :	rice (*MV*)	rice (*MV*)	maize+cow-pea (fodder)	jute
	Rabi :	potato	mustard	vegetable	oats
	summer :	green gram	onion	jute	rice bean
2nd	Kharif :	rice (*MV*)	rice (*MV*)	jute	*aus* rice
	Rabi :	mustard	potato	vegetable	Bengal gram
	summer :	black gram	vegetable	maize+cow-pea (fodder)	sugar-cane
3rd	Kharif :	rice (*MV*)	rice (*MV*)	*aus* rice	sugar-cane
	Rabi :	wheat	Bengal gram	vegetable	sugar-cane
	summer :	rice bean	maize+cow-pea (fodder)	onion	jute

*Crops in 1st, 2nd and 3rd years will be repeated in respective blocks or fields in 4th, 5th and 6th years.

Table 13.5 D:
Three-course rotation (two years' sequence)

Years and seasons		A	B	C	D
			Blocks or fields		
1st	kharif :	rice	rice	maize+ground-nut	jute
	rabi :	potato	mustard	wheat	cauliflower
	summer	vegetable	onion	rice-bean	*boro* rice
2nd	kharif :	rice	maize+groundnut	jute	rice
	rabi :	mustard	wheat	cauliflower	potato
	summer :	onion	rice bean	*boro* rice	vegetable

*Crops in 1st and 2nd years will be repeated in respective blocks or fields in 3rd and 4th years.

be any one among the Table Nos. 5 : A to E Minor rotation is usually practised in small, steep sloped areas subject to climatic and/or biotic hazards. Under this system there may be a *base crop* (the crop occupying the highest percentage of the sown area in the year round sequence of cropping) and others may be *alternative crops* (that may be the substitutes for the base crop in the same season or as the crops which fit in with the rotation in the subsequent season(s) of the year under changing circumstances).

Table 13.5 E:
Two-course land rotational cropping (four years'
sequence) under rain fed conditions

Years seasons	Blocks or fields			
	A	B	C	D
1st Pre-*kharif* :	*aus* rice	rice (*MV*)	vegetable	sunnhemp
Post-*kharif* :	black gram	Bengal gram +rape	barley	sweet potato
2nd Pre-*kharif* :	rice (*MV*)	vegetable	sunnhemp	*aus* rice
Post-*kharif* :	Bengal gram +rape	barley	sweet potato	black gram
3rd Pre-*kharif* :	vegetable	sunnhemp	*aus* rice	rice (*MV*)
Post-*kharif* :	barley	sweet potato	black gram	Bengal gram +rape
4th Pre-*kharif* :	sunnhemp	*aus* rice	rice (*MV*)	vegetable
Post-*kharif* :	sweet potato	black gram	Bengal gram +rape	barley

Table 13.5 F:
Two-course minor rotational cropping (two years' sequence)

Years and seasons	Blocks or field			
	A	B	C	D
1st *kharif* :	maize	maize	groundnut	pigeon-pea
rabi :	wheat	barley	wheat	wheat
2nd *kharif* :	maize	maize	groundnut	pigeon-pea
rabi :	Bengal gram	berseem	barley	mustard

Table 13.5 G:
Three-course minor rotational cropping (two years' sequence)

Years and seasons	Blocks or fields			
	A	B	C	D
rabi :	wheat	wheat	wheat	wheat
1st summer :	green gram	sunflower	vegetable	black gram
kharif :	maize+ cow-pea (fodder)	rice (*MV*)	pigeon- pea	rice (*MV*)
rabi :	wheat	wheat	wheat	wheat
2nd summer :	sunflower	vegetable	black gram	green-gram
kharif :	rice (*MV*)	pigeon-pea	rice (*MV*)	maize+cow- pea (fodder)

In Table 13.5F, maize, groundnut, pigeon-pea and wheat are the base crops and barley, Bengal gram, berseem and mustard are the alternative crops of the respective blocks or fields.

Again, each block having four fields may be adopted with three-course minor rotational cropping (two years' sequence).

Table 13.5 H:
Three-course rotational cropping in a block (two years' sequence)

Years and seasons		Fields			
		A ·	B	C	D
1st	kharif :	jowar+ pigeon-pea	jowar+ pigeon-pea	jowar+ groundnut	jowar+ groundnut
	rabi :	berseem	oats	peas	onion
	summer :	maize	green gram	sunflowea	sesame+green gram
2nd	kharif :	jowar+ pigeon-pea	jowar+ pigeon-pea	jowar groundnut	jowar+ groundnut
	rabi :	wheat	berseem	Bengal gram	wheat
	summer :	guar	sunflower	ragi	rice bean

or

		A	B	C	D
1st	kharif :	jute (seed)	jute (seed)	jute (seed)	jute (seed)
	rabi :	tobacco (seed)	onion (seed)	oats (seed)	berseem (seed)
	summer :	black gram (seed)	green gram (seed)	cow-pea (seed)	guar (seed)
	kharif :	jute (seed)	jute (seed)	jute (seed)	jute (seed)
2nd	rabi :	onion (nursery)	tobacco (nursery)	vegetable (nursery)	potato (seed)
	summer :	cow-pea (seed)	oats (seed)	guar (seed)	green gram (seed)

In India, there are a number of base crops grown in different agro-climatic conditions. Among the *kharif* crops, rice, *jowar*, *bajra*, maize, groundnut and cotton and among the *rabi* crops, wheat, together with barley and oats, gram and very recently mustard are the predominant base crops. In some areas sugar-cane, jute, potato, tobacco, chillies, coriander and onion are considered base crops.

13.2.4.B. Rotational use of varieties

Like rotational cropping the rotational use of varieties also
provides certain yield advantages. Agricultural varieties may
be considered agro-ecotypes where ecotype may be a group
of biotypes especially adapted to a specific environmental
niche. There may be climatic, edaphic and biotic ecotypes
(Wilsie and Shaw, 1954).

It is an established fact that the new plant types with
shorter stratures are not only dwarf above ground, they are
also dwarf below ground. Dwarf varieties have a shallower
root system as compared to medium and tall varieties. Very
short duration varieties have a shorter field duration and they
absorb nutrients from the soil for a shorter period, therefore,
a considerable quantity of nutrients among slowly released
manures and fertilisers are left unutilised by them. In general,
dwarf varieties have a higher harvest index $\left(\dfrac{\text{economic yield}}{\text{biological yield}}\right)$.
The nutrients needed for the structural development of non-
economic parts of these varieties are proportionately less than
that of varieties with poor harvest index values. It has been
found that by rotating varieties or even rotating the seed-
source of the same varieties that are grown in the same field
with the same cultural practices, there are yield advantages for
crops such as, potato and rice. On the other hand, when seeds
of a cultivar of one season (cv. ratna grown as *boro* crop) was
grown in another season (cv. ratna grown as *kharif* crop) or
vice-versa, there was a marginal reduction in yield, probably
due to the seasonal effect. When the seeds of the cultivars were
used for cultivation in the same season (*boro*-harvest-*boro*
grown or *kharif*-harvest-*kharif*-grown) of the next year there
were no such differences. In the case of crops with vegetative
propagules such as potato or sugar-cane records of lower
yields are not uncommon when the home-grown seed materials
are used continuously for a few years. This is more apparent
in ratoon crops.

Another interesting feature is that all the adaptable varieties do not respond equally in every year probably because
of weather effects. Some varieties do better in certain years
while others in other years. It is very difficult to apprehend

even during seeding which variety will perform better in the forthcoming season. Therefore, it is better to grow a number of varieties of the same crop by providing an equal chance to each variety. These varieties should also include short, medium and late and other problem-based varieties so that the failure of one variety is compensated by the best performing one. In addition there is on-farm seed production. Land rotation with these varieties will reduce the cost and trouble of procuring reliable good seed for each crop.

Strict (major) rotational cropping (fixed sequences of cropping) is no longer technically essential or financially desirable for profitable crop production. The factors of changing the sequence of cropping include market price, prevailing weather and soil conditions, availability of inputs in time and their prices. The length of either arable or grass-break or both can be adopted to suit current demand and economic advantages.

13.2.4.C. Prerequisites of rotational cropping

To follow suitable rotational cropping there are some prerequisites which are as follows:

1. The farm should have either equal or a multiple number of blocks (consisting of a number of fields which may be of various sizes and shapes according to topography, nature and extent of draft power, source of irrigation and other related conveniences) or fields of equal size as the number of years for which rotation is adopted.

2. Each block or field area should be relatively levelled and each field of the blocks should have equal physical facilities such as, soil conditions, irrigation and drainage, approach facilities, chemical capacities such as, fertility, pH and biological potential such as, microflora and fauna and the yielding ability of crops.

3. The area devoted to each crop should be uniform. The fields of minor rotation should be such that ploughing and other operations are not being impeded.

4. There should be some sorts of elasticity within the cohesiveness of rotational cropping. Crop substitution especially when one crop fails due to unavoidable circumstances (biotic or abiotic or both) or when the crop becomes more remunera-

tive, the inclusion of a non-traditional crop with respect to season and site, shifting of varieties etc., may be made whenever and wherever feasible.

13.2.4 D. Choice of crops in rotational cropping

In planning a cropping sequence under rotational cropping some important aspects are to be taken into consideration such as, soil type, local weather, crops grown surrounding the fields and blocks, the availability of inputs and their prices, demands and market outlets, soil fertility, the health of the soil and the crops, crop yield and quality, post-harvest opportunities, knowledge and skill of the farmers, crops in succession, specific problem of the agro-climatic situation, cropping methods and cultural systems prevalent, along with other factors such as the socio-economic conditions and traditions as well as the technical considerations of maintaining as well as increasing yield potential per ha per day by increasing use-efficiency of all the resources.

Though soil is the site of crop growth and yield yet, each soil type is suitable only for a group of crops or in other words all types of crops are not adaptable to a particular soil. Successful crop production basically depends on the choice of the right type of crop on the right type of soil. Efficient soil management (tillage, irrigation and drainage, leaching, plant food supply, application of bulky organic manure, checking erosion, correction of soil properties by judicious application of amendments and conditioners, land shaping) helps to modify the basic nature of the soil to some extent to make it suitable as a site for the growth of desirable crops.

The type of soil partly determines the crops that can be grown and the farming system because of (a) the suitability of the crop to the soil type and (b) the effect of the farming system on the soil and its fertility. A fertile soil should have certain properties (a) it should supply, in suitable forms and quantities, plant food, water and air, throughout the growth period of the crop concerned because a soil may be fertile for one crop but not for another; (b) it should be stable and free from harmful concentrations of toxic substances and pests (soil-borne ones); (c) it should be worked freely under a range of weather conditions.

An agricultural soil differs from soils for other uses such as, brick making, pottery and road building because of the presence of an adequate amount of organic matter along with the suitable nature of mineral matter. The organic fraction (both living and non-living) of the soil is important in maintaining and improving the physical, chemical and biological properties of the soil and bears an important role in soil formation and protection. In general, topsoil is richer in organic matter than subsoil because of the presence of plant roots and depopits of plant and animal residues as well as applied manures and organic mulches. This provides for the freer infiltration and aeration of the surface soil and permits it to warm up more quickly than the subsoil. Soil organisms flourish in surface soil with its plentiful supply of food, air and moisture. Soil organic matter acts as a buffer stock of all the essential elements and it helps to receive, retain and release nutrients slowly but steadily compared to purely mineral soils. It also releases antibiotics, vitamins, enzymes, hormones and organic acids which help in crop production. The modifying factors such as, tillage, irrigation and drainage, the application of major plant nutrients through fertilisers, liming, weed management, growing wide spaced row crops, non-leaf-shedding crops, bare fallowing and permitting erosion, aeration and heat dissipation deplete the soil organic matter. On the other hand, growing closed spaced or sod crops, leaf-shedding crops, zero or minimal tillage crops, adopting stubble farming, mulching, warping, pasturing, intercropping, ratooning, green manuring, biofertilising and the application of bulky and concentrated organic manures or tank silts helps to build up suitable soil conditions for crop production.

Each type of soil has its associated crops, for instance, light soils are better for potato, groundnut, tobacco, sweet potato, jute, onion, black gram and green gram; heavy soils are better; for cereals, grasses, sugar-cane, linseed, lentil and lathyrus; wet soil is better for rice, para grass *dhaincha* and jute; dry soil is better for cotton, pulses, oil-seeds and millets.

There is an axiom "climate decides the crop, weather decides the yield". The soil and weather of a locality determine what type of crop is to be grown in different seasons of the year. Rainfall, its intensity, duration and distribution with respect to time and space, determines its effectiveness on crop production and thus partially dictates the selection of crops for a season and sequence. Like rainfall, other elements of weather such as, temperature, light, humidity and wind have a considerable influence on the selection of crops.

Local weather may be modified 'y the presence of shelter belts of trees. In frosty weather, colder dense air flows downhill as it is heavier than the surrounding warmer air. If this colder air meets a barrier like a wall or hedge, it may collect and form a 'frost pocket'. Crops susceptible to frost damage such as, pigeon-pea should be grown in fields where the risk of a frost pocket is minimal. If susceptible crops are grown on the upper slopes of fields which are not broken by barriers, the frost may flow past them and settle on lower ground without damage to the crop. For crops grown during summer specially in areas prone to be parched, borders with wind brakes protect the crops from wind-burn.

The crops grown in surrounding fields or blocks also determine the crop to be chosen. Crops should be such that the cultivation of one crop does not hinder the production of another crop, for example, full grown tall crops such as, jute and sugar-cane provide shade to low crops or any crop at an earlier stage of growth grown to the north and west of it. The cultivation of a water impounded crop such as *boro* rice should not be adjacent to crops susceptible to waterlogging such as, maize or mustard except when they are grown at sufficiently high elevations. The crops that are collateral hosts of insects, pests and pathogens should not be grown closely and simultaneously and in quick succession.

The availability of quality seed, fertiliser, irrigation water, draft power, implement and pesticides, their prices, labour and fund partially regulate the choice of crops in sequence.

Some crops may be needed by the family for use or sold quickly at a remunerative rate in the local market Sometimes seedlings, saplings and stecklings are more in demand particularly when they are raised earliest in the season. If crops are chosen anticipating the demand in the locality, the farmer does not have to wait and worry for their disposal but rather can enjoy immediate and greater benefits from the produce.

The fertility status of a soil has a role in the choice of crops for rotation. The inclusion of fertility restorative crops in crop rotation has immense importance. Less fertile soils are not suitable for crops such as potato, maize and cauliflower. On the other hand if sesame is grown in highly fertile soil, it continues flowering at the apex and the lowermost capsules may ripen and dehisce. The crops susceptible to lodging should not be grown in highly fertile soils.

All crops do not require a number of plant nutrients in the same proportion and when different crops are grown in rotation, the fertility of the soil is utilised more evenly and effectively and the soil is not depleted regularly of particular nutrient elements. Growing cereal crops after legumes economises the nitrogen application of cereals.

Each crop leaves certain parts in the soil. If these residues carry pathogens or invite insect-pests and pathogens or release some toxins, the crops in succession will be affected. Again, in the roots and rhizosphere of legumes and some non-legumes there are some associated organisms which help to build up soil fertility. If host plants of such beneficial organisms are not grown for years, they may not survive and will need fresh inoculation when such crops will be grown.

The higher yielding crops generally remove a higher amount of nutrients but to achieve the targetted higher yield, higher quantities of manures and fertilisers are needed to be applied to the crop. A considerable quantity of these remain as residue which helps the succeeding crop. Succeeding crops should be such that they utilise these residual nutrients more intensively, for instance, onion or *boro* rice after potato, or wet rice after jute. Again if a crop fails or performs poor, due to unfavourable climatic or biotic factors, it is unable to utilise the resources fully. The succeeding crop should be such that it will utilise considerable amounts of residual as well as current resources fully.

With the introduction of very short duration varieties, hybrids and mutants such as, Jyoti (potato), Baisakhi mung (green gram), Bankura local (tori), Sattari (rice), T (black gram), Sathi, Teenpakhia, Kathri (maize), C-152 (cow-pea), CSH 5, SPV-245 (sorghum) or fodder crops, it becomes possible to raise a crop within 60 to 75 days. At the same time with the improvement in potential irrigation facilities it is possible to grow crops at the right time: the second fortnight of June is found to be ideal for sowing/transplanting modern varieties of rice, *jowar*, *bajra*, groundnut, maize, cotton, pigeon-pea, green gram, black gram, cow-pea, rice bean, Dinanath grass and napier as *kharif* crops. Among these very short duration varieties become ready for harvest at the end of August when monsoonal rains still continue. The removal or incorporation of stubble and the preparation of land for the successive crop determines what type of crop should be chosen to grow under these conditions. They should also fit in the post-monsoon. A number of catch crops can be raised successfully without affecting the yield of *rabi* crops. Under dryland situations if rainfall is received towards the end of August or early September, the early sowing of mustard (Baruna), safflower (N-62-8) and gram (C-235) may be done. In irrigated farming, the case for crops harvested early in the winter and summer seasons is similar.

It is also possible to obtain a crop season by circumventing sowing and harvesting times. Now there are many varieties of pigeon-pea such as 20 (105), Pusha-1, DA-6 which on sowing in the post-monsoon period (Sept.-Oct.) produce substantial yield and are harvested simultaneously with pre-monsoon sown pigeon-pea varieties.

In adopting suitable rotational cropping under an intensive cropping pattern in intensive farming systems, farmers should be sound in literacy, numeracy and fabracy. The farmer is the ultimate adopter of any technology suitable for his farm. He is the ultimate decision maker at the field level. Agriculture deals with lives, the factory of converting natural resources comprising both living and non-living entities into productive resources that provide life and longevity in this luminous world. The knowledge and skill of the farmer are of prime importance in acquiring, adapting and adopting newer ideas and also inventing newer ideals appropriate to his farming.

In agribusiness farmers of India enjoy the major share of income through crop yields. Therefore, it is of paramount importance to choose the right crop, in the right season, in an area considering its suitability for the cropping system so that a maximum profit is achieved from the entire system, not from a single crop which is a component of a cropping system and has to be grown in some crop sequence. In selecting a crop for a particular season both pre- and post-season crops are to be considered. Similarly, in selecting crops for a cropping system, crop planning and budgeting including the cropping programme, with cultural systems are to be chalked out beforehand in the cropping scheme. If a farmer wants to grow *boro* rice he has to plan and follow it from the beginning of the *kharif* season. He may grow a crop of early potato or toria after early rice to get the land free in time for planting *boro* rice. If he wants to grow wheat in the plains of West Bengal, he may grow a crop of black gram as *paira* crop after early rice but if he wants to grow potato or rape it is not possible to grow any post-monsoon crop in between rice and rape or potato. Even under such conditions the rice variety should be early-maturing otherwise it is not possible to get the land free for the second crop such as, mustard or potato to be grown in succession. Again, if pototo is grown after wet rice, considerable time and energy is lost in the turn around period for good land preparation where as, if potato is grown after *kharif* maize or groundnut or any other crop requiring well drained aerobic soil conditions, the tillage requirements will be less. If wet land rice is grown after the harvest of jute, maize, *jowar*, cotton or sugar-cane all of which leave stiff stubble, there will be difficulty and delay in field operations. On the other hand, any crop grown after the harvest of crops by digging, may require minimal or zero tillage. Thus crops to be grown in succession are partially dictated by the present crop and its position in the cropping system.

Indian agriculture is said to be a gamble with weather. Successful agriculture is that which integrates all the crop production and protection factors into the ultimate object of quality yield after a certain interval without much deterioration of soil, the site of agriculture. During this interval a series of changes occur in the soil and crops that are distributed in an uncontrolled outdoor area. These areas may be of different toposequences of which the higher ones are prone to erosion and the lower ones are prone to waterlogging and siltation. Selected crops should be suitable to those toposequences. For example, upland rice, maize, *jowar*, pigeon-pea, cow-pea etc., for high lands; transplanted rice, sugar-cane, jute etc., in medium lands and rice, *dhaincha*, jute (capsularis), para grass, Lathyrus etc., in low lands.

In moderate to steep slopes erosion promoting and erosion resisting crops are to be grown in alternate strips.

In precarious rain fed areas with normal to excess rainfall with intermittent drought or early and late onset of monsoons, crops to be selected should have a wide range of tolerance (euryoecious) and be compatible to inter or mixed cropping with crops of various durations, early, medium and late and they may be used for one or more purposes according to prevailing conditions. For example, maize+black gram; if conditions permit for harvesting full season crops, they will be used for grain and seed; if black gram is capable of producing seed but not the maize, then maize will be used as fodder; if none of the crops are capable of producing mature crops, they may be used either as fodder or green manure crops. A combination of a *sure crop* such as, castor, cotton, cassava, pigeon-pea, *bajra*, safflower plus a *risk bound* crop such as, green gram, black gram, cow-pea, groundnut etc., may be grown. Again, the crops which have vegetative organs as the economic yields such as, fodder and onion or the crops with indeterminate growth such as, mustard, cotton, pigeon-pea, peas, grams, beans, and sweet potato that have a longer span for producing economic yield non-synchronously, a certain quantity of yields will be obtained if they can pass about one half of their life period successfully. All these crops, should be thrifty in their use of moisture, and should conserve it for the completion of the life cycle.

In low lying flood-prone areas the crops should be tolerant of water stagnation and flood of varying depths and duration. In salt affected areas, crops should adjust to the varying concentrations of salts with the passage of time. Similar is the case for other factors of soil and climatic conditions for which crops and their varieties must be adaptable.

The selection of crops partially depends upon the cropping methods and cultural systems followed. In the multiple cropping system, crops should be of shorter to intermediate duration, responsive to high inputs, with minimum to zero residual hazard and rather high residual benefits. Sole, inter or mixed cropping, direct seeding or transplanting or crops with specific purposes such as, ware crop, truck crop, seed crop and fodder crop need special measures and on that basis, crops and their varieties are to be chosen for single course or many course rotational cropping in mono to poly-cropping systems in specialised or mixed or integrated farming systems.

13 2.4. E. Criteria to select crops in sequence for rotational cropping

There is a specific need to rotate crops either in immediate sequence or in the same season of the following year or years. There are a number of factors that are to be considered in adopting rotational cropping under various conditions. Some of them are listed below for easy understanding:

1. Non-leguminous crops are to follow leguminous crops and vice-versa such as, green gram—wheat/maize. If preceding crops are legume or non-legume grown as intercrops or mixed crops, the succeeding crop may be legume or non-legume or both.

2. Restorative crops should follow exhaustive or non-restorative crops such as, sesame/linseed/sunnhemp—cow-pea/green gram/black gram/groundnut.

3. Leaf shedding crops should follow non-leaf shedding or less exhaustive crops such as jute/pulses/cotton—wheat/rice.

4. Green manuring crops should be followed by grain crops such, as *dhaincha*—rice, green gram/cow-pea—wheat/maize.

5. Highly fertilised crops should be followed by non-fertilised crops such as, potato/maize—black gram/gourds.

6. Perennial or long duration crops should be followed by seasonal/restorative crops such as, napier/sugar-cane—groundnut/cow-pea/rice bean/green gram.

7. Fodder crops should be followed by field or vegetable crops such as, maize+cow-pea—wheat/potato/cabbage/onion.

8. Multicut crops should be succeeded by the seed crops such as, oats/berseem—green gram/maize.

9. Ratoon crops should be followed by deep rooted restorative crops such as, sugar-cane/*jowar*--pigeon-pea/lucerne/cow-pea.

10. Fouling crops should be followed by cleaning crops such as, *jowar*/maize—potato/groundnut.

11. Cleaning crops should be followed by nursery crops

such as, potato/colocasia/turmeric/beet/carrot—rice nursery/ onion nursery, tobacco nursery, vegetable nursery.

12. Deep rooted crops should be succeeded by shallow rooted crops such as, cotton/castor/pigeon-pea—potato/lentil/ green gram etc.

13. Deep tillage crops should be followed by zero or minimal tillage crops such as, potato/radish/sweet potato/sugar-cane—black gram/green gram/green manuring crops.

14. Dicot crops should be followed by monocot crops such as,   potato/mustard/groundnut/pulses—rice/wheat/sugar-cane/ oats/barley/*jowar* or dicot+fmonocot crops should be followed by dicot+monocot or either dicot or monocot crops.

15. Stiff stubble leaving crops should be followed by minimum intercultivation requiring crops such as, sugar-cane/*jowar*/ cotton/jute/pigeon-pea—fodder crops.

16. The crops of wet (anaerobic) soil should be followed by the crops of dry (aerobic) soil such as, rice-Bengal gram/Lathyrus/pulses/oil-seeds. The tendency to build up difficult-to-control weeds becomes less in such rotation than in continuous wet land rice culture.

17. The crops that are susceptible to soil-borne pests and nathogens should be followed by tolerant/break/trap crops such as, sugar-cane—marigold for pathogenic nematodes (IISR, 1982) tomato/brinjal/tobacco/potato—rice/pulses for *Orobanche, jowar* –castor for *Striga* and berseem—oats for *Cuscuta*.

18. The crops with problematic weeds (weeds that are difficult to distinguish at any one stage of crop, may be seedling or seed stage) should be followed by cleaning crops/multicut crops/other dissimilar crops or varieties such as wheat—wet rice for *Phalaris minor*, berseem—potato/*boro* rice/oats for *Cichorium intybus*, mustard—early potato for *Cleome viscosa*, rice—jute/sugar-cane / vegetables / maize + cow-pea for *Echinochloa causgalli*, jute—multicut fodder/vegetable for *Corchorus acutangulus*.

19. Pasture crops should be followed by fodder or seed crops such as, para grass—maize+cow-pea/cow-pea/rice bean/ tetrakalai/oats for seed.

20. Silage/hay/cleaning crops should bi followed by seed crops such as, maize/groundnut—onion, juie, tobacco, oats/cow-pea/*jowar* for seed crops.

21. Crops with the same symbiotic/associated microbes should be followed by common host crops, such as,

Rhizobium melilote	lucerne, sweet clover, fenugreek
R. trifolli	berseem, Persian clover
R. ieguminosorum	peas, lentil, Lathyrus
R. phaseoli	beans, green gram, pillipesara, black gram
R. lupini	Lupines
R. japonicum	cow-pea, pigeon-pea, *guar*, sunnhemp, Bengal gram, soybean, *kudzu*.

22. Crops in dryland or drought-prone areas should follow fallowing (chemical, bastard-, tilled-, legume-, stubble-, mulch- or clean-fallow) to conserve soil moisture, check the growth of weeds, pathogenic microbes and harmful insects and rodents.

The rotational use of crop varieties, and cultural practices in addition to rotational cropping provides more and assured benefits than that of adopting only crop or land rotation.

13.2.4.F. Objectives and advantages of rotational cropping

1. Comparatively higher yields are obtained without incurring extra expenditure.

2. Better maintenance of soil fertility and fair dealings to soil and its conditions is achieved by encouraging microbial activities for atmospheric nitrogen fixation and mineralisation, avoiding accumulation of toxins, checking erosion and protecting soil from acidity or alkalinity.

3. Better utilisation of both native and applied manures and fertilisers.

4. The soil below the plough layer may be made workable after the inclusion of very deep rooted crops in rotation.

5. Adequate time is available for good land preparation, care and management of crops because of less competition due to less seasonality of produce and managerial capacity becomes more easy and effective.

6. Slow but steady income is gained as well as low expense is incurred which is beneficial to small and marginal farmers.

7. Widens the scope of development of complementary enterprises as cultivation of fodder crops in cereal farming.

8. Introduction of newer crops to non-traditional seasons and belts is possible and under most of the cases, they produce bumper yields for a few years until the nutrients of the soil are depleted and/or associated pests and parasites develop to reduce the yield considerably.

9. Higher possibilities of continued supply of food, feed, fibre, fuel as well as funds for the family and farm animals or to feed market.

10. Greater possibilities of the best uses of residual moisture, manures, micro-organisms and management practices.

11. Greater scope for improving the physical, chemical and biological properties of the soil by the incorporation of more organic matter from various sources and repeated disturbances of the soil.

12. Greater scope to include crops which have a greater demand in the local market for getting comparatively higher advantages.

13. Greater insurance against natural devastation which may not be able to upset the total production. Farmers with a variety of crops have a fair chance of having at least one or two good crops in most seasons which has great importance both to the livestock and to arable land farmers, it helps to spread the risk of poor price of the produce marketed at one time in huge quantities.

14. Helps to maintain the health of the soil and crop by keeping the soil free from troublesome diseases and pests, permitting ecological weed control by the shifting agro-ecological situation, cultivation at different times of the season and year and the inclusion of smother, clean and competitive crops. Rotational cropping is very effective to eradicate monophagous pests and parasites and slowly spreading soil inhabiting organisms.

15. Provides distribution of labour, power and capital use throughout the year in contrast to peak demands during seasons of sowing and harvesting.

13.2.4.G. Disadvantages of rotational cropping

Merits do not come alone. All meritorious activities are mingled with some demerits. Some of the dark sides of rotational cropping may be listed:

1. Growing different crops in rotation needs the preservation and/or procurement of various seeds, special types of implements and equipment and these may not be available in the local market. Special care and management may not be done satisfactorily because of the unavailability of skilled labour. Marketing of the produce may be difficult even though the crops are well adapted to the agro-climatic situation and cropping patterns.

2. It is difficult to shift to crops based on priority and prevailing demands.

3. Repetition of the same rotational cropping for years may nullify the long-term benefits of such cropping and may develop pests including weeds in association with the system thus they will be difficult to manage.

13.2.5 Sequential cropping

Sequential cropping depends on several factors. The most important are the availability of water, the agro-climatic situation of the locality, the farmer's preference and requirements for the family and farm animals and birds and the price of the produce as well as the suitability of raising crops one after another even in turn around periods with constant energy requirements.

Among the field crops, sugar-cane is the most energy intensive, while oil-seeds and legumes are the least energy intensive. Determining the economic optimum level of various energy inputs to achieve the pre-determined yield level may be the more realistic and rational approach (Pal *et al*, 1985). Agronomic productivity, physiological energy output and economic return are fairly high in a complete cereal cropping system where the yield levels often exceeded 9-10 t/ha under optimum input conditions. However, the inclusion of a pulse crop helped to raise the low production of proteins in a complete cereal cropping system. In the wheat growing areas of north and north-east India, the rice-wheat-green gram approach seems to be ideal, while in many parts of Maharashtra the maize-wheat-groundnut or *jowar*-wheat-green gram system gave the best results. The inclusion of a tuber crop such as potato boosted the food output in the alluvial soils. In the south, rice-rice-groundnut, rice-rice-cow-pea system in the lateritic coastal areas and rice-rice-rice system in the red sandy soils in Tamil Nadu are most promising. In eastern India potato-jute-rice, rice-potato-rice are the most energy intensive, highly productive and profitable crop sequences.

Energy relationships in cropping spstems vary and constitute a dependent function of the total energy content of the crop: knitted in a sequence consisting of the yield level, the nature o power used, the soil type, energy input and the agro-climate Efforts should be made to include legumes, oil-seeds and roo crops in the systems for favourable monitoring of energy rela tionships as far as possible (Pal *et al*, 1985).

Climatic conditions of India are such that cropping through out the year can be practised in almost all the arable areas pro vided water is available for crop growth. Indian soils are said to be more thirsty as the availability of soil water for rais ing crops is more scarce than that of other inputs and the supply of quality water in adequate amounts and at the correct times determines the cropping pattern and the succes: or failure of individual crops in sequence.

According to water supply Indian cropping may be broadly grouped into exclusively rain fed, rain fed but with limited irri- gation and irrigated. Rain fed areas may have a regular topo- sequence such as upland, medium land and low land with the supply of pluvial, phreatic and fluxial water respectively and they have a strong influence on crop planning. Areas under fluxial water are situated in between the natural levees and become inundated at different depths for different periods of time and are periodically eroded or silted and formed due to meandering, braiding and course changing of rivers, on the onrush of flood or back waters and are subject to impeded drainage in most situations. Such lands may have shallow or deep water which may be lentic or lotic and also fresh or saline water.

Rain fed upland and medium lands are also subject to various levels of soil moisture (from dry to saturated conditions) with periodic changes according to the rainfall pattern. All these conditions have a great influence on land preparation and crop planning.

Efficient sequential cropping provides input use-efficiency, production, income, employment and land utilisation. Some model sequential croppings may be listed considering water supply systems:

A. *Rain fed areas*:

1. *Pluvial:*
    a) *Regions with 350—1000 mm rainfall*:
    Rice—rape/radish—sesame/fallow.
    Rice—gram+rape/safflower/niger/black gram—fallow.
    Rice+pigeon-pea—*ragi*/black gram/green gram.
    Jute—*toria*/rape/maize/green gram/radish/greens—cow-pea/rice bean/tetrakalai/sesame.
    Maize—gram+rape/black gram/sesame—sunnhemp (green manure).
    Maize+pigeon-pea—fallow.
    Maize+groundnut—toria/rape/black gram/radish/*taramira*—fallow.
    Jowar/ragi—*guar/jowar/bajra*/oats—cow-pea.
    Millets+pigeon-pea—sesame+green gram.
    Groundnut—*jowar/bajra*/cotton/safflower—fallow.
    Castor/sunnhemp/sunflower/soybean—black gram.
    Castor+pigeon-pea—fallow.
    Sunnhemp—radish/black gram/green gram/sesame—fallow.
    Cotton+groundnut/soybean—*bajra*—fallow.
    Maize+cow-pea (fodder)/*jowar*+cow-pea/rice bean—oats/berseem+*sarson*—Dinanath grass/*bajra*.
    Jowar+cow-pea—ratoon—fallow.
    Dinanath—oats/berseem+*sarson/kulti*—black gram/tetrakalai/rice bean.
    Napier/Guinea grass/marvel grass/anjan grass—ratoon+berseem.
    Sweet potato/colocasia/corm/ginger+lady's finger/turmeric/tapioca/lady's finger/brinjal/chillies+colocasia/sweet gourd + amaranth/snake gourd/ridged gourd/bitter gourd/cucumber/ash gourd/pumpkin/*palwal*/little gourd—cow-pea/black gram/*kulti/methi*/spinach/greens—fallow.

Colocasia+lady's finger+brinjal+amaranth—fallow.
Gourds+ginger+lady's finger—fallow.
Turmeric—fallow
Jute (for seed)—fallow.

b) *Regions with 1000 mm and above rainfall*:

Rice—wheat/maize/mustard/coriander/cumin—black gram/green gram/sesame/sesame+green gram.

Rice/jute/sunnhemp—rice/cauliflower/tobacco/tomato/cotton/chillies/radish/sweet potato/gram/peas/beans/onion/coriander/cumin/black cumin/*methi*/*jowar*/*bajra*—summer vegetable/*dhaincha* (green manure)/maize/sunnhemp/groundnut+maize/groundnut.

Sugar-cane+vegetables/greens/pulses—ratoon.

Sugar-cane—rice/*jowar*/maize/cotton/groundnut/sesame.

Rice nursery—rice—*boro* rice/summer vegetable/*dhaincha* (green manure).

Maize+black gram/groundnut—*jowar*/*bajra*/berseem/*kulti*—fallow.

Rice—gram/berseem+*sarson*/oats—maize/tetrakalai/rice bean/teosinte+cow-pea/*jowar*+cow-pea.

Jute—black gram/*kulti*/berseem/Lathyrus/cotton/tobacco/winter vegetable—fallow.

Jowar+cow-pea—ratoon *jowar*—bersem/*senji*/*guar*—Dinanath grass/rice bean.

Kharif vegetables—pigeon-pea—sesame/summer vegetables.

2. *Phreatic*:

Rice/jute—wheat/*kulti*/Lathyrus—sesame + green gram/maize.

Rice—pigeon-pea—sesame/summer vegetable/green manure.

Sugar-cane—ratoon/sesame/maize/*jowar*+cow-pea/*dhaincha* (green manure).

Rice nursery—rice—berseem/*boro* nursery—*boro* rice.

Rice—onion nursery/tobacco nursery—onion/tobacco.

Soybean—sugar-beet/winter vegetable—pulses/fodder.

Napier/para grass/Guinea grass—ratoon+berseem—ratoon.

3. *Fluxial:*
   Rice/jute/sugar-cane/*dhaincha* (for seed)—fallow—summer pulses/oil-seeds/vegetables.

B. *Rain fed but with limited irrigations*:

   Rice/jute/maize/groundnut/cotton/*jowar*—wheat/potato/ sugar-beet/mustard/tobacco/maize/cabbage/cauliflower/ brinjal/chillies/tomato/sweet   potato/coriander/onion/ cumin/black   cumin/*jowar*/*randhuni*/fenugreek/carrot/ knolkhol/turnip/*bajra*/peas/gram/oats/berseem/beans— gourds/maize/sesame/black gram/green gram/soybean colocasia/greens/rice bean/tetrakalai.

   Rice nursery/*dhaincha* (green manure)—rice+Azolla—potato —summer gourds.

   Rice—onion (nursery)/tobacco (nursery)/vegetable (nursery) —black gram + sesame/green gram/cow-pea/maize/ groundnut/cotton.

   Rice/groundnut/sunnhemp—sugar-cane+ potato/sugar-cane +greens—ratoon.

   Sugar-cane—sesame+green gram/sunnhemp/*dhaincha* (green manure).

   Jute (for seed)—onion (for seed) – sunnhemp.

   Napier/Guinea grass/marvel grass/para grass/anjan grass— ratoon+berseem+*sarson*/ratoon+*guar*—ratoon.

   Jowar+cow-pea—*jowar* (ratoon)—berseem/*guar*/oats—tetrakalai/rice bean/Dinanath grass.

C. *Irrigated*:

   Rice—*toria*/potato—*boro* rice—*dhaincha* (green manure).

   Rice—potato—summer vegetables.

   Rice—bersem+*sarson*—colocasia.

   Rice—cabbage/cauliflower—*boro* rice—sunnhemp.

   Maize—potato—groundnut/summer gourds.

   Sugar-cane+green gram + vegetables—sesame/sunnhemp/ green gram/black gram/cow-pea/*jowar*+cow-pea/maize cotton/groundnut/summer gourds.

   Maize+groundnut/groundnut—sugar-cane + potato/sugar- cane+berseem/sugar-cane+wheat.

Jowar—*bajra*—groundnut/black gram/green gram.

Maize/*jowar*+cow-pea/rice   bean--berseem+*sarson*/oats—
Dinanath/*jowar*+ cow-pea/tetrakalai.

Napier/Guinea grass/para grass—ratoon+berseem+*sarson* ratoon+rice bean/ratoon+tetrakalai.

Rice/maize/groundnut—berseem (for seed)/oats (for seed).

The above sequences may be adjusted with favourable conditions of soil and atmosphere for sowing suitable crop varieties and their duration should be befitting to the sequence, purpose of cultivation including seed production, seedling raising and green manuring in specific locations of the land and considering other facilities available to the farmer. Due consideration should also be given to soil health, input use-efficiency, the utilisation of land and other resources, energy output and economic return per unit time, area and input. Other considerations should be on the production of food, feed, fibre, fuel, fund, feeding and feed-back materials for and from factories, flood control and the availability of a free, fair and fresh environment.

CHAPTER 14

Cropping Pattern

A cropping pattern is the yearly sequence and spatial arrangement of crops on a given land area (a district, a part of a state, a state, or a part of the country) or even a farm (an operational holding may be consolidated and enclosed and used for agricultural production).

Individual crops are the components of a given cropping pattern in a yearly sequence. The plot-wise record of crops grown in each season of a year gives a complete picture of the cropping pattern of a particular locality or geographical areas. This determines the multiple cropping index (MCI) under intensive cropping and the relative distribution of crops under extensive cropping.

Depending upon the prevailing atmospheric temperatures and the occurrence of rains three distinct agricultural seasons are generally recognised. They are *kharif* (sowing—mid June to mid August and harvesting—September to December), *rabi* (sowing—October to December and harvesting—February to May) and summer or *zaid* (sowing—February to March and harvesting—May to June). With the introduction of modern short duration varieties a short cropping season—a post-monsoon but pre-*rabi* season is coming up in some eastern parts of India.

India is said to be a subcontinent (the seventh largest country in the world) with a total geographical area of 3287782 km^2 and a total population of 68,51,84,692 (including the estimated population of Assam—according to the 1981 census) which lives in 5,78,842 settlements or villages. The country measures about 3200 km from north to south and about 2700 km from east to west. India has a land frontier 13120 km long and a coastline of 5600 km. Its altitude varies from below the mean sea-level to the highest mountain ranges of the world.

14.1 TOPOGRAPHIC CLASSIFICATION OF ARABLE LAND

According to the topography of land the country may be divided into three distinct regions. They are:

a) *The great mountain zone of Himalayas*: The Himalayas are a series of three almost parallel ranges interspersed with large plateaus and valleys some of which, like the Kashmir and Kulu valleys, are fertile and extensive.

b) *The Indo-Gangetic Plain*: The plain with the Himalayas in the north and the tableland of peninsular India in the south extends over about 2400 km from the western border of Bangladesh to the eastern border of Pakistan. Agriculturally, this is the most important region of India. It is supplied with a perennial supply of water from the Ganges and its tributaries, the Yamuna, the Gomati, the Ghagra and the Gandhak. Portions of Assam are watered by the Brahmaputra and West Bengal by the Damodar, the Mayurakshi, Kangsabati and those of Punjab by the Ravi, Beas, and Sutlej

c) *The southern tableland in the Peninsula*: This tract is marked off from the Indo-Gangetic plains by a mass of hill ranges, varying from 500 to 1330 m in height. The peninsula is flanked in one side by the Eastern Ghats and on the other by the Western Ghats. This region is rocky and uneven. The plateau is traversed by the rivers Narmada and Tapti which fall into the Arabian Sea and the Mahanadi, the Krishna and the Cauvery which drain into the Bay of Bengal.

14.2 MAJOR SOIL GROUPS IN INDIA

The soils in India are classified in accordance with their origins into the following groups:

14.2.1 Black soils

These are also known as black cotton soils or *regur* and are specially suited for cotton cultivation. They occupy the greater part of Maharashtra, the western part of Madhya Pradesh, the Hyderabad region of Andhra Pradesh and some parts of Tamil Nadu. Some dryland farming regions in black soils comprise

Akola, Bellary, Bijapur, Indore, Kovilpatti, Rajkot, Rewa, Solapur and Udaipur. The black soils are deeper, clay to clay-loam and are characterised by low permeability and high water holding capacity. These soils contain a high percentage of clay (more than 30 per cent) and the clay and silt fractions together constitute more than 80 per cent of the soil. The dominance of montmorillonite type of minerals and the consequent cracking property due to enormous shrinkage on drying are other features associated with this group. They have a low infiltration rate, high plasticity and stickiness, low organic matter content, and are generally deficient in nitrogen and phosphate but rich in potash and lime, with a high cation exchange capacity (CEC), a calcareous nature and slightly alkaline reaction. All this poses problems of management practices. Vertisols, when kept fallow during *kharif*, are exposed to soil erosion hazards. On deep black soils with low infiltration rates and on flat lands with less than 0.3 per cent slope, waterlogging is a more serious problem than erosion. The soils of the Deccan Trap region are usually undulating and shallow, medium deep and deep black with certain unfavourable physical characteristics. Small showers of rain do not penetrate to great depth and are of not much use. Most of the soil moisture is held back by the soils due to the high moisture content at wilting point.

The soil character plays a great part in deciding the regional cropping.

14.2.2 Red soils including red loams and yellow earths

These soils comprise practically the whole of Tamil Nadu, Karnataka, South-east Maharashtra, eastern Hyderabad and the tract running along the eastern part of Madhya Pradesh to Chhotanagpur and Orissa Red soils are light textured, shallow to medium in depth and usually underlain by compact subsoil, fairly porous and with a low water holding capacity. Soils are prone to erosion and surface crusting. Alfisols are light and drought prone. Kaolinite is the dominant clay mineral and as such the capacity of retention of cations such as, potassium, calcium and magnesium are very small. Such soils are generally deficient in organic matter, nitrogen, phosphate and in some cases zinc.

Such soils are suitable for various crops both when rain fed and irrigated.

14.2.3 Laterite and lateritic soils

These are found on the summits of the hills of the Deccan Madhya Pradesh, Rajmahal and Eastern Ghats and certain parts of Orissa, West Bengal, Maharashtra, Malabar and Assam. Soils on slopes are susceptible to severe sheet erosion. A regular toposequence is a characteristic feature. The upland soils are more acidic, shallow, lighter and prone to crust formation. On the slopes, they are medium deep, moderately fertile. In low lands soils are deep, clayey, relatively fertile and slightly acidic. Such soils are in general deficient in organic matter, nitrogen, phosphate, calcium and megnesium. Some soils show deficiency of boron and zinc.

Various crops are grown in these soils.

14.2.4 Alluvial soils

Agriculturally such soils are the most important. These soils cover extensive areas in the northern and north-eastern parts including the states of Punjab, Uttar Pradesh, Bihar, West Bengal and parts of Assam and Orissa and also in the coastal regions of southern India including the deltas of the mouths of the rivers. The whole of the Indo-Gangetic plain is comprised in this area. The soils are deficient in nitrogen, phosphate in some locations and humus but generally rich in potassium and lime. These soils are fairly level, deep and light to medium in texture with favourable physical characteristics and good permeability. Small showers are useful and there is the utilisation of most of the water held by the soils due to the low moisture content at wilting point. The range of soil reaction varies from slightly acidic to saline and alkaline. A distinct toposequence of soils exist in some locations. Crusting is a problem in uplands.

On the whole, they are fertile and well-drained. Due to the presence of perennial rivers and the level topography of the land this tract has extensive irrigation systems of canals and tubewells. Various types of crops are grown in this region.

14.2.5 Sierozemic soils

Sandy, loamy sands and sandy loams, very deep loams are found in Jodhpur, Hissar and neighbouring areas. Low soil moisture storage, instability of soil structure and poor fertility are the characteristic features of such soils. High wind velocity leads to severe wind erosion. Soil drifting leads to soil and nutrient losses. Surface crust formation is a problem. They are light, loose and low in bulk density which leads to deep percolation and high evaporation of rain water. These soils are deficient in organic matter and nitrogen.

14.2.6 Sub-montane soils

Such soils are distributed in the dry sub-humid environment of Hoshiarpur in Punjab, and Rakh Diansar in Jammu and Kashmir and the humid tract of Dehradun. The lands are sloping and undulating with a regular toposequence. Soils are light, poor in organic matter and shallow and prone to be eroded. Soil crusting, low moisture retention and low fertility status restricts the choice of crops.

Besides the above soil groups of normal arable soils a large area of land needs to be reclaimed and developed for bringing into cultivation. Such areas are:

a) Lands infested with shrubs and bushes (1.04 million ha)
b) Ravines (3.67 million ha)
c) Waterlogged lands (6.0 million ha)
   (1) Due to surface flooding (3.4 million ha)
   (2) Due to high water-table (2.6 million ha)
d) Lands affected by salinity and alkalinity (0.7 million ha)
e) Riverine lands.

14.3 RAINFALL PATTERN

The rainfall pattern of a locality not only regulates the supply of natural water but also dictates to a greater extent the temperature, relative humidity, transparency of atmosphere and wind velocity.

Temperature and other climatic conditions being favourable throughout the year over most parts of the country, it is possible to grow more than one crop a year provided water is available. In some parts of the country the rainy season is spread over a long enough period to provide ample scope for double cropping. The area under irrigation is not expected to be more than 42 per cent of the total cropped area in 2000 AD. Judicious utilisation of direct rainfall and irrigation water, singly and in combination, will have to be thought of for increasing production.

Farming technology has so advanced that it is possible to increase crop yields even under rain fed conditions, but the choice of crops would have to depend upon the amount and distribution of the prevailing rainfall. Additionally, it will be necessary that the maximum possible quantity of rain water is conserved in ponds and pools situated either within the farm area or elsewhere, in soil profiles and underground storage so that the same could be readily used to save crops in time of stress. Not only in rain fed farming but even under irrigated farming one will have to plan for the most economic and efficient use of water to derive the maximum possible benefit from rainfall reducing dependence on irrigation so that the advantage of availability of water could be extended to as large area as possible. This necessitates a close study of the existing cropping patterns vis-a-vis rainfall patterns, aimed at determining the nature of changes needed in cropping patterns to make the maximum use of rain water. The cropping patterns depend primarily on soils and climatic factors, but as they evolved, also represent the integrated effect of the requirements, local habits and economic factors through time. For increasing production, it is necessary to examine them from the scientific angle and discover possible alternative patterns which have a higher potential. The other relevant factors to study to facilitate an integrated assessment are orography, land use pattern, human and livestock populations, power availability for cultural operations, socio-economic, employment and educational status, soils and climate.

The distribution of rainfall during periods of crop growth has a close relation to the crop and its yield potential. Some limits have been identified. They are:

a) Rainfall of greater than 30 cm per month for at least three consecutive months would be suitable for rice whose water need is very high;

b) 20 to 30 cm per month for not less than three consecutive months would be suitable for crops whose water need is high but less than that of paddy such as maize and black gram;

c) 10 to 20 cm per month for at least three consecutive months is considered suitable for crops requiring much less water such as, *bajra* and small millets;

d) 5 to 10 cm per month is just sufficient for crops which are low water requiring such as, moth (*Phaseolus aconitifolius*) and ephemeral grasses; and

e) rainfall less than 5 cm per month is not of much significance for crop production (NCA, 1976).

The period of moisture availability in dry farming regions determines the length of the cropping season and cropping pattern on the basis of weather-soil-crop environment for each soil group (Randhawa and Singh, 1983).

a) Less than 20 weeks   sole cropping
b) 20 to 30 weeks   intercropping
c) More than 30 weeks   sequence cropping

Cropping patterns as governed by rainfall and soil character under rain fed conditions are suggested (Chatterjee and Maiti, 1984).

Rainfall (mm)	Moisture storage capacity of soil (mm)	Cropping pattern
350—625	100	Single crop in *kharif*
650—750	100	Intercropping can be attempted.
780—900	150	Sequential cropping is possible.
900 and above	200	Sequential cropping is assured.

A threshold of 10 to 15 cm water is required for rapeseed, mustard, *taramira* and sesame; 15 to 20 cm for Bengal gram, lentil, niger and safflower; 20 to 25 cm for barley, linseed, green gram and black gram; and 25 to 30 cm for wheat and oats.

Climatic water availability is independent of the soil in low rainfall conditions (less than 250 mm). In mid rainfall (250 to 500 mm) seasons, the soil type plays a definite role and in relatively high rainfall (500 to 550 mm) seasons again stress-free periods for crop growth are independent of the soil type (Chatterjee and Maiti, 1984).

The southwest monsoon months, June to September form the principal rainy season and outside it there are hardly a few areas and months with rainfall of even 5 to 10 cm per month (except for areas along the west coast, the extreme northeast and some hill areas where the monthly rainfall seldom exceeds 40 to 50 cm) during the post-monsoon months of October to January and the pre-monsoon months (February to May). The year is divided into three periods of four months each. The rainiest month(s) in each season are well known; July and August are usually the rainiest in the June to September period; May is the rainiest month between February and May and October between October and January (NCA, 1976).

Even though the above pattern is said to be normal considering the mean values for 50 years or more, such a pattern of rainfall is quite variable with respect to years and localities.

If an identical distribution of rainfall occurs over two or more adjacent taluks, the distribution is designated as a pattern and the area covered by it is distinguished as a zone and indicated suitably by numbers. Although rainfall is a highly variable element, it is seen that differences in monthly, seasonal and annual amounts are small over short distances. The coefficient of variation of monthly rainfall is as high as 40 to 50 per cent even in the rainiest month of July over most of central, northern and eastern India. In the south, excluding the west coast, the coefficient of variation is 60 to 100 per cent. The variability of weekly or fortnightly rainfall is many times greater. This is a major difficulty in using weekly and fortnightly rainfall as dependable indicators of rainfall distribution. Isolines for the same or nearly the same type of distribution of monthly rainfall can be drawn. These isolines will not necessarily follow the boundaries of taluks but intersect them.

1) where variations are small, isolines follow the taluk boundaries;

2) where essential, isolines intersecting taluk boundaries

have been retained; and

3) when listing taluks included in different rainfall zones, any area less than a quarter of a taluk is omitted.

The study of the cropping structure of the country is the area distributions of crops in the taluks. Numerous crops are grown in a taluk but most of them occupy a very small area as a per cent of the gross cropped area. In general, the crops which occupy ten per cent or more of the gross cropped area of the taluk are taken to constitute the crop distribution of the taluk. In this process, several crops have often got excluded, which may be of local importance such as, vegetables, potato and tobacco. The minimum upper limit of the total areas of crops in a distribution or a pattern is fixed at 70 per cent. The maximum number of crops needed to reach the minimum of 70 per cent area limit, seldom exceeds four to five. Thus, a combination of crops, each occupying not less than ten per cent of the gross cropped area of the taluk and a total of not less than 70 per cent of the gross cropped area of the taluk is considered as constituting a cropping pattern provided it is the same over two or more adjacent taluks (NCA, 1976).

Rain fed agriculture will continue to dominate Indian agriculture, almost as a permanent feature. The distribution of gross irrigated areas shows wide variations among the states and among districts and taluks in the states. This is causing a great disparity among taluks, districts and states with respect to crop production and productivity and thus, socio-economic conditions which reflect the sociui, political and other conditions of the peasantry who depend directly and indirectly on them.

14.4 SUGGESTIONS FOR FUTURE CROPPING PATTERNS

Five steps are visualized in the formulation of cropping. These are:

1) Delineation of rainfall patterns;

2) Identification of the existing cropping pattern;

3) An idea of the area needed for each ciop for national self-sufficiency and the ideal location for its distribution;

4) Juxtaposition of (3) over (2) and studying them together with (1);

5) Consideration of related factors such as soil, irrigation, pressure of population, needs of livestock, proportion of forest vegetation, cropped area and then arriving at the future cropping patterns on the basis of (4) (NCA, 1976).

For substituting the existing less efficient crops by more efficient crops the poins to be considered are:

a) quantity and distribution of rainfall;

b) soil type, altitude and topography;

c) nature, extent and quantities of irrigation water available;

d) existing cropping patterns, needs of the farmers, locality and the country as a whole;

e) available research data, profitability and adaptability to the means of the farmers of the locality;

f) experience and observations of scientists;

g) existing agricultural development programmes;

h) characteristics of the more efficient crops on physical, chemical, biological and technological aspects on soil, crop and its environment for the current season crop and crop seasons in sequence.

India exhibits many varieties of climate. The most important feature of the Indian climate is the alteration of monsoons, the seasonal winds whose direction more or less reverses twice during the year and on which depends the occurrence of rains in diffierent regions. The presence of land of different topographies in the country accounts for variations in the climate due to the differences in the prevailing temperatures at places situated at different elevations. The tropic of cancer divides the country into two halves. The northern part presents a semitropical to temperate climate while the southern one a tropical climate. A long coastline is responsible for great variations in climates of regions near the sea or away from it. The cumulative effect of all these is that Assam and its adjoining areas (Khasi-Jaintia hills of Meghalaya) in the east and Rajputana in the west present a striking contrast of dampness (with more than 10,000 mm of average rainfall and an average potential evaporation of 660 mm per year with less than 10 cm of average rainfall and more than 3,400 mm of potential evaporation per year respectively). Punjab and its adjoining areas have a typical continental climate with extreme summer heat

(45-50°C) and are almost freezing cold in winter, while Travancore (Kerala) in the south has an equal maritime climate with an almost unvarying temperature and higher relative humidity that prevails throughout the year. At places during the rainy season the relative humidity may be 100 per cent while in hot months in upper India it is frequently less than ten per cent. As India is located in the north hemisphere the day length increases from December 22 and reaches a maximum on June 21; thereafter day length decreases and reaches the minimum on December 21. The duration of both days and nights are equal on March 21 and September 21. Bright sunshine hours vary from zero on overcast or rainy days to more than 13 hours in summer months with a high solar incidence of 500 cal cm^2. Some areas regularly have devastating floods with a colossal loss of soil, nutrients and lives with wanton waste of water while some locations have a wind velocity with desiccating winds causing a high rate of evapo-transpiration, wind erosion of soil and nutrients and sometimes the drifting of sand. Extensive climatic hazards such as, weather aberrations with drought, flood (with an intensity of more than 400 mm of rainfall in 24 hours), frost, fog, gale (wind speed exceeding 150 knots), hailstone (upto 10 cm in diameter), cyclone (with a minimum momentary wind speed of 250 knots), snowfall, dewfall (exceeding 30 mm in the month of January), heat and cold waves, dust storms and thunder storms (exceeding wind speed to 100 knots) also take place irregularly in several non-traditional areas causing a serious threat to man and his agriculture. Making allowance for the local variations, the climate of India is essentially of the monsoon tropical type. The wide variation of climate, soil and human behaviour both spatially and temporally, make it possible to grow a large number of crops requiring diverse agro-ecological conditions.

Considering the rainfall pattern—its onset, intensity, duration, distribution and retreat, farming areas are grouped into four categories: rain fed, seasonally dry, drought-prone and dry areas.

A rain fed crop can be grown in areas receiving about 200 mm of rainfall per month for a minimum of two consecutive months and the gap between two effective rainfalls should not exceed seven to ten days. Supplemental or life-saving irrigation by harvesting excess rainfall (pluvial water of the same site) during the period helps to mitigate certain intermittent droughts. Rain fed farming (without irrigation with water from other sources or areas) is practised in areas receiving 400 to 3,000 mm or more rainfall per year. One to three or more crops per year can be grown according to the total precipitation and its evenness of distribution.

Some areas receive more than 1,125 mm rainfall per year but have a distinct dry season when no crop can be raised without irrigation. These areas are known as seasonally dry areas. Cropping in such areas is restricted to the rainy period, and only one or two short duration crops are grown.

In some sub-humid rain fed areas the occurrence of drought (when a seasonal rain fed crop suffers from water stress because of a deficit in rainfall during the period) is a regular feature. Areas subject to 25 per cent deficit of normal rainfall one in five times (20 per cent or more situations) are designated as drought-prone areas. Careful cropping with drought tolerant and ever-usable crop varieties, water harvesting, adoption of location specific agro-techniques are the possible means of arable cropping in addition to the rearing of domestic animals, birds and insects and tree farming under the integrated farming systems.

Such areas that receive annual precipitation ranging from 300 to 1,125 mm under Indian conditions and with precarious and erratic distribution (drought frequency of 80 per cent) with respect to space and time are called dry areas. Ordinary farming in these areas is more or less a gamble. Farming practices tend to maintain water in soil and so combat the lack of rainfall and these practices are adaptable in areas with very good soil conditions but poor pluviometric conditions where irrigation is out of the question. Three types of agriculture are possible in such areas: crop production (one or two crops per year or one in two years according to the rainfall pattern) with millets, pulses, oil-seeds or cotton, animal husbandry (sheep, goat, camel, cows) including the management of grazing land and agro-forestry.

Crop patterns of different regions of a country reveal no sharp division lines between commodity regions. There exists much overlapping in the crops being grown in different regions.

With the introduction of photononsensitive modern varieties suitable for various agro-ecological conditions and the development of potential irrigation facilities as well as newer agro-technologies, it becomes possible to grow different crop varieties in non-traditional areas, non-traditional seasons and even in all the different seasons of the year. Such crop varieties and available technologies (chemical, hydrological, mechanical, biotic and cultural) very often substitute or complement the traditional crop and cropping patterns. Even so the distribution of crops and cropping patterns followed are not at all haphazard and the characteristic crops of a region are profitable to the cultivators who grow them. There are certain definite physical (climate, topography of the land, conditions of the soil, irrigation and drainage facilities), economic (relative demand of the produce and price, wage and hire rates of labour and other unit specific capitals (such capital should be a definite unit of other capitals for its economic utilisation—for instance, if a farmer owns a tractor he should hold a minimum of five ha of land), supply of inputs including credit and their rates and interests, transport facilities through land, water and air and freight rates), social (historical, racial, caste and creed, pressure of population, food habits, socio-economic conditions, knowledge, skill and education of the farmer, research and developments based on local problems, agro-based industries, processing and preservation) political (national pricing and procurement policies, import-export policies, subsidy on inputs), biological (natural foes such as, weeds and pests, pathogens, parasites and predators and friends such as, pollinating insects, rhizobium and predators on pests) and technological (degree of management, skill in increasing efficiency of non-monetary, less-monetary and monetary inputs, degree of utilisation of natural resources such as, land, light, air, water, temperature by intensive cropping or intercropping) factors which govern the production of any particular crop in a locality.

The prevalent cropping systems in a particular locality are the cumulative results of past and present decisions by individuals, communities, governments and their agencies. These decisions are usually based on experience, tradition, expected profit, personal preferences and resources, social and political pressure (Singh, 1980) over and above the agro-climatic situatiou and the farming systems followed.

A broad picture of the major cropping patterns in India can be presented by taking the major crops into consideration. To begin with, the southwesterly monsoon crops (*kharif*) are considered. They include rice, sorghum, *bajra*, maize, *ragi*, groundnut and cotton. Among the post monsoon crops (*rabi*), wheat, sorghum, maize, mustard, potato and gram can be considered to be the base crops. Among pre-monsoon crops (*zaid*), jute, upland (*aus*) rice may be considered as base crops of different cropping patterns. Sugar-cane is a major crop in many regions but as it is a crop of nine to 18 months it does not fit into the yearly sequence of cropping.

To describe the cropping pattern of a region the crop occupying the highest percentage of the sown area in a particular season or year of the region is taken as the base crop and a major emphasis is bestowed upon the same crop. while all other possible alternative crops which are sown in the region either as substitutes for the base crop in the same season, or as the crops which fit in with the rotation in the subsequent season(s). or supplementary crops which are grown in addition to base crops in the same season as intercrops are considered in the pattern. Also these crops have been identified as associating themselves with a particular type of agro-climate, and certain other minor crops with similar requirements are grouped under cne category. For example, wheat, barley and oats are taken as one category. Similarly, the minor millets are grouped with sorghum or *bajra* (Singh, 1980).

Based on similarities within the diversities of several factors stated earlier the country may be divided into five agricultural regions since the most important element of farming in India is the production cf grains on which both man and his domestic animals and birds survive.

1. The rice region extending from the eastern part to include a very large part of the northeastern and the southeastern India, with another strip along the western coast.

2. The wheat region, occupying most of the northern, western and central India.

3. The millet-sorghum region, comprising Rajasthan, Madhya Pradesh and the Deccan Plateau in the centre of the Indian Peninsula.

4. The temperate Himalayan region of Kashmir, Himachal Pradesh and Uttar Pradesh and some adjoining areas. Here potatoes are important as cereal crops (mainly maize and rice) and in other areas the tree fruits form a large part of agricultural production.

5. The plantation-crops region of Assam and the hills of southern India where good quality tea is produced. The high quality coffee is produced in the hills of the western peninsula of India. Rubber is mostly grown in Kerala. Several spices are also grown in Kerala and parts of Karnataka and Tamil Nadu (Singh, 1980).

14.4.1 The Kharif season cropping pattern

A. *The rice-based cropping patterns*

Rice is a facultative hydrophyte and generally semi-aquatic in nature. Rice is essentially a short-day and C_3 type of plant. It requires warm and humid ecological conditions. It is said to be a unique crop as it grows throughout the year on all types of soils including saline and alkaline soils though the best rice soils are heavy soils with high water retention capacity. Rice cultivation depends more on the conditions under which the crop is grown than upon the nature of the soil itself. It grows as a rain fed upland crop in areas receiving 400 to 1,000 mm of rainfall during its field duration. It grows as a rain fed wet land crop (in lowland, deep water and flooded rice) in areas receiving more than 1,000 mm of rainfall. It grows well as an irrigated crop. It is an exclusive crop in some sub-humid to humid areas (Tripura, Manipur, Mizoram) and also in areas receiving supplemental flood irrigation from canals or lift irrigation causing prolonged submergence of the soil during monsoon months.

In different states of India, the base crop rice during *kharif* is alternated with a number of crops at different stages. Some of them may be listed as follows:

States	*Alternate crops to rice*
Andhra Pradesh	cotton, pulses, groundnut, *jowar*, maize, sugar-cane, tobacco, vegetables.

States	Alternative crops to rice
Assam	jute.
Bihar	pulses, jute, maize, sugar-cane, oil-seeds, cotton, vegetables.
Karnataka	*ragi*, *bajra*, cotton, groundnut, *jowar*, maize, vegetables.
Kerala	tapioca.
Madhya Pradesh	small millets, pulses, groundnut.
Maharashtra	*ragi*, pulses, groundnut, sugar-cane, oil-seeds, fruit crops like banana, papaya.
Orissa	jute, pulses, *ragi*, oil-seeds, maize, small millets, vegetables, sugar-cane.
Uttar Pradesh	pulses, groundnut, sugar-cane, *bajra*, *jowar*, maize.
West Bengal	jute, sugar-cane, vegetable, maize, spices.

Rice is also grown as an intercrop with other crops such as, pigeon-pea and orchards particularly in their early stages of establishment. Dual cropping of rice plus fish is also practised in some humid regions.

B. *The Kharif cereals other than rice*

Maize, *jowar* and *bajra* form the main *kharif* cereals, whereas, *ragi* and small millets come next and are grown on a limited area.

1. *The maize-based cropping pattern*:

Maize is a warm weather plant with the C_4-dicarboxylic acid pathway of photosynthesis at later stages. The rate of development of maize from planting to anthesis depends almost entirely on the temperature experienced by the growing point over the whole period.

It is an important crop in the warm temperate regions as well as in subtropical and tropical zones with the water-table well below 80 cm. It is not a satisfactory crop in regions with a semi-arid climate. Maize grows well on a variety of soils but it prefers fertile, deep, well-drained loam soils. Water stagnation for more than four hours is extremely harmful to the crop. With irrigation maize can be grown throughout the year. It can be grown as a rain fed crop in regions receiving only 250 mm to as high as 5,000 mm rainfall per year. It is a successful crop in areas receiving 200 to 300 mm of rainfall per month in at least two consecutive months.

A number of cropping patterns are followed considering maize as the crop during *kharif*. In some areas of Rajasthan, Uttar Pradesh and Bihar maize is grown as an exclusive crop. Different states have different alternative crops of *kharif* with maize as the base crop. Some of them are as follows:

States	Alternate crops to kharif maize
Andhra Pradesh	rice, jowar, oil-seeds.
Bihar	rice, *bajra*, groundnut, sugar-cane, wheat (rabi), *ragi*, pulses.
Gujarat	rice, groundnut, cotton.
Himachal Pradesh	*jowar*, rice, groundnut.
Punjab	groundnut, fodder crops, *bajra*, rice, wheat (*rabi*).
Rajasthan	small millets, pulses, groundnut, wheat (*rabi*).
Uttar Pradesh	rice, groundnut, sugar-cane, *bajra*, *ragi*, pulses, wheat (*rabi*).

Maize is also intercropped with green gram, pigeon-pea, groundnut, kidney bean, black gram and soybean. Maize is also grown as a subsidiary crop with cotton and pigeon-pea.

2. *The kharif jowar-based cropping pattern*:

Jowar ranks third after rice and wheat among cereals in India. *Kharif jowar* is mainly grown where rainfall distribution ranges from 10 to 12 cm per month at least for two to three months of the southwesterly monsoon. It is also grown on a limited scale in sloping areas as a fodder crop in moderate to high rainfall zones. Medium and black soils are predominantly suitable for this crop. *Kharif jowar* is also grown on well drained light soils on a limited scale. It can be grown throughout the year with irrigation. It is grown both for grain and fodder.

Jowar is a warm weather crop initially exhibiting C_4 photosynthesis which later changes to C_3.

A number of cropping patterns are followed in the medium rainfall zones of the country taking *kharif* jowar as the base crop. A number of alternative crops are also adopted in these zones. They are:

States	Alternate crops to kharif jowar
Andhra Pradesh	groundnut, pulses, cotton, oil-seeds.
Gujarat	*bajra*, cotton, groundnut.
Karnataka	cotton, groundnut, *ragi*.
Madhya Pradesh	cotton, wheat, fodder.
Maharashtra	cotton, pulses, groundnut, small millets.
Rajasthan	wheat (*rabi*), cotton, *bajra*, maize.

Jowar is also intercropped with pigeon-pea, soybean and groundnut. A ratoon crop of grain can successfully be harvested after a reasonable yield of succulent fodder.

The bajra-based cropping patterns:

Bajra is the fourth important food crop in India. It is a more drought tolerant crop grown for grain and green fodder. It is preferred in low rainfall areas and on well drained light soils. It has a wide range of adaptations under different day lengths, temperatures and moisture stresses. Under irrigation conditions three to four *bajra* crops can be harvested in a year. A larger area in this crop is covered during the south-westerly monsoon period.

Different *bajra* growing zones follow different cropping patterns. Different states adopt different alternative crops:

States	Alternate crops to kharif bajra
Gujarat	pulses, groundnut, oil-seeds, jowar, cotton, tobacco.
Maharashtra	pulses, wheat, *jowar* (*rabi*), cotton, groundnut.
Rajasthan	pulses, groundnut, oil-seeds, *jowar*.
Uttar Pradesh	maize, rice, wheat (*rabi*).

Bajra is also intercropped with crops such as pigeon-pea, cotton, and groundnut. A good ratoon crop is also harvested both for grain and green fodder.

The groundnut-based cropping patterns

Groundnut is a leguminous vegetable oil-seed crop that can be grown throughout the year with irrigation. The rainfall in the groundnut growing areas ranges from 20 to 30 cm per month in one of the monsoon months and much less in other months (50 to 125 cm per year). In some cases the rainfall is even less than 10 cm per month during the growth of the crop. It is grown on a variety of soil types. However, the crop does best on light loamy soils and in black soils with good drainage.

It is grown as an exclusive crop in some parts of Gujarat. Several alternative crops are grown in different states:

State	Alternate crops to kharif groundnut
Andhra Pradesh	rice in irrigated areas; *bajra, jowar,* small millets, cotton, pulses in rain fed areas.
Gujarat	*bajra, jowar,* cotton pulses.
Karnataka	jowar, cotton, tobacco, sugar-cane, wheat (*rabi*).
Maharashtra	both *kharif* and *rabi, jowar,* small millets.
Tamil Nadu	rice in irrigated areas; *bajra, jowar,* small millets, cotton, pulses in rain fed conditions.

As mixed or intercrops pigeon-pea, blackgram, green gram, *jowar, bajra* and castor are grown with groundnut.

The cotton-based cropping patterns

Cotton is an important seed fibre crop of C_4 photosynthesis. It is a tropical and subtropical (temperature ranges being 21 to 43°C) crop. Most of the cotton areas in the country are under medium to high rainfall (50 to 500 cm) zones. Long days, warm days and cool nights with diurnal variations are conducive to fruiting, good boll and fibre development. It requires a rainfall of 50 cm during crop growth with proper distribution but dry weather during the flowering and fruiting stages is beneficial. Under irrigation two to three crops can be raised in a year. Ratoon crops also provide considerable yield.

Cotton is grown in a variety of soil types yet the soils with good drainage facilities are favoured as it cannot tolerate water-logging. It is grown mainly as a dry crop in the black and medium black soils and as an irrigated or rain fed supplemented by irrigation in alluvial, red and lateritic soils.

It is a wider spaced and proportionately long duration (four to nine months) annual or perennial crop plant. Different alternate crops to cotton are grown in different states for instance, in Andhra Pradesh, Maharashtra, Madhya Pradesh and Karnataka cotton is alternated with *jowar* (both *kharif* and *rabi*), groundnut, small millets, pulses, wheat (*rabi*) under rain fed and rice and sugar-cane under irrigated conditions. In Gujarat rice, tobacco, maize, *jowar*, *bajra* are the alternate crops to cotton.

Mixed or intercropping with maize, *jowar*, sesamum, pulses, vegetables, *ragi* and groundnut are also practised in rain fed cottons in particular.

14.4.2 The *Rabi* Season Cropping Patterns

Among the *rabi* crops, wheat together with barley and oats, *jowar* and gram, potato, rape and mustard are the main base crops among the *rabi* season cropping patterns.

The wheat—and—gram-based cropping patterns

Wheat is the second important food crop in India. Throughout the country spring wheats are grown though they are raised in winter (*rabi*). In the southern hills of Tamil Nadu two crops of wheat are grown in a year. The first crop is raised in October to April and the second crop in May to September.

The cool winters and the hot summers are very conducive for wheat. Both irrigated and rain fed crops are raised in different parts of the country. Wheat is grown in the areas receiving an annual rainfall ranging from 12.5 to 100cm. During the crop duration rainfall ranges between 3-7 cm. As an irrigated crop only one to six irrigations are scheduled according to the availability of water for irrigation. Late crops are often damaged by early hailstorms and desiccating winds.

Wheat is grown under a wide range of soil conditions. The best wheat soils are well drained loams and clayey loams. Based on agro-climalic conditions, the country is broadly divided into five wheat zones: (1) The northwestern plains zone; (2) The northeastern plains zone; (3) The central zone; (4) The peninsula zone and (5) The northern hills zone.

Gram is an important soil improving, nutrient rich pulse crop. It is used as food and feed. Gram is grown in identical agro-climatic conditions of wheat It is grown on more moisture retentive soils but with partial irrigation or in areas receiving low to moderate rainfall with mild cold weather.

The *rabi jowar, bajra,* groundnut, maize, cotton, berseem, mustard, potato, *boro* rice, pulses and oil-seeds, spices (*rabi*) are the alternative crops to wheat, barley and gram. In Maharashtra and Madhya Pradesh heavy black soils are left fallow during *kharif* because of operational difficulties and wheat is grown after the cessation of rains with stored moisture. *Rabi jawar* is an alternative crop in such areas

Wheat is intercropped with mustard, lentil, safflower. Wheat is also grown as a companion crop with autumn planted sugarcane. Gram is mixed or intercropped with linseed, mustard and safflower.

The rabi jowar-based cropping patterns

Rabi jowar is grown exclusively in Maharashtra. They are grown on residual soil moisture and also as a rain fed crop. *Bajra*, pulses, oil-seeds and tobacco are grown as alternate crops. In Karnataka, small millets, groundnut, *bajra*, pulses and oil-seeds form alternate crops to *rabi jowar*. Cotton and tobacco are also grown in some parts of the *rabi jowar* area of Karnataka. In Andhfa Pradesh, short duration pulses, small millets, rice and oil-seeds form the main alternative crops in the *jowar* area.

The potato-based cropping patterns

Potato is a wholesome food crop which can be harvested from 60 days onward in the plains. It is an irrigated tuber crop of temperate to semi-tropical climates. Two successive crops can be raised on the same land where a mild winter continues for four or more months. In the Nilgiri hills, three crops are raised in succession in April, August and January. In the hills of the north the crop is sown in March and April.

Potato flourishes well on light loam, well drained soils and on soils of high natural fertility. Alkaline soil conditions favour the developmemt of scab disease of potato. Incessant rain for more than 24 hours, night temperature not below 10°C and higher relative humidiiy is the inclement weather which causes development of late blight of potato.

The alternative crops for potato are vegetables, mustard, wheat, maize, pulses, groundnut, spices (*rabi*). Potato is also grown as a companion crop with autumn planted sugar-cane. Summer gourds are intercropped with potato by staggering the time of sowing.

The Rape and mustard-based cropping patterns

Rapeseed and mustard are the most important edible vege-table oilseed crops of the tropical as well as of the temperate zones. They require a relatively cool temperature for satisfactory growth. They are crops of the *rabi* season from September-October to February-March. They keep the fields yellow for a considerable period—thus they are regarded as the component of the yellow revolution in agriculture.

They are grown as crops of post-monsoon residual soil moisture, rain fed, rain fed with partial irrigation and as irrigated crops. They prefer well drained deep, light to heavy loam soils.

The alternative crops for rape and mustard are wheat, barley, potato, gram, coriander, maize, *jowar*, berseem and oats.

Rape and mustard are grown mixed or intercropped with wheat, barley, gram, linseed, lentil and coriander. They are also grown as companion crops with autumn planted sugar-cane.

The jute-based cropping patterns

Jute is an important bast (bark) fibre and one of the most important cash crops of eastern India. Jute was said to be a crop of the 80's (i.e. of 80 inch rainfall, 80°F temperature and 80 per cent relative humidity) but now it has become a crop of the 100's (i.e. of 100 cm rainfall and 100 days duration) and is a crop of a warm humid climate. Jute is a rain fed crop. Alternate sunshine and rainy days are most conducive to the growth of jute.

Capsularis varieties can stand waterlogging to some extent during their later stage of growth. *Capsularis* varieties are sown early during February-March while *Olitorius* varieties during April-May with pre-monsoon rains.

The new alluvial soils of good depth and areas receiving new silts from floods is the best for jute. Sandy to clay loams are also good for jute.

Jute is an exclusive crop in some districts. Upland *aus* rice is the main alternative crop to jute. Other alternative crops are green gram, black gram, sesame, maize, sugar-cane and vegetables.

Some other important crops are grown exclusively in certain regions. They are tobacco, chillies, coriander, onion and a number of fodder crops. A number of crop varieties are grown as alternative crops to these crops.

14.4.3 Relative Yield Index (RYI) of Crops

The performance of crops in different parts of the country varies greatly. Instead of using absolute yield values they have been expressed as a per cent of the whole of India. The yield of a crop in a district expressed as a percentage of the whole of India is called the Relative Yield Index (RYI) i.e.

$$RYI = \frac{\text{Mean yield of a crop in a district}}{\text{Mean all India yield of that crop}} \times 100.$$

For example, in Karnataka yields of maize are the best in the country with RYI values ranging from 200 to 400 per cent. With respect to rice only about 30 districts have RYI exceeding

150 per cent and they are confined to Tamil Nadu, Karnataka, Punjab, Jammu and Kashmir and Andhra Pradesh where irrigated rice is grown. Ludhiana has the highest RYI (exceeding 240 per cent) for wheat. On the contrary, Karnataka and Andhra Pradesh have only 13 to 34 per cent RYI for wheat. The RYI for cotton is 340 to 350 per cent in Punjab and Tamil Nadu whereas it is nearly 50 per cent in Maharashtra (NCA 1976).

The All India Co-ordinated Agronomic Research Project (ICAR) clearly noted that although wheat is the most remunerative *rabi* crop in the alluvial soils of north and north-east India, gram, mustard and lentil appeared to have better stability in the alluvial plains under conditions of limited availability of fertiliser and irrigation. In black soil areas *jowar* is the principal *kharif* crop, gram and safflower appeared to be promising *rabi* crop under constraints of fertiliser use and irrigation. In regions of low rainfall with *bajra* as the principal *kharif* crop, gram and wheat are the most productive crops in desert or red sandy soils. In these areas in the arid western plains as well as in semi-arid lava plateaus, gram appears to be more suitable for marginal farmers. In the latrite soils of Bhubaneswar in region V, groundnut may be the most remunerative *rabi* crop following *kharif* rice or *jowar*.

Research on intercropping revealed that *jowar*+pigeon-pea, maize+pigeon-pea and *jowar*+gram in black soils (Vertisols) and *bajra*+groundnut and pigeon-pea+groundnut in red soils (Alfisols) are profitable intercropping systems. In the Bellary region of Karnataka, intercropping sugar-cane with onion yielded excellent results. Intercropping castor with green gram S.K. Nagar (Gujurat), *jowar* with green gram at Parbhani (Maharashtra) and *jowar* with groundnut at Rahuri (Maharashtra) are some of the successful intercropping systems developed by the respective centres in castor growing areas.

Mixed cropping of wheat+mustard in the alluvial soils at Kalyani and in the black soils of Navasari, wheat+gram in the black soils of Jabalpur and Parbhani and groundnut+*jowar* at Rahuri are some of the successful mixed cropping practices.

In the cafetaria system for the drylands black gram gave better results than pigeon-pea or *jowar* in the drylands of Rajasthan. Even delayed sowing did not seriously affect the yield of gram under dryland conditions as compared to *jowar*.

Appendix I

LIST OF FIELD CROPS

Botanical Name	Common Name
Abelmoschus esculentus (L) Moench	Lady's finger, *Okra Bhindi*
Agave sisalana Perrine	Sisal
Allium cepa L.	Onion
A. sativum L.	Garlic
A. Porrum L.	Leek
Amaranthus spp.	Amaranth
A. tricolor L.	*Lalsag*
A. blitum L.	*Sada Notey*
Amorphophallus campanulatus Blume Ex Dcne.	Elephant's foot yam, Corm.
Arachis hypogaea L.	Groundnut, peanut
Asparagus officinalis L.	Asparagus
Atylosia scarabaeoides (L.) Benth	Atylosia
Avena sativa L.	Oats
Basella alba L.	Climbing spinach, *poi*
Benincasa hispida	Ash gourd
Beto vulgaris L.	Sugar beet
Boehmenia nivea (L.) Gaud.	Ramie
Brachiaria mutica Stapf.	Para grass
Brassica alba Boiss.	White mustard
B. campestris L.	Rape
B. campestris var *rapa*	Turnip
B. caulorapa L.	Knalkhol
B. juncea (L.) Czern and coss	Brown mustard
B. nigra Koch.	Black mustard
B. oleracea var *botrytis* L.	Cauliflower
B. oleracea var *capitata* L.	Cabbage
B. pekinensis (Lour) Rupr.	Chinese cabbage

Botanical Name	Common Name
Cajanus cajan (L.) Millsp.	Pigeon-pea, Red gram, Arhar, Tur
Capsicum annum L.	Chillies, Red pepper
C. frutescens L.	Chillis
Carthamus tinctorius L.	Safflower
Carum roxburghianum	*Randhuni*
Cenchrus cillaris L.	Anjan or Buflel grass
Centrosema pubescens Benth.	Centrosema
Chloria gayana Kunth.	Rhodes grass
Chrysopogon fulvus	*Dhaula*
Cicer arietinum L.	Bengal gram, Gram, Chick pea
Citrullus vulgaris Schred.	Watermelon
C. vulgaris var. *fistulosus*	Round gourd, *Tinda*
Coccinia indica Wight and Arn.	*Kundru*
Colocasia esculenta (L.) Schott.	Arum
Corchorus capsularis L.	Capsularis (bitter) jute
C. olitorius L.	Olitorius (sweet) jute
Coriandrum sutivum L.	Coriander
Crotalaria juncea L.	Sunnhemp, Bombay hemp, Banaras hemp
Cucumis melo L.	Musk-melon
C. melo var *utilissium* Dulhie & Fuller	Long melon, *Kakri*
C. melo var *reticulatus*	Sweet melon, *Kharbuza*
C. melo var *momordica* Duthie & Fuller	Snapmelon, *Phoot*
C. sativus Linn.	Cucumber
Cucurbita maxima Ducn.	Winter squash
C. moschata Duch ex poir	Pumpkin
C. pepo L.	Summer squash
Cuminum cyminum	Cumin, *Jiro*
Curcuma longa L.	Turmeric
Cyamopsis tetragonoloba	*Guar*, Cluster bean
Cymbopogon flexuosus Stapf.	Lemon grass
C. martinii var *motia* Stapf.	Palmarosa oil grass
C. winterianus Jowitt.	Citronella grass
Cynodon dactylon (L.) Pers.	Bermuda grass

Botanical Name	Common Name
C. plectostachyum (Sebum) Pilger.	Bermuda grass
Dactylis glomerata L.	Orchard grass
Daucus carota L.	Carrot
Dichanthium annulatum (Forsk.) Stapf.	Marvel grass
Dioscorea alata L.	Yam
Dolichos biflorus L.	Horse gram
D. lablab. L.	Lablab bean
D. lablab var lignasus	Field bean, carpet legume
Echinochloa frumentacea (Roxb.) Link.	Barnyard millet
Eleusine coracana Gaertn.	Fipger millet, *Ragi.*
Eruca sativa Mill.	*Taramira*
Euchlaena mexicana Scbrad.	Teosinte
Foeniculum vulgare Mill.	Fennel, Anise
Glycine max Merr.	Soybean
Gossypium arboreum L.	Cotton (Old world)
G. barbadense L.	Cotton (New world)
G. herbaceum L.	Cotton (Old world)
G. hirmtum L.	Cotton (New wond)
Guizotia abyssinica L. f. cass.	Niger
Helianthus annuus L.	Sunflower
Heteropogon contortus (L.) Beauv. ex Roem and Scbult.	Spear grass
Hibiscus cannabinus L.	Roselle
H. sabdariffa var altissima	*Mesta*, Kenaf.
Hordeum vulgare L.	Barley
Indigofera tinctoria L.	Indigo
Ipomoea batatus (L.) Poir.	Sweet potato
Lactuca sativa L.	Lettuce
Lagenaria siceraria (Mol) Standi.	Bottle gourd
Lathyrus sativus L.	Chickling-pea, Lathyrus, *Khesari*
Lens esculenta Moench.	Lentil
Linum usitatissimum L.	Linseed, Flax.
Luffa acutangula Roxb.	Ridge gourd
L. cylindriea (L) Roem.	Sponge gourd
Lycopersicon escultntum Mill.	Tomato

Botanical Name	Common Name
Manihot esculenta Crantz	Tapioca, cassawa
Medicago sativa L.	Lucerne, Alfalfa
Melilotus parviflora Desv.	*Senji*, Indian clover
Mentha sp.	Mint, Podina
Momordica charantia L.	Bitter gourd
Nicotiana rustica	Tobacco
N. tabacum L.	Tobacco
Oryza sativa L.	Asian rice
O. glaberrima Steud.	African rice
Panicum antidotale Retz.	Blue panicum
P. maximum Jacq.	Guinea grass
P. miliaceum L.	Common millet
P. miliare Lamk.	Little millet, Cheena
P. repens. L.	Torpedo grass
Paspalum dilatatum Poir.	Dallis grass
P. notatum Fluegge	Bahia grass
P. scorbiculatum L.	Kodo millet
Pennisetum Pedicellatum Trin.	Dinanath gra ss
P. purpureum Schumach	Napier grass, Elephant grass
P. typhoides (Burm f.) S & H	Pearl millet, *Bajra*
Phaseolus aconitifolius Jacq.	Moth Bean, *Meth, Matki*
P. aureus Roxb.	Green gram
P. calcaratus	Rice bean
P. lunatus L.	Double bean
P. mungo Roxb.	Black gram, *Urd, Mash*
P tetraploid	Tetrakalai
P. trilobus (L.) W. T. Ait.	Pillipesara
P. vulgaris L.	French bean
Piper nigrum L.	Black pepper
Pisum sativum Walh.	Peas
Portulaca oleracea L.	Portulaca
Pueraria hirsuta	Kudzu vine
P. phaseoloides (Roxb.) Benth	Kudzvuvine
Raphanus sativus L.	Radish
Ricinus communis L.	Castor
Sacharum officinarum L.	Sugar-cane

Botanical Name	Common Name
Sehima nervosum	Sain
Sesamum indicum L.	Sesame, *sesamum*, *Gingelly*, *Til*
Sesbania aculeata Poir	*Dhaincha*
Setaria italica (L) Beauv.	Italian millet, Kaun
Solanum melongena L.	Brinjal, Egg plant
S. tuberosum L.	Potato
Sorghum bicolor Moench.	*Jowar*, Sorghum
S. vulgare Pers.	*Jowar*, Sorghum
S. sudanense (Piper) Stapf.	Sudan grass
Spinacia oleracea L.	Spinach
Stylosanthes gracilis	Stylo
S. hamata Taub.	Stylo
S. humilis	Stylo
Trachyspermum ammi	*Jowan*
Trichosanthes anguina	Snake gourd, *Chichinga*
T. dioica Roxb.	Pointed gourd, *palwal*
Trifolium alexandrinum	Berseem, Egyptian clover
Trigonella foenumgraecum	*Methi*, Fenugreek
Triticum aestivum L.	Wheat
T. dicoccum Shubl.	Wheat
T. durum Desf.	Wheat
Vicia faba L.	Broad bean
Vigna sinensis	Cow-pea
Zea mays L.	Maize
Zingiber officinale Rosc.	Ginger

Appendix II

LIST OF SOME IMPORTANT WEEDS

Botanical Name	Common Name	Family
Abutilon indicum (L.) Sweet	(Petari)	Malvaceae
Acalypha indica L.	(Muktajhuri)	Euphorbiaceae
Achyranthes aspera L.	(Apang)	Amaranthaceae
Aeschynomene aspera L.	(Shola)	Papilionaceae
A. indica L.	(Shola)	,,
Ageratum conyzoides L.	(Sarhand/ Chagrona)	Asteraceae
Alhagi camelorum Fisch.	(Jawasa)	Papilionaceae
Allium vinedle L.	(Jangli rasoon)	Alliaceae
Amaranthus spinosus L.	(Kanta note)	Amaranthaceae
A. viridis L.	(Jangli note)	,,
Ammania baccifera L.	(Dadmari)	Lathyraceae
A. multiflora Roxb.	,,	,,
Anagallis arvensis L.	(Krishnanil)	Primulaceae
Andrographis paniculata Nees.	(Kalmegh)	Acanthaceae
Argemone mexicana L.	(Satyanasi/ Sealkanta)	Papaveraceae
Asphodelus tenuifolius Gav.	(Jangli piaz)	Liliaceae
Avena fatua L.	(Jangli jai)	Poaceae
A ludoviciana Dur.	,,	,,
Azolla pinnata R. Br.	(Pana)	Salviniaceae
Blumea lacera Burm f. Dc.	(Kukursonga)	Asteraceae
Boerhavia diffusa L.	(Santb)	Nyataginaceae
Calotropis gigantes (Willd.) Dry and ex. W.T. Ait	(Akand)	Asclepiadaceae

Botanical Name	Common Name	Family
Celosia argentea L.	(Morogphul)	Amaranthaceae
Centelia asiatica (L.) Urb.	(Thalkuri)	Apiaceae
Centipeda minima (L.) A Br. & Aschers	(Hanchiphal)	Asteracae
Chara zeylanica Willd.	(Sewla)	Characeae
Chenopodium album E.	(Bathua)	Chenopodiaceae
C. murale L.	,,	,,
Chrysopogon aciculatus Reti. Trin.	(Cborkanta)	Poaceae
Cladophora sp.	(Sewla)	Cladophoraceae
Colocasia antiquorum Schott	(Ghet kachu)	Araceae
Commelina benghalensis L.	(Kansira)	Commelinaceae
Convolvulus arvensis L.	(Hirankhuri)	Convolvulaceae
Corchorus acutangulus Lam.	(Jangli pat)	Tiliaceae
C. fascicularis Lam.	,,	,,
Croton bonplandianum Bail.	(Bonmircha)	Euphorbiaceac
C. sparsiflorus Morong.	,,	,,
Cuscuta reftexa Roxb.	(A mar bel)	Convolvulaceae
Cymbopogon citratus (Dc.) Stapf.	(Dhannantari)	Poaceae
Cynodon dactylon (L.) Pers.	(Durba)	,,
Cyperus difformis L.	(Mutha)	Cyperaceae
C. rotundus L.	,,	,,
Dactyloctenium aegyptium [L.] Richt	(Makra)	Poaceae
Datura stramonium L.	(Dhutura)	Solanaceae
Daucus carota L.	(Jangli gajar)	Apiaceae
Dichanthium annulatum (Forsk.) Stapf.	(Ghas)	Poaceae
Digit aria sanguinalis (L.) Scop.	(Suina ghas)	Poaeeae
Dopatrium junceum (Roxb.) Buch.-Ham ex Benth.	(Jabri)	Scrophalaria-ceae
Echinochloa eolonum (L.) Link	(Shyama ghas)	Poaceae
E. crus-galli (L.) Beauv.	(Jangli dhan)	,,
Eclipta alba (L.) Hasak.	(Kesut)	Asteraceae

Botanical Name	Common Name	Family
Eichhornia crassipes (Mart.) Solms	(Kachuripana)	Pontederiaceae
Eleusine indica (L.) Gaertn.	(Sawan)	Asteraceae
Eragrostis cynosuroides (Retz.) Beauv.	(Kus)	Poaceae
Eriocaulon sieboldianum Sieb et Zuce	(Chata phuli)	Eriocaulaceae
Euphorbia hirta L.	(Dudhi)	Euphorbiaceae
Fimbristylis miliacea (L.) Vahl.	(Jawane)	Cyperaceae
Fumaria parviflora Lam.	(Gajri)	Fumariaceae
Gnaphalium indicum L	(Jangli palang)	Asteraceae
G. purpureum L.	,,	,,
Gomphrena celosioides Mart.		Amaranthaceae
Heliotropiutn indicum L.	(Hatisur)	Boraginaceae
Hydriila verticillata (L.f.) Royle	(Jhanji)	Hydrocharita-ceae
Hydrolea zeylanica (L) Vahl.	(Nila)	Hydrophylla-ceae
Imperata cylindrica (L.) Beauu.	(Ulu ghas)	Poaceae
Inula indica L.	(Haldi)	Asteraceae
Ipomoea carnea Jacq.	(Bera kalke)	Convolvulaceae
Kyllinga erecta Schum	(Mutha)	Cyperaceae
Lantana camara L.	(Lantana)	Verbenaceae
Lathyrus aphaca L.	(Jangli mator)	Papilionaceae
Lolium temulentum L.		Poaceae
Ludwigia adscendens (L.) Hara	(Jalkalmi)	Onagraceae
L parviflora Roxb.	(Labangaphul)	,,
L. octovahis (Jacq.) Raven	,,	,,
Martynia annua L.	(Baghnakh)	Martyniaceae
Medicago denticulata Willd.	(Jangli rijhaka)	Papilionaceae
Melilotus alba Desr	(Methi/Senji)	,,
M. indica (L.) All	,,	,,
Mimosa pudica L.	(Lajyabati)	Mimosaceae
Mucuna pruriens (L.) Dc.	(Alkusi)	Papilionaceae
Nicotiana trigonophylla Dun.	(Jangli Tamak)	Solanaceae

Botanical Name	Common Name	Family
Nitella hyalina (Dc.) Ag.	(Sewla)	Characeae
Ocimum sanctum L.	(Tulsi)	Lamiaceae
Opuntia dillenii Haw.	(Pheniraonsa)	Cactaceae
Orobanche indica Buch-Ham.	(Baniabao)	Orobancha-ceae
Oxalis repens Thunb.	(Amrul)	Oxalidaceae
Paederia foetida L.	(Gandhal)	Rubiaceae
Panicum repens L.	(Ghas)	Poaceae
Parthenium hysterophorus L.	(Partha bish)	Asteraceae
Paspalum distichum L.	(Ghas)	Poaceae
Phalaris minor Retz.	(Phala ghas)	,,
Phragmites karka (Retz.) Trin.	(Gubnal)	,,
Phyllanthus niruri L	(Hazardana)	Euphorbiaceae
P. simplex Retz.		,,
Physalis minima L.	(Boa tepary)	Solanaceae
Pistia stratiotes L.	(Pana)	Araceae
Portulaca oleracea L.	(Nuniasak)	Portulacaceae
Pteridium aquilinum (L.) Kuhn.		Dennstaedtia-ceae
Ranunculus aqualilis L.		Ranunculaceae
Rullia prostrata Poir.	(Phatphati)	Acanthaceae
Rumex acetosella L.	(Tok palang)	Polygonaceae
Saccharum munjo	(Sar)	Poaceae
S. spontaneum L.	(Kash)	,,
Sagittaria sagitifolia L.	(Tirmukhi)	Alismataceae
Salsola foetida Delie		Chenopodia-ceae
Scirpus supinus L.	(Chatchati)	Cyperaceae
Sida rhombifolia L.	(Berala)	Malvaceae
Solanum nigrum L	(Makoy)	Solanaceae
Sorghum halepense (L) Pers.	(Johnson ghas)	Poaceae
Spergula arvensis L.	(Satganthia)	Caryophylla-ceae
Sphaeranthus indicus L.	(Chagalnadi)	Asteraceae
Sphenoclea zeylanica Gaertn.	(Hansi)	Sphenoclea-ceae

Botanical Name	Common Name	Family
Striga lutea Lour.	(Agiya)	Scrophuleria-ceae
Suaeda maritima	.	Chenopodia-ceae
Tephrosia purpurea (L.) Pers.	(Bon nil)	Papilionaceae
Tragia involucrata Linn.	(Bichuti)	Euphorbiaceae
Trapa natans L.	(Panifal)	Trapaceae
Trianthema monogyna L.	(Patharchata)	Aizoaceae
T. portulacastrum L.	(Punarnava)	
Tribulus terrestris L.	(Gokbru)	Zygopbylla-ceae
Tridax procumbens L.	(Phanaphuli)	Asteraceae
Typha angustata Bory & Chaub.	(Hogla)	Typhaceae
T. elephantina Roxb.	,,	,,
Vallisneria spiralis L.	(Patasewla)	Hydrocharita-ceae
Vernonla cinerea (L.) Less.	(Jhurjhuri)	Asteraceae
Vicia sativa L.	(Entke)	Papilionaceae
Xanthium strumarium L.	(Bichu)	Asteraceae
Zizyphus rotundifolia Lam.	(Jbarber)	Rhamnaceae
Zornia diphylla (L.) Pers.		Papilionaceae

Selected Questions

1. a) Give a brief outline of the past and present forms of agriculture.
   b) How is agriculture related with the basic sciences ?
   c) What is agriculture ? How does agriculture play an important role in the socio-economic conditions of the Indian people ?
   d) How is agriculture gaining importance in global development ?
   e) What are the roles of agriculture in the regional economy ?

2. a) What is agronomy ? How is agronomy related with other disciplines of agriculture ?
   b) What are the basic principles of agronomy ?
   c) How does agronomy influence increasing crop production ?

3. a) What are the principles of classifying crops ?
   b) What are the different methods of classifying crops ?
   c) What is meant by the agrarian classification of crops ? Cite with examples eight important classes under this classification.
   d) Illustrate with examples the taxonomic classification of field crops.
   e) How are crops grouped under agronomic classification ?
   f) Differentiate between:
      1) Cropping system and farming system;
      2) Commercial and economic classification of crops;
      3) Catch crop and cash crop;
      4) Restorative and exhaustive crop;
      5) Smother and cover crop;
      6) Nurse and guard crop;
      7) Parallel and *paira* crop;

8) Ware and truck crop;
9) Cleaning and soiling crop;
10) Sole and cole crop;
11) Silage and sod crop;
12) Companion and contingent crop;
13) Trap and turf crop;
14) Spices and beverage crop;
15) Cereal and industrial crop;
16) Legume and pulse crop;
17) Avenue and strip crop:
18) Contour and riparian crop;
19) Plant and ratoon crop;
20) Seed and grain crop.

4. a) What is a seed ? What are the characteristics of a good seed ?

   b) What are the different types of pure seeds ? What are the implications of sowing impure seeds ?

   c) Why is the improved sowing quality of seeds desirable? How can a farmer procure quality seeds ?

   d) What is meant by the real value of a seed ?

   e) What is seed dormancy ? How can seed dormancy help in plant survival ? Why does a farmer prefer seed dormancy under certain circumstances ?

   f) What are the different types of seed dormancy ? What are the mechanisms of seed dormancy ?

   g) What are the endogenous chemicals responsible for seed dormancy ? What are the different forms of seed dormancy ?

   h) What is meant by seed viability ? How does it differ from seed dormancy ? How does a seed become non-viable ?

   i) Define seed treatment. What are the objectives of seed treatment ?

   j) Differentiate between:
      1) Germination and emergence;
      2) Hypogeal and epigeal germination;
      3) Embryo and endosperm;
      4) Hybrid and composite seed;
      5) Stratification and scarification;
      6) Microbiotic and mesobiotic seed;

7) Induced and enforced dormancy;
8) Certified seed and breeder seed.

5. a) What are the common methods of sowing seeds ?
   b) Explain the merits and demerits of broadcasting and line sowing of crop seeds.
   c) What are the merits and demerits of dibbling seeds ?
   d) What are the different methods of drilling seeds ?
   e) What are the factors responsible for variations in depth and density of sowing crop plants ?
   f) What is the ideal field condition for sowing seeds ?
   g) How are the seed rates determined for direct seeding and transplanting ?
   h) What are the characteristics of a good seedling ?
   i) What are the considerations in selecting a site for a nursery bed ?
   j) What should be the areas of different types of rice nursery beds as compared to the transplanted area ?
   k) What are the advantages of transplanting over direct seeding ?
   l) What are the steps to be taken in preparing a wet nursery ?
   m) What are the precautions to be taken during the pre- and post-transplanting periods to get maximum recovery of cole crop seedlings ?
   n) How can seedling recovery be improved from a wet bed nursery ?
   o) What are the care and management practices to be followed in raising seedlings in the nursery ?
   p) Describe the method of hand transplanting.
   q) Differentiate between:
      1) Planting and transplanting;
      2) Rose end and heel end;
      3) Seed-bed and nursery bed;
      4) Unidirectional and bi-directional sowing;
      5) Inoculation and incubation of seed materials.

6. a) Define soil. What are the major components of soil ? How are soils formed ? What are the transported soils?
   b) What is parent material ? What are the major factors of weathering ? How do soils develop ?
   c) Define soil profile. What is solum? What are the differen

ces in soil profiles of arable lands, forest lands, eroded lands and recent flood plains ? What is horizonation ? How does it differ from one location to other ?

d) What are the important physical properties of the soil responsible for its productivity ? What is an organic soil ? Where do organic soils form ?

e) What are the different types of soil separates? What are the specific roles played by them ?

f) What are the important minerals constituting the soil ? What are the properties of three main clay minerals ?

g) What is soil colloid ? What are the properties of soil colloids ?

h) Define soil texture. What are the different textural classes? What is particle density? How is it related with porosity ?

i) What is pore space ? What are the roles played by pore spaces in soil ?

j) What is plasticity ? How is it related to the workability of soils ?

k) What are the factors influencing soil temperature ? What are the important factors governed by soil temperature ?

l) What are the effects of soil aeration ? How can soil aeration be improved ?

m) What are the different forms of soil water ? What is available water ? What is wilting point ?

n) What are the major soil structures ? What are the influences of soil structure ? How are soil structures disturbed in arable lands ? How can soil structure be improved ?

o) What are the principal minerals found in mineral soils? What are the important mineral nutrients present in the soil ?

p) Define ion exchange. What are the seats of ion exchage ? What are the effects of ion exchang ?

q) What is CEC ? What is base saturation ? What are the effects of a base-unsaturated soil ?

r) Define soil pH. What are the different soil classes based on soil reaction ? What are the effects of soil pH ?

s) What are the roles played by organic matter in the soil ? What are the factors responsible for organic matter content in the soil ?

t) What are the different types of soil organisms that inhabit the soil ?

u) What are the land capability classes ?

v) What are the different orders in modern classification systems of soils ? How are they related with earlier classification system ?

w) Briefly describe tho major soil groups of India.

x) What are problem soils ? What is an acid soil ? How does a soil become acidic ? How can it be reclaimed ?

y) What is an alkali soil ? What are the principal causes of soil salinity and alkalinity ? How can alkaline soils be reclaimed ?

z) How can a soil be managed for its sustained use ?

7. a) What is tillage ? How does it differ from tilth ?

b) What are the merits and demerits of tillage ?

c) What are the different types of tillage ?

d) What is meant by a zero tillage system ?

e) What is primary tillage ? What are the prerequisites of primary tillage ?

f) What is preparatory tillage ? What are the various factors influencing preparatory tillage ?

g) What is secondary tillage ? What are the different secondary tillage implements ?

h) What is intertillage ? What are the benefits of inter-tillage ?

i) What are the merits and demerits of post-harvest tillage ?

j) What is fallow tillage ? How does it differ from mulch tillage ?

k) What are the benefits of levelling during preparatory tillage ?

l) What are the advantages and disadvantages of puddling the rice field before transplanting ?

8. a) What is a plough ? Differentiate between indigenous and improved ploughs.

b) What are the different methods of ploughing ?

c) What are the prerequisites for a ploughing operation ?

d) What are the qualities of a good ploughing operation ?

e) What is harrowing ? What are the advantages of harrowing ?

f) What is intercultivation ? What are the objectives of intercultivation ? What are the implements used for such operations ?

g) Explain how preparatory tillage differs with crop variety and soil type ?

h) What should be the preparatory tillage for an upland nursery bed ?

i) What are the benefits of earthing up to rhizomatous and tuber crops ?

j) Differentiate between:

1) Intercultivation and intercropping;

2) Dry and wet land tillage;

3) Harrow and cultivator;

4) Furrow slice and furrow sole;

5) Splitting and gathering;

6) Soil turning and soil stirring plough;

7) Rolling and sliding type of plough;

8) Tyne and blade;

9) Push and pull type tillage implement;

10) Scrapper and roller.

9. a) What is meant by the essential elements of plants ? What are the essential elements and how are they classified ?

b) What are the functions of N and K and what are the symptoms produced by plants when B and Zn are deficient ?

c) What are the forms of nutrients absorbed by the plants? How are the nutrients absorbed by the plants ?

d) How are the plant nutrients lost from the soil ? List the sources of plant nutrients in the soil.

e) What is meant by soil fertility. How does it differ from soil productivity ?

f) What are the different sources of soil organic matter ? What is green manuring ? What are the roles played by organic matter in the soil ?

g) What is meant by commercial fertilisers? How are they classified ?

h)   What are the different methods of nutrient application to the soil ? What are the other methods of application of nutrients ?

i)   What are the principles governing the selection of the proper time and method of application of manures and fertilisers ?

j)   How are the doses of nutrients decided ?

k)   What is meant by integrated nutrient management ?

10. a)   What is meant by water management ? Why is water management necessary in successful crop husbandry ?

b)   How does water move in the plant system ? What do you mean by soil-plant-environment continuum ?

c)   Justify the statement "Irrigation is an important practice for higher production".

d)   What is irrigation efficiency ? How can it be improved? What should be the characteristics of an irrigable land and soil ?

e)   What are the problems associated with irrigated agriculture ?

f)   How does water move in the soil ? How does soil moisture deplete in the crop field ?

g)   What are the soil moisture conditions determining the schedule of irrigation ? How does water retention and release capacity vary with soils ?

h)   What are the different sources of irrigation water ? How can these sources be exploited for irrigation throughout the year?

i)   What is meant by the combined use of water for irrigation ? How can it be adopted ?

j)   What are the different types of water conveyances ? What are the benefits of lining earthen conveyances?

k)   What are the factors determining the type of layout of farm fields for irrigation ?

l)   What is meant by quality water for irrigation ? How is it possible to use water of less satisfactory quality ?

m)   What is meant by organic irrigation ? How is it practised ?

n)   What are the basic principles of irrigation ?

o)   What are the factors to be considered in scheduling

irrigation ?   What is meant by the critical stages of irrigation ? How do they differ from crop to crop ?

p) What are the different systems of irrigation ? What are the different methods of surface irrigation ?   What are the crops suitable for adopting such methods ?   What are the merits and demerits of these methods ?

q) What is sprinkler irrigation ?   How can it be adopted ?

r) What are the different water lifting devices ?   What are the capacities of each such device ?

s) What are the different types of drainage systems ? Why is a good drainage system desirable ?   How can drainage and irrigation channels be combined together in fields ?

11. a) Define dry farming.   What is drought ?   What are the different forms of drought ?

b) Define aridity index and moisture index. What are the drought and chronic drought areas ?

c) What are the roles played by the soil condition in dry farming ?

d) What are the types of agriculture possible in dry farming areas ?

e) Discuss the importance of dry farming in India.

f) Give a brief account of the rainfall pattern in India.

g) How are dry farming zones divided and distributed in India ?

h) What are the aberrant weather conditions faced by the farmers in rain fed farming areas in India ?

i) Define MAI, OMAI, AAR and WUE.

j) What are the major constraints associated with dryland farming areas?

k) What should be the characteristics of the crop varieties adoptable under dry farming conditions ?

l) Enlist some important crop varieties of cereals, pulses and oil-seeds suitable for *kharif* and *rabi* cropping.

m What are the major management aspects to be taken into consideration for dry farming ?

n) What are the effects of water stress on crop plants ?

o) What are the suitable cropping systems in dry farming zones ?

p) What are the steps to be taken to improve production and productivity in dry farming zones ?

q) What do you mean by water harvesting ? Suggest appropriate methods and specify the conditions for water harvesting *in situ* and interplots.

r) What are the different agro-chemicals used in dry farming ?

s) What are the land management practices needed to improve water use-efficiency in dry farming ?

t) Give a brief outline of the nutrient management and yield stability in dry farming.

u) What are the effects of weeds and other crop pests in dry farming ?

v) Write short notes on:
1) Antitranspirants; 2) Sealants; 3) Effective rainfall; 4) Water use-efficiency; 5) Drought hardening; 6) Stress injury; 7) Soil mulch; 8) Xerophyte.

w) Briefly discuss:
1) Mixed cropping as a means of survival in dryland farming; 2) Moisture conservation practices for dryland farming regions; 3) Adjusting the planting pattern to limited moisture supply; 4) Procedure to increase run off from the catchment area of a water reservoir.

12. a) Define weeds. Classify the weeds.

b) What are the characteristics of weeds ? How do weeds perpetuate under field conditions ?

c) What are the harmful effects of weeds in crop fields ?

d) Why are weeds not desirable in surface water courses and water bodies ?

e) What are the beneficial effects of weeds ?

f) How can weeds be utilised for economic purposes ?

g) What is a crop-weed association ? What are the effects of crop-weed competition ?

h) What is meant by weed management ? What the basic principles of weed management ?

i) What are the different methods of weed management ? Why is the eradication of some weeds necessary ? How can preventive methods be effective under field conditions ?

j) What are the different methods of weed control ? Classify herbicides based on methods of application and mechanisms of action.

k) How can biological agents be used for controlling weeds ? What is meant by integrated weed management ?

l) Briefly discuss:
1) Weeds are the guardians of the scil
2) Dissemination of weeds
3) Selectivity is a unique property of herbicides
4) Adaptation of weeds
5) Allelopathic effects of weeds
6) Problematic weeds,

m) Write short notes on:
1) Pre-emergence application of herbicides
2) Aquatic weeds
3) Facultative weeds
4) Competitive crops
5) Parasitic weeds.

13. a) What are the potentials and limits of multiple cropping ?

b) What are the advantages and disadvantages of intercropping ?

c) What are the criteria in selecting suitable intercrop components ?

d) What are the different competitive relationships between intercrops ?

e) What is LER ? How is LER calculated ?

f) What is PER ?

g) What is multi-storeyed cropping ? How can it be adopted under field crop cultivation ?

h) What is crop rotation ? What are the benefits of crop rotation in multiple cropping ? Give an example of a four year rotation practice of a farm located far from a city.

i) What is rotational cropping intensity ? How does it differ from cropping intensity ?

j) What is minor rotational cropping ? How does it differ from major rotational cropping ?

k) What is land rotational cropping ? What are the prerequisites for adopting land rotational cropping ?

l) What are the benefits of rotational use of varieties ?

m What are the important considerations made in selecting crops in sequence in rotational cropping ?

n) What is meant by sequential cropping ? Cite some examples for different pluviometric conditions.

4. a) What is a cropping pattern ?

b) Give a short account of the rainfall pattern in India.

c) How is the cropping pattern governed by rainfall and soil characters under rain fed conditions?

d) What is the crop potential in dry farming zones ?

e) What is meant by a drought prone area ? What are the potential cropping patterns in such areas ?

f) What are the major cropping patterns prevalent in India ?

g) Differentiate between:
   1) Base crop and substitute crop;
   2) Traditional crop and non-traditional crop;
   3) Supplementary and augmenting crop;
   4) Principal crop and alternate crop.

h) What are the suitable cropping patterns with respect to the period of moisture availability in low rainfall areas ?

i) What are the suitable cropping patterns in irrigated areas ?

References

Agarwala, S.C. and C.P. Sharma, 1976. Plant Nutrients—Their Functions and Uptake. [In] Soil Fertility-Theory and Practice. *Ed*. J.S. Kanwar, ICAR 7-64.

Ahlawat, I.P.S., Masood Ali, M. Pal and A. Singh, 1985. Research Needs and Directions on Pigeonpea based Cropping Systems. Abs. of Papers, Sympo. on 'Cropping Systems' Indian Soc. of Argon. CSSRI, Karnal, 3– 5 April, 8—9.

Aiyer, A.K.Y.N.. 1949. Mixed Cropping in India. Indian J. of Agri. Sci. 19: 439-543.

All India Coordinated Maize Improvement Project, Sabour Centre. Progress Report, 1975 – 84. ICAR, Sabour.

All India Coordinated Rice Improvement Project, 1984. Technology for Increasing Rice Production. Extension Pub. No. 12, April, 1984. AICRIP, ICAR. 21.

Allen, L.H., T.R. Sinclair and E.R. Lemon, 1975. Radiation and Micro-climate Relationship in Multiple Cropping Systems. [IN] Multiple Cropping Sympo. (Proceedings), American Soc. of Agronomy Annual Meeting, Knoxville, Tennessee. 24-29th May, 191-200.

Amen, R.D. 1968. A Model of Seed Dormancy. Bot. Rev. 34: 1—31.

Anderson, J.D., 1970. Physiological and Biochemical Differences in Deteriorating Barley Seed. Crop. Sci. 10: 36—39.

Arnon, D.I., 1954. [In] Trace *Elements in Plant Physiology. *Ed*. T. Wallace. Chronica Botanica, Waltham, Massachusetts. 31 – 39.

Arnon, I., 1972. Mixed Cropping. [In] Crop Production in Dry Regions. London, England, Leonard Hill. Vol. 1: 475-476.

Arnon, I., 1975. Physiological Principles in Dryland Crop Production. [In] Physiological Aspects of Dryland Farming. *Ed*. U.S. Gupta. Oxford & IBH Pub. Co. 3—145.

Baeumer, K. and W.A.P. Bakermans, 1973. Zero Tillage. [In] Advances in Agronomy. *Ed.* N.C. Braday. Academic Press. Vol. 25: 75-123.

Bailey, L.H., 1950. The Standard Cyclopedia of Horticulture. Vol. III. Macmillan, New York.

Baker, E.F.I., 1974. Research on Mixed Cropping with Cereals in Nigerian Farming Systems —a System for Improvement. [In] International Workshop on Farming Systems (Proceedings) ICRISAT, Hyderabad, 18-21 Nov. 287-301.

Baker, E.F.I. and Y. Yusuf, 1976. Research with Mixed Crops at the Institute for Agricultural Research, Samaru, Nigeria. [In] Symposium on Intercropping in Semi-Arid Areas (Proceedings) Morogoro, Tanzania, 10—12 May.

Biddulph, O., 1959. Translocation of Inorganic Solutes. [In] Plant Physiology. *Ed.* F.C. Steward. Aca. Press. N.Y.

Biswas, G.C., P.C. Mitra and B.N. Chatterjee, 1965. Productivity of Potato-Jute-rice Sequence under Different Fertilizer Management. Abs. of Papers. Sympo. on Cropping Systems. Indian Soc. of Agronomy. CSSRI, Karnal. 3-5 April. 20.

Brady, N.C., 1974. The Nature and Properties of Soils. A College Text Book of Edaphology. Eurasia Pub. House (Pvt) Ltd., New Delhi.

Buol, S.W., F.D. Hole and R.J. Mc. Cracken, 1980. Soil Genesis and Classification. The Iowa State Univ. Press. Ames.

Central Arid Zone Research Institute, 1982. Annual Report, CAZRI, 1982, Jodhpur, ICAR.

Chatterjee, B.N. and S. Maiti, 1984. Cropping System (Theory & Practice). Oxford & IBH Pub. Co.

Crafts, A S. and W.W. Robbins, 1973. Weed Control: A Text Book and Manual. Tata McGraw-Hill Pub. Co. Ltd.

Currie, J.A., 1973. The Seed-soil System. [In] Seed Ecology. *Ed.* W. Heydeker. The Butterworth Group. England. 463-480.

Curtis, H.J., 1963. Biological Mechanisms underlying the Ageing Process. Science. 141, 686.

Dastane, N.G., 1980. Water Management in Crop Production. [In] Handbook of Agriculture, ICAR. 158-202.

De, G.C., 1979. Produce more Rice by Adequate Water Management. Farmer and Parliament XIV (3). 17.

De, G.C., 1980a Effect of Row Orientation and Different Proportions of Sesame and Blackgram under Intercrop Situation. Madras Agric. J. 67 (10) 695-697.

De, G.C., 1980b. Harmful Effects of Weeds. Sabujsona 3(16) & (17).

De, G.C., 1981. Characteristics of Submerged Rice Soils. Science Reporter. CSIR. March-April, 1981. 159-60.

De, G.C., 1984. Growth Analysis and Production of Field Crops (a mimeograph). Palli Siksha Bhavana, Visva-Bharati.

De, G.C., 1986. The Concept of Yield Advantages and Price Equivalent Ratio (PER) in Sole and Intercropping Situations. Abs. of Papers. Sympo. on "Alternate Fmg. Systems". Indian Soc. of Agron. ICAR. 30.

De, G C. and S.K. Mukhopadhyay, 1984. Weed flora in Sub-humid Lateritic Belt of West Bengal, India. Indian J. Weed Sci. 16(2) 101-15.

De, G.C , S.K. Mukhopadhyay and D. Jayaram, 1986. Extent of Germinable Weed-seed Population in Different Field Soils under Different Land Situations. Abs. of Papers. 73rd Session of the Indian Sci. Cong. 38.

De, R., 1983. Some Aspects of Nutrient Supply in Drylands. [In] Dryland Agril. Res. in India—Thrust in the eighties. Ed. R.P. Singh and V.B. Subramanian. AICRP for Dryland Agric. Hyderabad. 61-67.

De, R. and S.P. Singh, 1983. Cropping System under Adverse Weather Conditions. Ferti. News. 28(7): 49-58.

De Datta, S.K., 1981. Principles and Practices of Rice Production. John Wiley & Sons. New York.

Devlin, R.M., 1975. Plant Physiology. Affiliated East-West Press Pvt. Ltd.

Donald, C.M., 1961. Competition for Light in Crops and Pastures. Symposia of the Society for Exptl. Bio. XV. Mechanisms in Biological Competition (Proceedings), 282-313.

Drake, M., 1964. Soil Chemistry and Plant Nutrition. [In] Chemistry of the Soils. Ed. F.E. Bear. Reinhold Pub. Corporation. New York. 2nd Ed. 395-444.

Elliot, J.C., 1967. The Sowing of Seeds in Aqueous Fluid, Part of Work in Progress on 'Grassland' by the Agronomy Sec. of the Dept. of Weed Control, Weed Res. Organisation, 2nd Report. 1965-66: 31-32.

Fisher, R.A. and R.M. Hagan, 1965. Plant-water Relations, Irrigation Management and Crop yield. Exptl. Agric. 1: 161-77.

Francis, C.A., C.A. Flor and S.R. Temple, 1975. Adapting Varieties for Intercropped Systems in the Tropics. [In] Multiple Cropping Sympo. (Proceedings). American Soc. of Agron. Annual Meeting, Knoxville, Tennessee, 24-29th August.

Friesen, G.H , 1983. Integrated Weed Management for Dryland. [In] Dryland Agril. Res. in India—Thrust in the eighties. Ed. R.P. Singh and V.B. Subramanian. AICRP for Dryland Agric. Hyderabad. 45—53.

Fryer, J.D. and R.J. Makepeace, 1977. Weed Control Handbook. Vol. 1. Blackwell Scientific Pub. London.

Gates, C.T., 1968. Water Deficits and Growth of Herbaceous Plants. [In] Water Deficits and Plant Growth. Vol. II. Ed. T.T. Kozlowski, Aca. Press, New York. pp. 135-90.

Gautam, O.P., 1987. Agronomy in the Present Context of Agricultural Development. The first H.R. Arakeri Memorial Lecture Delivered at the Nat. Sympo. on Alternate Fmg. Systems, held at IARI, New Delhi.

Gooding, H.J., 1962. The Agronomic Aspects of Pigeonpea. Field Crop Abs. 15: 1-5.

Goswami, N.N., 1986. Optimising Soil Environment for Sustained Agricultural Production. Presidential address. Section of Agril Sciences. 73rd Session of the Indian Sci. Cong.

Gregory, P.J. and M.S. Reddy, 1982. Rootgrowth in an Intercrop of Pearlmillet/Groundnut. Field Crops. Res. 5: 241-252.

Gupta, O.P. and P.S. Lamba, 1978. Modern Weed Science. Today and Tomorrow's Printers and Publishers. New Delhi.

Harper, J.L., 1957. The Ecological Significance of Dormancy and its Importance in Weed Control (Proceedings), 4th Int. Cong. Crop Protection, Hamburg. Vol. 1. 415-20. Braunschweig, Hamburg.

Hartt, C.E., 1967. Effect of Moisture Supply upon Translocation and Storage of 14 C in Sugarcane. Pl. Physiol. 42: 338-46.

Herbert, Hauson, 1962. Dictionary of Ecology. Peter Owen Ltd. London.

Holliday, R., 1963. The Effect of Row Width on the Yield of Cereals. Part I, Field Crop Abs: 16: 71-81.

Hoque, M.Z., 1984. Cropping Systems In Asia. On-farm Research and Management. IRRI. Philippines.

Indian Council of Agricultural Research, 1979. 50 years (1929-79) of Agricultural Research and Education. ICAR.

Indian Society of Agronomy, 1987. Agronomic Terminology. I.S.A., Div. of Agron. IARI, New Delhi.

Indian Institute of Sugarcane Research, 1982. Annual Report. IISR. ICAR. Lucknow.

International Centre for Agricultural Research in the Dry Areas. 1981. Annual Report, ICARDA, Beirut, Lebanon.

Israelsen, O.W. and V.E. Hansen, 1962. Irrigation Principles and Practices. Wiley International ed. John Wiley and Sons. Inc. N.Y., London, Toppan Co. Ltd., Tokyo, Japan.

Jodha N.S., 1976. Resource Base as a Determinant of Cropping Pattern. [In] Sympo. on Cropping Systems Research and Development for the Asian Rice Farmer. IRRI, Philippines. 21-23 Sept.

Kanwar, J.S , 1976. Micronutrients. [In] Soil Fertility—Theory and Practice. *Ed.* J.S. Kanwar, ICAR. 229-300.

Kassam, A.H , 1973. In Search for Greater Yields with Mixed Cropping in Northern Nigeria—a Report on Agronomic Work. Report, Inst. for Agri. Res. Samaru, Nigeria.

Kaul, R., 1965. Effect of Water Stress on Respiration of Wheat. Can. J. Bot. 44: 623-32.

Levitt, J., 1956. Significance of Hydration of the State of Protoplasm. [In] Handbuch der Pflanzenphsiologie. *Ed.* W. Ruhland. Springer Verlag, Berlin, Band III. 650-51.

Maguire, J.D. , 1972. Physiological Disorders in Germinating Seeds Induced by the Environment. [In] Seed Ecology. *Ed.* W. Heydecker. Butterworth & Co. 289-310.

Marshall, B. and R.W. Willey, 1963. Radiation Interception and Growth in an Intercrop of Pearlmillet/Groundnut. Field Crops Res. 7: 141-160.

Matsuki, G , 1956. *Beisaku zoshu no gijutsu* (Techniques for Improving Rice Cultivation). Asakura Pub. Co. Tokyo.

Matsushima, S. 1967. Crop Science in Rice—Theory of Yield Determination and its Application. Fuji Pub. Co. Ltd. Tokyo.

May, L.H. and F.L. Milthrope, 1962. Drought Resistance of Crop Plants. Field Crop Abs. 15: 171-9.

Mazumdar, S.K., 1984. A New Method for Competition Free Intercropping. Abs. of Papers. Section of Agri. Sci. 71st Session of the Indian Sci. Cong. Ranchi. 20.

Mead, R. and R.W. Willey, 1980. The Concept of a Land Equivalent Ratio and Advantages in Yields from Intercropping. Expl. Agric. 16: 217-228.

Melsted, S.W., 1954. New Concept of Management of Corn Belt Soil. Advances in Agronomy. Aca. Press New York. Vol. 6: 121-142.

Michael, A.M., 1978. Irrigation—Theory and Practice. Vna Educational Books. Vikas Pub. House. Pvt. Ltd. New Delhi.

Miranda, S.M., P. Pathak and K.L. Srivastava, 1983. Runoff Management on Small Agricultural Watersheds: the ICRISAT Experience. [In] Dryland Agril. Res. in India— Thrustin the eighties. *Ed*. R.P. Singh and V.B. Subramanian, AICRP for Dryland Agric. Hyderabad.

Misra, N.M. and D.P. Dwivedi, 1980. Effect on Presowing Seed Treatment on Growth and Dry Matter Accumulation of High Yielding Wheats under Rainfed Conditions, Indian J. Agron. (2) 230-4.

Mitra, P.C., C.S. Patel, T. Ghosh and S. Sanyal, 1972. Multiple Cropping with Jute in Tossa Belt. Indian Fmg. 21 (10): 31-32.

Mitrieva, N. and P.K. Pavlov, 1963. Ovzaimnom Vliyanti Kulturnykh Restenii Cherezpozhnivnye Ostatki (On the mutual influence of cultivated plants exerted through postharvest residues) 2. Selskestop. Nauka, 2, 9, 1161-1165.

Moore, R.F., 1962. Effects of Mechanical Injuries on Viability. [In] Viability of Seeds. *Ed*. E.H. Roberts. 94-113.

Murthy, R.S., 1980. Soils. [In] Handbook of Agriculture, ICAR. 20-72.

Natarajan, M. and R.W. Willey, 1980. Sorghum-pigeonpea

Intercropping and the Effects of Plant Population Density. J. Agric. Camb. 95: 51-58.

National Academy of Sciences, 1971. Principles of Plant and Animal Pest Control. Vol. 2. Weed Control: NAS. Washington D.C. 20418.

National Commission on Agriculture, 1976 Report of the NCA Vol. IV, V & VI, Govt. of India, Ministry of Agriculture and Irrigation, New Delhi.

Nelliat, E.V., K.V. Bavoppa and P.K.R. Nair, 1974. Multistoreyed Cropping for Cocoanut plantations. World Crops. Nov.-Dec. 262-266.

Newman, E.I., 1966. Relationship between Root Growth of Flax. (*Linum usitatissimum*) and Soil-water Potential. New Phytol. 65: 273-83.

Onions, C.T., 1956. The Shorter Oxford English Dictionary on Historical Principles. Oxford Univ. Press, Amen House, London, EC—4.

Osiru, D.S.O., 1974. Physiological Studies of Some Annual Crop Mixtures. Ph.D. Thesis, Makerere Univ., Kampala, Uganda, 260.

Ovcharov, K.E., 1977. Physiological Basis of Seed Germination. Amerind Pub. Co. Pvt. Ltd. New Delhi.

Pal, M. and K. Shiks, 1969. Multiple Cropping—Multiple Profit. Indian Fmg. XI (2): 29-32.

Pal, M., K.A. Singh, J.P. Saxena and H.K. Singh, 1985. Energetics of Cropping Systems. Indian J. Agron. 30 (2) i-Lxi.

Patnaik., N, 1980. Soil Fertility and Fertiliser use. [In] Handbook of Agriculture. ICAR. 203-47.

Pearson, L.C., 1973. Principles of Agronomy. East-West Student Edition. Affiliated East-West Press Pvt. Ltd., New Delhi.

Peter, Gray, 1967. The Dictionary of the Biological Sciences. Reinhold Pub. Corporation, N.Y.

Pollock, M. Bruce, 1972. Effects of Environment after Sowing on Viability. [In] Viability of Seeds. *Ed.* E.H. Roberts. Chapman and Hall Ltd. 150—171.

Quick, C.R., 1961. [In] Seeds—the Year Book of Agriculture. *Ed.* A. Stefferud, USDA, Washington D.C.

Rachie, K.O. and L.M. Roberts, 1974. Grain Legumes of the

Lowland Tropics. Advances in Agronomy. Academic Press. New York. 26: 32-44.

Raju, R.A., 1978. Kalium (K)—the Kingpin in the Production of Paddy. Indian Potash J. 3 (4).

Randhawa, N S. and R.P. Singh, 1983. Fertiliser Management in Rain-fed Areas: Available Technologies and Future Needs. Ferti. News. 28 (9): 17.

Rao, A.C.S. and S.K. Das. 1982. Soil Fertility Management and Fertiliser use in Dryland Crops. [In] A Decade of Dryland Agril. Res. in India. 1971 80. AICRP for Dryland Agric. Saidabad, Hyderabad, ICAR.

Rao, M.R. and R.W. Willey, 1983. Effects of Pigeonpea Plant Population and Row Arrangement in Sorghum-Pigeonpea Intercropping. Field Crops Res. 7: 203-212.

Rao, V.S., 1983. Principles of Weed Science. Oxford & IBH Pub. Co.

Rao, A.V. and B. Venkateswarlu, 1982. Biological Nitrogen Fixation Associated with the Roots of Graminaceous Plants. [In] Annual Report, Central Arid Zone Research Institute, Jodhpur (ICAR), 110-112.

Reddy, M.G.R.K. and N.M. Misra, 1986. Effect of Phenyl Mercuric Acetate (PMA) on Yield and Uptake of N, P and K of Wheat under Rainfed Conditions. Indian J. Agron. 31 (2): 158-61.

Reddy, M.S. and R.W. Willey, 1980. The Relative Importance of Above and Below-ground Resource use in Determining Yield Advantages in Pearlmillet/Groundnut Intercropping. Presented at the 2nd Sympo. on Intercropping in Semi-Arid Areas, Univ. of Dar-es-Salaam, Morogoro, Tanzania, 4—7 Aug.

Reddy, M.S. and R.W. Willey, 1981. Growth and Resource use Studies in an Intercrop of Pearlmillet-Groundnut. Field Crops Res. 4: 13-14.

Reddy, M.S. and R.W. Willey, 1982. Improved Cropping Systems for the Deep Vertisols of the Indian Semi-Arid Tropics. Expl. Agric. 18: 277-287.

Reddy, P.S., G. Saran and G. Giri, 1985. Research Needs and Directions on Groundnut based Cropping Systems. Abs. of Papers. Sympo. on Cropping Systems. Indian Soc. of Agron. CSSRI. Karnal, 3-5 April. 9.

Roberts, E.H., 1972 a. Loss in Viability and Crop Yields. [In] Viability of Seeds. *Ed.* E.H. Roberts. Chapman & Hall Ltd. London. 307-320.

Roberts, E H., 1972b. Dormancy: A Factor Affecting Seed Survival in the Soil. [In] Viability of Seeds. *Ed.* E.H. Roberts. Chapman & Hall Ltd. London. 321-357.

Russel, E.W., 1961. Soil conditions and Plant growth, Longmans, Green and Co. Ltd., London.

Salisbury, E., 1961. Weeds and Aliens. Collins, London.

Sarkar, R.P. and B.C. Biswas, 1980 Agroclimatic Classification for Assessment of Crop Potential and its Application to Dry Farming tracts of India. [In] Climatic Classification: A Consultants Meeting 14-16 April. ICRISAT, Hyderabad, 89-107.

Sen, S.P., 1985. [In] Welcome Address. Abs. of Papers. National Sympo. on Physiology of Crop Yield. Plant Physiology Lab. Dept. of Botany, Bose Institute, Cal. 9. 4-5 March.

Sharma, S.N., V.K. Rao and K. Vijayalakshmi, 1982. Soil and Moisture Conservation for Drylands. [In] A Decade of Dryland Agri. Res. in India. 1971-80. AICRP for Dryland Agric. Saidabad, Hyderabad. ICAR.

Singh, A., 1980. Cropping Patterns. [In] Handbook of Agriculture, ICAR. 248-262.

Singh, A. and B. Lal. 1976. Organic Matter in Soil and its Maintenance. [In] Soil Fertility— Theory and Practice. *Ed.* J.S. Kanwar. ICAR. 128-53.

Singh, A.I. and B.N. Chatterjee, 1980. Barley Production under Rainfed Condition with Pretreated Seeds. Indian J. Agron. 25(4) 600-07.

Singh, B. and O.P. Awasthi, 1984. Effect of time of Application of Antitranspirants on the Yield of Rainfed Wheat under Mid-hill Conditions of H.P. Indian J. Agron. 29(3) 363-66.

Singh, H. and S.L. Seth, 1983. Development Plant for Increasing Crop Production in Dryland Areas. Ferti. News. 28 (10).

Singh, M., 1983. Cropping Systems for Drylands. [In] Dryland Agril. Res. in India—Thrust in the Eighties. *Ed.* R.P. Singh and V.B. Subramanian. AICRP for Dryland Agric. Hyderabad.

Singh, S N., 1982. **Agrometeorological** Research for Drylands of India. [In] A Decade of Dryland Agril. Res. in India 1971-80. AICRP for Dryland Agric. ICAR.

Singh, S.S., 1985. Principles and Practices of Agronomy. Kalyani Publishers. New Delhi.

Slatyer, R O., 1967. Plant-water Relations. Academic Press. N.Y.

Sriram, C., G.C. Yadav, P.D. Gupta and J.S. Atwal, 1982. Tillage and Seeding Practices Relevant to Drylands. [In] A Decade of Dryland Agril. Res in India. 1971-80. AICRP for Dryland Agric. Saidabad, Hyderabad. ICAR. 140-52.

Subramanian, V.B., D.G. Rao, and C.H. Rao, 1982. Crops and Cropping Systems for Drylands. [In] A Decade of Dryland Agr l. Res. in India. 1971-80. AICRP for Dryland Agric. Hyderabad, ICAR. 26-47.

Takahashi, N., 1984. Seed Germination and Seedling Growth. [In] Biology of Rice. *Ed.* S. Tsunoda and N. Takahashi. Japan Scientific Societies Press. Elsevier. 71-88.

The Encyclopedia Americana, 1962. American Corporation, N.Y. Vol. 28.

Throne, G.N., 1966. Physiological Aspects of Grain Yield in Cereals. [In] The Growth of Cereals and Grasses. *Ed.* F.L. Milthrope and D.J. Ivnis. Butterworths, London.

Timson, J., 1965. Nature. 207, 216-217.

Tisdale, S.L. and W.L. Nelson, 1963. Soil Fertility and Fertilisers. The Macmillan Co. New York.

Toole, E.H., S.B. Hendricks, H.A. Borthwick and V.K. Toole, 1956. Physiology of Seed Germination. Ann. Rev. Pl. Physiol. 7: 299-324.

United States Department of Agriculture, 1964. A Manual on Conservation of Soil and Water: Handbook of Professional Agricultural Workers. Oxford & IBH Pub. Co.

Venkateswarlu, J., 1981. Fertiliser Management in Dryland Areas. Indian Fmg. Oct. 1981. p. 47.

Villiers, T.A., 1972. Seed Dormancy. [In] Seed Biology *Ed.* T.T. Kozlowski, Vol. II. Aca. Press. New York. 219-281.

Willey, R.W., 1979. Intercropping—its Importance and Research Needs. Part I. Competition and Yield Advantages. Field Crop Abs. 32 (1): 1-10.

Willey, R.W. and E.H. Roberts, 1976. Mixed Cropping [In]

Solar Energy in Agriculture. Joint International Solar Energy Soc. Conf. (Proceedings), Univ. of Reading, England.

Willey, R.W., M.R. Rao and M. Natarajan, 1980. Traditional Cropping Systems with Pigeonpea and their Improvement. International Workshop on Pigeonpeas. Vol. I. 15-19 Dec. 11-25.

Willey, R.W. and M.R. Rao. 1981. A Systematic Design to Examine Effects of Plant Population and Spatial Arrangement in Intercropping, illustrated by an Experiment on Chickpea/Safflower. Expl. Agric. 17: 63-73.

Wilsie, C.P. and R.H. Shaw, 1954. Crop Adaptation and Climate. [In] Advances of Agronomy. Aca. Press. Vol. 6. 199.

Withers, B. and S. Vipond, 1974. Irrigation: Design and Practice: B.T. Batsford Ltd., London.

Yawalkar, K.S. and J.P. Agarwal, 1962. Manures and Fertilisers. Agri. Horti. Pub. House. Nagpur.

Author Index

Subject Index